NONSTANDARD METHODS IN FUNCTIONAL ANALYSIS

lectures and notes

NONSTANDARD METHODS IN FUNCTIONAL ANALYSIS

lectures and notes

Siu-Ah Ng

University of KwaZulu-Natal, Pietermaritzburg, South Africa
ngs@ukzn.ac.za

World Scientific

NEW JERSEY · LONDON · SINGAPORE · BEIJING · SHANGHAI · HONG KONG · TAIPEI · CHENNAI

Published by

World Scientific Publishing Co. Pte. Ltd.
5 Toh Tuck Link, Singapore 596224
USA office: 27 Warren Street, Suite 401-402, Hackensack, NJ 07601
UK office: 57 Shelton Street, Covent Garden, London WC2H 9HE

Library of Congress Cataloging-in-Publication Data
Ng, Siu-Ah.
    Nonstandard methods in functional analysis : lectures and notes by Siu-Ah Ng.
    p. cm.
    Includes bibliographical references and index.
    ISBN-13 978-981-4287-54-8
    ISBN-10 981-4287-54-7
    1. Functional analysis. 2. Nonstandard mathematical analysis. I. Title.
    QA321 .N47 2010
    515'.7--dc22

                                                        2010004689

British Library Cataloguing-in-Publication Data
A catalogue record for this book is available from the British Library.

Cover design by Hua Yang.

Printed in Singapore.

To my dearest MOTHER

—my first Teacher

from whom I learnt

一山還比一山高

---

[0]Translation: *Any high peak is yet to be humbled by a higher peak somewhere.*

Preface

In the early 1960s, using techniques from the model theory of first order logic, Robinson gave a rigorous formulation and extension of Leibniz infinitesimal calculus. Since then the methodology has found application in a wide spectrum of areas in mathematics, with particular success in probability theory and functional analysis. In the latter, fruitful results were produced with Luxemburg's invention of the nonstandard hull construction.

There is so far no publication of a coherent and self-contained treatment of functional analysis using methods from nonstandard analysis. Therefore, this publication seeks to fill such a gap.

In a way, writing a book like this is tantamount to writing a fantasy novel on a plausible alternative evolution of mathematics: *What if rigorous nonstandard analysis were invented and become popularized before the development of Banach space theory?*

However, by adhering to such a theme too dogmatically, one misses out lots of excitement—as it is unwise to prescribe a methodology before investigating the problems. For that reason, the purpose of this book is simply to demonstrate how intuition and methods from both the classical camp and the nonstandard camp are brought together to create the fundamental concepts and results in functional analysis.

READERSHIP

This book aims at both senior/graduate level students and researchers in functional analysis. For the former, it can be used as a self-study aid or a textbook for a course in functional analysis. For the latter, the book can be used as a reference for techniques from nonstandard analysis applicable to functional analysis, as well as for directions to further research.

PREREQUISITE

Undergraduate level courses covering naive set theory, real analysis, complex analysis and preferably some basic measure theory.

SYNOPSIS

Chapter 1 (p.1–75): A brief introduction to the logical and set-theoretical framework needed for nonstandard analysis is first presented in §1.1. Then two constructions of the nonstandard universe are given in §1.2 and §1.3. In §1.4, as a warm-up exercise, elementary calculus is used as a testing ground for convincing the reader that nonstandard analysis is indeed a collection of simple, effective and intuitive mathematical tools. §1.5 and §1.6 continue to serve the same purpose and present in a nonstandard manner all measure-theoretical and topological background required in later chapters.

§1.1–1.4 can be skipped by those with requisite skills in nonstandard analysis, namely being fluent with the transfer principle, the saturation principle, internal sets and internal extensions. §1.5 and §1.6 can be skipped by those who already had a thorough introduction to measure theory and topology, although they may still enjoy browsing through the nonstandard treatment of such topics.

Chapter 2 (p.77–180): In §2.1 and §2.2 basic results concerning normed linear spaces, Banach spaces, linear operators and the nonstandard hull construction are given. In §2.3, Helly's Theorem is placed in the nonstandard context and regarded as the most fundamental result in Banach space theory. The nonstandard hull construction is no doubt the most central notion in this book. A version of this construction is applied in §2.4. to represent the bidual of a Banach space. Reflexive spaces are perhaps the most studied class of Banach spaces. They are dealt with in §2.5. In §2.6 Hilbert spaces are given just sufficient coverage for the operator theory in the next chapter. §2.7 consists of a selection of topics, including the invariant subspace problem which witnessed the success of nonstandard analysis in the early days of its development.

Chapter 3 (p.181–275): In §3.1 Banach algebras and spectra are introduced. The nonstandard hull construction is extended in this context. C^*-algebras, Gelfand transform and the GNS construction are handled in §3.2. The norm-nonstandard hull of a C^*-algebra is the topic of §3.3. As a prominent type of C^*-algebras, von Neumann algebras are featured in §3.4 with a study of the effect of various kinds of nonstandard hull constructions on them. §3.5 is about some aspects and usage of projections.

Chapter 4 (p.277–301): This chapter is made up of some new results in Banach space and Banach algebras, such as isometric identities for Hilbert space-valued integrals, fixed point theorems, representation of Arens product on a bidual and a noncommutative version of Loeb measures. Some tangible open problems and questions are listed with the intention of inviting the serious reader to make contribution to the advancement of nonstandard methods.

COURSE TOPICS

One-semester senior level course: Chapter 1 (§1.3 may substitute part of §1.2.2), §2.1–2.6.

Two-semester senior level course or *one-semester graduate level course*: The above, together with §2.7, §3.1–3.4 and any part of §3.5 and Chapter 4. Possibly with more detail on logic and ultraproduct supplemented by Chang and Keisler (1990).

Whenever possible, exercise problems should be done by using methods from nonstandard analysis.

ATTRIBUTIONS

Other than the author's negligence, the absence of ascribing credit for a result means that it is easy or folklore or due to the author.

ACKNOWLEDGEMENT

I am indebted to Hua and Yisun for their patience and tolerance during the writing of this book. I also thank Hua for the cover design and drawings.

Siu-Ah Ng
January 2010

x   *Nonstandard Methods in Functional Analysis*

Convention and Symbols

- A *Banach space, linear functional, etc.* is an object which is a Banach space, linear functional, *etc.* in the ordinary (*standard*) sense.
- An *internal Banach space, internal linear functional, etc.* is an *internal* object which is a *Banach space, *linear functional, *etc.*
- Notation evolves along the pages as topics become more specialized. For example, the generic notation for a linear operator on a Hilbert space changes from f to ϕ then to T and sometimes may reverse, depending on the context.
- In the first half of the book, the symbol $*$ is more rigorously attached to any internal extension. In the second half, this practice is eased out for better readability. So $\widehat{*\mathcal{B}(*H)}$ can be written as $\widehat{\mathcal{B}(H)}$.
- From p.58 onward, unless stated otherwise, all topological spaces are assumed to be *Hausdorff.*
- By $\forall i \in I \, (\cdots)$ holds, we mean $(\cdots)$ holds for every $i \in I$. We also use $\exists i \in I \, (\cdots)$ in a similar way. The conjunction $\bigwedge_{i \in I} (\cdots)$ is equivalent to $\forall i \in I \, (\cdots)$ and the disjunction $\bigvee_{i \in I} (\cdots)$ is equivalent to $\exists i \in I \, (\cdots)$. None of these are first order formulas if I is infinite.
- $X := (\cdots)$ means X is defined as $(\cdots)$.
- Respectively, $\subset$ and $\supset$ mean $\subseteq$ and $\supseteq$, but $<$ and $>$ mean $\lneq$ and $\gneq$.
- The image of a set X under a function f is denoted by $f[X]$.
- $\mathbb{N}$ = natural numbers, $\mathbb{Z}$ = integers, $\mathbb{Q}$ = rationals, $\mathbb{R}$ = reals and $\mathbb{C}$ = complex numbers. $\mathbb{N}^+ = \{n \in \mathbb{N} \mid n > 0\}$ and $\mathbb{R}^+ = \{r \in \mathbb{R} \mid r > 0\}$.
- $\mathbb{F}$ stands for either $\mathbb{R}$ or $\mathbb{C}$.
- In the second half of the book, all Banach algebras are over $\mathbb{C}$.
- The appearance of a terminology in boldface italic font is an indication that its **definition** is to be found nearby.

Sets:

$\emptyset$ (empty set),   1

$\mathcal{P}(X)$ (power set),   1

ω (first infinite ordinal),   6

$\aleph_0$ (first infinite cardinal),   6

$\aleph_1$ (first uncountable cardinal),   6

$\prod_U X_i$ (ultraproduct),   19

$A \backslash B$,   38

$A \triangle B$ (symmetric difference),   38

Nonstandard analysis:

$^*X$ (nonstandard extension),   12

κ (saturation cardinality),   13

$V(\mathbb{R})$, $V(^*\mathbb{R})$ (universes),   10,12,16

$\approx$ (infinitely closed),   23,58,56

$\mu(x)$ (monad),   23,54,56

$^\circ x$ (standard part),   23,58

Fin$(\cdot)$ (finite part),   23,80

$|\cdot|$ (internal cardinality),   28

$\mathbb{T}$ (hyperfinite timeline),   28

SL^p (integrability),   47

ns$(\cdot)$ (nearstandard part),   54

st$(\cdot)$ (standard part mapping),   58

$\widehat{X}$ (nonstandard hull),   81

$\widehat{x}$ (element of $\widehat{X}$),   81

$\widehat{X}^w$ (general nonst. hull),   121

$\approx_w$, $\mu_w(\dots)$, Fin$_w(\cdot)$,   121

$\widehat{^*X^w}$ (weak nonst. hull),   123

Fin$_q(\cdot)$,   124

$\widehat{\mathcal{M}}^\tau$ (tracial nonst. hull),   258

$\widehat{x}^\tau$ (element of $\widehat{\mathcal{M}}^\tau$),   257

$HL_2(\Omega, \widehat{X})$ (hyper-int. func.),   284

Measures:

$\sigma\mathcal{B}$ (σ-algebra),   33

Leb (Lebesgue measure),   38

$\overline{\mu}$ (outer measure),   38

$\underline{\mu}$ (inner measure),   38

$L(\mu)$ (Loeb measure),   40

a.e. (almost everywhere),   43

$\alpha(\cdot)$ (intersection number),   48

$\beta(\cdot)$ (measure number),   48

Topology:

$\mathcal{T}$ (topology),   54

$\overline{A}$ (topological closure),   54

$B(x,r)$ (open ball),   55

$S(x,r)$ (sphere),   55

X^ϵ (expansion by ϵ),   68

$\prod_{i\in I} X_i$ (Tychonoff product),   69

int$(\cdot)$ (interior of a set),   70

βX (Stone-Čech comp. of X),   72

Banach space:

dim$(\cdot)$ (dimension),   77

$\|\cdot\|$, $\|\cdot\|_X$ (seminorm, norm),   77

$x + I$ (coset),   79

X/Y (quotient space),   79

dist(A,B) (distance function),   79

Lin$(\cdot)$ (linear span),   84

$\overline{\text{Lin}}(\cdot)$ (closed linear span),   84

$\|\cdot\|_\infty$ (maximum norm),   87

$\|\cdot\|_\infty$ (supremum norm),   87

$\|\cdot\|_\infty$ (essential sup norm),   90

ℓ_p (sequence space),   87

c, $c(\mathbb{N})$, c_0 (sequence space),   87

$\mathbb{M}(\Omega)$,   88

$C_0(\Omega)$, $C_b(\Omega)$, $C(\Omega)$,   88

$ba(\mathbb{N})$, $ca(\mathbb{N})$,   89

$L_p(\mu)$, $L_p(\Omega)$ (Lebesgue space),   89

$L_p(\Omega, \mathcal{B}, \mu)$ (Lebesgue space),   89

$W^{k,p}(\Omega)$ (Sobolev space),   90

$\oplus_p X_i$ (direct sum),   91

Ker$(\cdot)$ (kernel),   94

B_X (open unit ball),   94

$\bar{B}_X$ (closed unit ball),   94

S_X (unit sphere),   94

$\mathcal{B}(X,Y)$ (bdd. lin. operators),   95

X' (dual space),   95

Convention and Symbols   xiii

Abraham Robinson (1918 - 1974)

List of Theorems

Method of proof: nn = nonstandard   ss = standard   ns = mixed
       zz = proof not included
Abbreviations: nsh = nonstandard hull   nst = nonstandard

Eduard Helly
(1884 - 1943)

Stefan Banach
(1892 - 1945)

John von Neumann
(1903 - 1957)

Israel M. Gelfand
(1913 - 2009)

Contents

Contents   xxi

Chapter 1

Nonstandard Analysis

1.1   Sets and Logic

*Mathematical objects are sets and mathematical theories are
part of the set theory.*

Intuitively, all mathematical objects can be collected to form what we call
a *universe* U. More formally, members of U are termed **sets** and there is
a system of written rules, *i.e.* axioms, specifying certain objects to be basic
sets and establishing some acceptable set-building machineries.

1.1.1   Naïve sets, first order formulas and ZFC

As a minimum requirement, U must contain $\emptyset$, the empty set—a set with
no members, and an infinite set, such as $\mathbb{N}$, the set of natural numbers.
Moreover, if $X, Y \in U$, then U must also contain sets such as

- $\{X, Y\}$, the pairing of X, Y;
- (X, Y), ordered pair, identifiable with $\{\{X\}, \{X, Y\}\}$;
- $X \times Y$, Cartesian product, *i.e.* $\{(A, B) \mid A \in X \text{ and } B \in Y\}$;
- $\bigcup X$, union, *i.e.* $\{A \mid A \in Z \text{ for some } Z \in X\}$;
- $\bigcap X$, intersection, *i.e.* $\{A \mid A \in Z \text{ for every } Z \in X\}$;
- $\mathcal{P}(X)$, power set that collects *all* subsets of X, *i.e.* $\{Z \mid Z \subset X\}$;
- $\{Z \in X \mid Z \text{ satisfies } \phi\}$ for certain admissible property ϕ;
- $F[X]$, image under a function $F \subset X \times Y$; *i.e.* F is such that for
  any $A \in X$ there is a unique $B \in Y$ such that $(A, B) \in F$;
- Y^X, *i.e.* $\{F \mid F \subset X \times Y \text{ and } F \text{ is a function}\}$.

The most widely accepted axiom system for sets is ZFC, the **Zermelo-
Frankel** set theory together with the **Axiom of Choice**. ZFC is formu-

2   *Nonstandard Methods in Functional Analysis*

lated in a formal language based on the following:

- Logical symbols (whose intuitive meanings are listed underneath):

$$\neg, \quad \wedge, \quad \vee, \quad \Rightarrow, \qquad \forall, \qquad \exists, \qquad = ;$$

  not, and, or, implies, for all, exists, equals;

- variables: $x, y, z, x_0, x_1, \ldots, y_0, y_1, \ldots$;
- a binary relation symbol: $\in$;
- brackets for making expressions unambiguous.

(To economize on notation, $=$ and $\in$ are used both as symbols in the language and as relations between sets in U.)

Valid expressions in this language are called **formulas**.

Beginning with two basic types of formulas:

$$x = y \quad \text{and} \quad x \in y,$$

where x, y are variables, other formulas are built up by iterating the following operations finitely many times:

$$\neg\phi, \quad \phi \wedge \theta, \quad \phi \vee \theta, \quad \phi \Rightarrow \theta, \quad \forall x\, \phi, \quad \exists x\, \phi,$$

where ϕ and θ are formulas that have been built already. $\forall$ and $\exists$ are called **quantifiers** and in the above, the variable x is bounded by them.

Some frequently used abbreviations:

- $x \neq y$, and $x \notin y$ stand for $\neg(x = y)$ and $\neg(x \in y)$ *resp.*;
- $\phi \Leftrightarrow \theta$ stands for $(\phi \Rightarrow \theta) \wedge (\theta \Rightarrow \phi)$;
- $\forall x, y$ and $\exists x, y$ stand for $\forall x \forall y$ and $\exists x \exists y$ *resp.*;
- $\forall x \in y\, \phi$ means $\forall x\, (x \in y \Rightarrow \phi)$;
- $\exists x \in y\, \phi$ means $\exists x\, (x \in y \wedge \phi)$;
- $\forall x \subset y\, \phi$ means $\forall x\, (\forall z\, (z \in x \Rightarrow z \in y) \Rightarrow \phi)$;
- $\exists x \subset y\, \phi$ means $\exists x\, (\forall z\, (z \in x \Rightarrow z \in y) \wedge \phi)$;
- $\exists! x\, \phi$ means $\exists x\, (\phi \wedge \forall y(\theta \Rightarrow (x = y)))$, where θ is obtained by replacing all occurrences of x in ϕ by y.

Formulas given above are more formally called **first order formulas** in the language consisting of a single binary relation symbol $\in$, *i.e.* in the language of set theory. This language is denoted by $\mathcal{L}$. So $\mathcal{L} = \{\in\}$.

A formula ϕ is written as $\phi(x_1, \ldots, x_n)$ if the variables in ϕ which are not bounded by any quantifiers $\forall$ or $\exists$ are among $x_1, \ldots, x_n$. When $x_1, \ldots, x_n$ in $\phi(x_1, \ldots, x_n)$ are replaced by $y_1, \ldots, y_n$ the resulted formula is written as $\phi(y_1, \ldots, y_n)$. Formula in which all variables are bounded by some quantifiers is called a **sentence**.

The axioms of ZFC consist of first order sentences describing the following set existence with the corresponding titles:

- (*empty set*) the existence of an empty set;
- (*extensionality*) sets are uniquely determined by membership;
- (*pairing, union, power set, comprehension, replacement*) the existence of those sets listed on p.1;
- (*infinity*) the existence of an infinite set;
- (*regularity*) no set contains a decreasing chain of membership;
- (*AC*) the axiom of choice.

To illustrate, the axiom of empty set is given by the sentence

$$\exists x \, \forall y \big(y \notin x \big)$$

while the axiom of extensionality is written as

$$\forall x, y \big(\forall z \, (z \in x \Leftrightarrow z \in y) \Leftrightarrow x = y \big).$$

The axiom of regularity is just a technicality needed to give U a hierarchical structure and is not relevant to most ordinary mathematics. The ϕ in $\{Z \in X \mid Z \text{ satisfies } \phi\}$ on p.1 is required to be a first order formula. The collection of first order sentences asserting the existence of such set, one for each such ϕ, constitutes the axiom scheme of comprehension. Likewise, the existence of the image under each function constitutes the axiom scheme of replacement. Not everyone on that list needs to be explicitly given an axiom, as the existence of some of them is derivable from others.

There is a natural and rigorous deduction system for first order formulas whose details are skipped over here. For example, for our purpose here, it is sufficient to see by an informal argument that $\text{ZFC} \vdash \exists! x \, \forall y (y \notin x)$, *i.e.* the uniqueness of the empty set. ($\vdash$ is the symbol for logical deduction.) A constant symbol $\emptyset$ is then introduced and a sentence like $\exists x \, (\emptyset \in x)$ abbreviates $\exists x, y \big(\forall z \, (z \notin y) \wedge (y \in x) \big).$

Note that $\emptyset$ also stands for a particular object in U, the set having no members. When the symbol $\emptyset$ is interpreted as the empty set $\emptyset$ in U (and symbols $=, \in$ interpreted as the relations $=, \in$ on U *resp.*), the set $\{\emptyset\}$ is a realization of the axiom of empty set. That is, the symbol $\emptyset$ is just a name for the empty set in U. Other common abbreviation symbols are also used. For example, $\subset$ is interpreted as the subset relation in U.

More generally, if a *theory* T (*i.e.* a collection of sentences) is realized by some $X \subset U$ and some binary relation E on X under some interpretation of the language, we say (X, E) is a *model* of T. This is written as

$$(X, E) \models T.$$

It follows for instance that $(U, \in) \models \exists x \, (\emptyset \in x)$ under the natural interpretation. Given a formula $\phi(x)$ and $A \in U$, the expression $(U, \in) \models \phi(A)$ means that in the natural interpretation with x interpreted as A, the formula ϕ is realized in U. We also say the first order property ϕ holds of A in U. So, if $\phi(y)$ is $\exists x \, (y \in x)$, then $(U, \in) \models \phi(\emptyset)$.

1.1.2 First order theory and consistency

We say that a theory T is consistent if no falsehood is deductable from it, *i.e.* $T \nvdash \exists x \, (x \neq x)$.

First order logic is simply the interplay between formulas, deduction and models. The crucial connection between them is the following:

Theorem 1.1. (Gödel's Completeness Theorem) *A theory is consistent iff it has a model.* $\qquad\square$

A deduction in first order logic requires only finitely many steps. Consequently, we have:

Theorem 1.2. (Compactness Theorem for First Order Logic) *If every finite subset of a theory has a model, then the theory has a model.* $\qquad\square$

To continue with the discussion about ZFC, the following is the axiom of pairing:

$$\forall x, y \, \exists z \, \big(\forall u \, (u \in z \Leftrightarrow (u = x \vee u = y)) \big).$$

Then ZFC $\vdash$

$$\forall x, y \, \exists z \, \big(\forall u \, [u \in z \Leftrightarrow ((\forall w(w \in u \Leftrightarrow w = x)) \vee (\forall w(w \in u \Leftrightarrow (w = x \vee w = y))))] \big)$$

i.e. the existence of the ordered pair (x, y). Hence the existence of an ordered pair is logically deduced from ZFC.

Given a formula $\phi(z, u_1, \ldots, u_n)$, the following is included in the comprehension axiom scheme:

$$\forall x \, \forall u_1, \ldots, u_n \, \exists y \, \forall z \, \big(z \in y \Leftrightarrow (z \in x \wedge \phi(z, u_1, \ldots, u_n)) \big)$$

stating the existence of the set $\{ z \in x \mid z \text{ satisfies } \phi \}$. We leave as exercises the formulation of other ZFC axioms. Viewed in an intuitively manner, starting from an empty set, all other sets in U are built up by applying ZFC axioms. We will return to some more examples of ZFC axioms in a moment.

It is sensible to claim that $(U, \in) \models$ ZFC. However we have:

Theorem 1.3. (Gödel's Incompleteness Theorem) *Unless ZFC is incon-sistent, the consistency of ZFC is not provable under ZFC.* $\square$

Hence no set model of ZFC can be produced. In particular U is not a set, *i.e.* $U \notin U$. (In fact $U \in U$ contradicts the regularity axiom as well.) Pragmatically, mathematicians work only in a large enough universe of sets and are content with the trust that mathematics is consistent—hence the consistency of ZFC.

In any case, we regard U as the collection of *all* sets and $\in$ the *true* membership relation on sets.

1.1.3   Infinities, ordinals, cardinals and AC

Recall that a binary relation $\leq$ on a set X is a *partial ordering* if it is reflexive, antisymmetric and transitive. *i.e.* the following are satisfied:

- $\forall x \in X \, (x \leq x)$;
- $\forall x, y \in X \, ((x \leq y \leq x) \Rightarrow (x = y))$;
- $\forall x, y, z \in X \, ((x \leq y \leq z) \Rightarrow (x \leq z))$.

We write $x < y$ for $(x \leq y) \wedge (x \neq y)$. Sometimes it is convenient to formulate partial ordering using $<$ with the obvious adjustment to the above.

A partial ordering $\leq$ on X is called a *linear ordering* if everything in X is comparable under $\leq$. *i.e.* $\forall x, y \in X \, ((x \leq y) \vee (y \leq x))$. A linear ordering $\leq$ on X is called a *well-ordering* if every nonempty subset of X has a least element *w.r.t.* $\leq$. We say that $a \in X$ is a $\leq$-*maximal element* if $\forall x \in X \, (a \leq x \Rightarrow a = x)$.

The pair $(X, \leq)$ is called a partial/linear/well order if $\leq$ is respectively a partial/linear/well- ordering on X.

In U, natural numbers are identified with sets:

$$0 := \emptyset, \quad 1 := \{0\}, \quad 2 := \{0, 1\}, \ldots, \; n + 1 := n \cup \{n\}, \quad \ldots$$

and $\mathbb{N}$ is the union of all such sets. Then $\mathbb{N}$ and every $n \in \mathbb{N}$ is linearly ordered by $\in$. In fact $\in$ is a well-ordering on them. More generally, we call a set in U an *ordinal* if it is well-ordered by $\in$.

Note that if α is an ordinal then so is $\alpha \cup \{\alpha\}$, *i.e.* $\alpha + 1$. Ordinals of the form $\alpha + 1$ are called *successor ordinals*. A nonzero ordinal α such that $\beta + 1 \in \alpha$ for every $\beta \in \alpha$ is called a *limit ordinal*. The collection of ordinals is denoted by Ord. Then $Ord \notin U$ for reasons similar to why $U \notin U$. On Ord, it is common to write the ordering as $<$ instead of $\in$. There is a natural arithmetic on Ord. For example, $\alpha + \beta$ is the unique ordinal order-isomorphic to the one given by adjoining β to the tail α. Multiplication and exponentiation are defined as well.

The ***infinity axiom*** of ZFC states that a limit ordinal exists:

$$\exists x \left(\emptyset \in x \wedge (\forall y \in x(y + 1 \in x)) \right).$$

Then the existence of $\mathbb{N}$ as a set is provable from ZFC using the well-ordering of $\in$. To emphasis its well-ordering structure, as an ordinal, we write ω instead of $\mathbb{N}$.

ZFC allows one to perform ***transfinite induction*** and ***transfinite recursion*** on Ord. This is a natural generalization of the usual induction and recursion on $\mathbb{N}$, as a consequence of the well-ordering.

Then the following objects are definable in ZFC by transfinite recursion:

$$U_0 := \emptyset, \quad U_\alpha := \bigcup_{\beta \in \alpha} \left(U_\beta \cup \mathcal{P}(U_\beta) \right), \quad \alpha \in Ord.$$

Note that $U_\alpha \in U$ and $U_\alpha \subset U$. Moreover, one regards

$$U = \bigcup_{\alpha \in Ord} U_\alpha,$$

hence giving a hierarchy of sets. (In particular the axiom of regularity is satisfied.) For $X \in U$, the rank $\mathrm{rk}(X)$ is therefore defined as the least α such that $X \in U_\alpha$.

It is easy to check that $\mathrm{rk}(\omega) = \omega + 1$.

Let $X \in U$. If

$$(U, \in) \models \left(\text{there exists a bijection between some} \, n \in \omega \text{ and } X \right),$$

we say that X is ***finite***. (Such a bijection is then in U.) If there is an injection from ω to X then X is ***infinite***. If there is an injection from X to ω then X is ***countable***, otherwise we say that it is ***uncountable***. So ω is the least infinite ordinal and is a countable one. We write the ***cardinality*** of X as $|X|$, this is the least ordinal so that there is a bijection between them.

By ***cardinals***, we mean $\kappa \in Ord$ such that $|\kappa| = \kappa$. So the cardinality of a set is always a cardinal. As a cardinal, ω is written as $\aleph_0$, the countably infinite cardinal. Then $\aleph_1$ is the least cardinal greater than it, the first uncountable cardinal, and so on. It can be proved that cardinals can be well-ordered and so infinite cardinals can be denoted as $\aleph_\alpha$ for some $\alpha \in Ord$.

In order to be able to define the cardinality of an arbitrary set $X \in U$, the following axiom is required:

$$\forall x \, \exists y \left(y \text{ is a well ordering on } x \right).$$

This is known as the **Axiom of Choice** (AC). Among all axioms of ZFC, it may look like an unusual one. As a matter of fact, until quite recently most mathematicians were suspicious of AC for its nonconstructive nature. However, AC is accepted nowadays for pragmatic reasons such as the existence of cardinality or the claim that every vector space has a basis.

AC is **independent** of ZF—the rest of ZFC. That is, ZF $\nvdash$ AC and ZF $\nvdash \neg$AC. Another famous result is that CH, the Continuum Hypothesis, the assertion that $|\mathbb{R}| = \aleph_1$, is independent of ZFC. In a sense ZFC cannot be effectively completed.

The above form of AC is normally referred to as the **well-ordering principle**. There are two other forms, all three are equivalent under ZF:

The existence of a choice function:

$$\forall x \ \exists f \subset x \times \cup x \left(f \text{ is a function } \wedge \ (\forall y \in x \, (y \neq \emptyset \Rightarrow f(y) \in y))\right).$$

Zorn's Lemma: Suppose X is partially ordered by $\leq$ with the property that every subset of X linearly ordered by $\leq$ has an upper bound. Then X has a $\leq$-maximal element.

With AC, one can be assured the existence of some sets without explicit prescription of their construction. The practice of nonstandard analysis can be characterized as an aggressive and efficient program of exploiting AC. Moreover, this is done within ZFC. In particular, results proved using methods from nonstandard analysis are never independent of ZFC.

1.1.4 Notes and exercises

The axiomatization of sets is analogous to Euclid's axiomatization of geometry. The main difference is the use of a formal language such as the first order language in ZFC axioms, which makes the statements and deduction more precise. Moreover, the syntax is given meaning via interpretations in models. Gödel's Completeness Theorem give a crucial correspondence between the syntax and semantic. There are other formal logics stronger in descriptive power, such as the second order logic, but poorer in model theory. First order logic appears to have the right balance.

In practice, mathematical theories are axiomatized in a specially tailored language, such as the first order language of groups (consisting of one binary function symbol and one constant symbol), that of rings, *etc.* But they can

8   *Nonstandard Methods in Functional Analysis*

be embedded as part of set theory by treating the relevant relation, function, constant symbols as the first order formulas defining them in set theory.

For basic first order logic, the reader is referred to textbooks such as [van Dalen (2004)]. For a complete list of the ZFC axioms and a more rigorous presentation of set theory, see [Enderton (1977)], [Jech (2003)] or [Kunen (1983)]. The last two give detailed treatment of independence results based on the forcing construction of models.

A standard reference and a complete treatment of model theory for first order logic is [Chang and Keisler (1990)].

AC was formulated by Zermelo who proved the equivalence between the choice function version and the well-ordering version. It is mainly the latter that causes uneasiness among mathematicians in the early days. The relative consistency with ZF was proved by Gödel and the independence was proved by Cohen. The history and variants of AC can be found in [Moore (1982)].

EXERCISES

(1) Write down as first order sentences in $\mathcal{L}$ the axiom of union, the axiom of power set and the axiom scheme of replacement.
(2) Write down a first order formula expressing the property that x is well-ordered by $\in$.
(3) Write down a first order sentence specifying the unique existence of ω.
(4) Show using ZFC axioms how $\mathbb{R}$ is constructed from $\emptyset$. Find $\mathrm{rk}(\mathbb{R})$.
(5) Write down in full details the first order sentence for AC.
(6) Show that a nonzero ordinal is either a successor or a limit.
(7) Explore the natural way to do arithmetic on ordinals. Investigate the order topology on ordinals.
(8) Show that ZFC $\vdash \forall x \left(|x| < |\mathcal{P}(x)| \right)$.
(9) Show under ZFC that AC is logically equivalent to the other two alternative formulation.
(10) Prove that every vector space has a basis.
(11) Prove that there is a subset of $\mathbb{R}$ which is not Lebesgue measurable and discuss how the proof relies on AC.

1.2   The Nonstandard Universe

Infinite mathematical objects can be idealized with all finitary information coded within.

In the practice of nonstandard analysis, a mathematical object X is idealized to an extension denoted as $^*X$. Moreover, this is done uniformly to all mathematical objects under consideration. By idealization, we mean an enrichment so that any finitary approximation property about X is actually realized in $^*X$. However, $^*X$ shares the same first order properties with X, hence $^*X$ can be regarded as an idealized copy of X. The situation is analogous to $\mathbb{C}$ taken as the algebraic closure of $\mathbb{R}$. In comparison, $^*\mathbb{R}$ is the "first order closure" of $\mathbb{R}$.

1.2.1   *Elementary extensions and saturation*

We continue to regard U as the collection of all sets and $\in$ the true membership relation on sets. Depending on the context, $\in$ also stands for the corresponding binary relation symbol from $\mathcal{L}$.

Given an infinite $X \in U$, think of $(X, \in)$ as a model for a first order theory in the language of set theory $\mathcal{L}$. Then there is a set $^\circ X \supsetneq X$ and a binary relation $^\circ\!\in$ on $^\circ X$ such that for for every formula $\phi(x_1, \dots, x_n)$ and any $A_1, \dots A_n \in X$, we have

$$(X, \in) \models \phi(A_1, \dots, A_n) \quad \text{iff} \quad (^\circ X, {}^\circ\!\in) \models \phi(A_1, \dots, A_n).$$

This is called an **elementary extension** and written as

$$(X, \in) \preceq (^\circ X, {}^\circ\!\in).$$

Note that $^\circ\!\in$ needs not be the true membership relation on the set $^\circ X$, only its restriction to X is the true membership relation $\in$.

For an infinite X, as a consequence of the Compactness Theorem, X has an elementary extension of size greater than an arbitrarily given cardinal. However, if X is finite, there is no proper elementary extension of X, *i.e.* $^\circ X$ is always the same as X. This is because one can write down a first order sentence characterizing the finite size of X. Therefore nonstandard analysis would not lend any extra power to finite mathematics.

Moreover, for reasons which become clear later on, for a fixed large enough uncountable cardinal κ (normally $\kappa = \aleph_1$ is enough), one looks for an elementary extension $(^\circ X, {}^\circ\!\in)$ satisfying a property called κ-saturation which we discuss now.

First some notation. Let $C \subset {}^{\circ}X$. We expand $\mathcal{L} = \{\in\}$ to the language $\mathcal{L}_C$ that includes a constant symbol c for each $c \in C$. Then the symbol c is naturally interpreted in $\left({}^{\circ}X, {}^{\circ}\in\right)$ by the element $c \in {}^{\circ}X$, so $\left({}^{\circ}X, {}^{\circ}\in\right)$ is also regarded as a model for the language $\mathcal{L}_C$. (More precisely, this model in the expanded language $\mathcal{L}_C$ should be written as $\left({}^{\circ}X, {}^{\circ}\in, c\right)_{c \in C}$.)

The above-mentioned property is the following:

κ-**saturation:** *Given arbitrary* $C \subset {}^{\circ}X$ *with* $|C| < \kappa$ *and a family* $\mathcal{F}$ *of* $\mathcal{L}_C$*-formulas in free variable* x, *if for any finite* $\mathcal{F}_0 \subset \mathcal{F}$,

$$\left({}^{\circ}X, {}^{\circ}\in\right) \models \exists x \bigwedge_{\phi(x) \in \mathcal{F}_0} \phi(x),$$

then there is $a \in {}^{\circ}X$ *such that* $\left({}^{\circ}X, {}^{\circ}\in\right) \models \phi(a)$, *for every* $\phi(x) \in \mathcal{F}$.

i.e. in $\left({}^{\circ}X, {}^{\circ}\in\right)$, *if* $\mathcal{F}$ *is finitely satisfiable then it is satisfiable.*

For infinite X and any cardinal κ, the existence of a κ-saturated elementary extension $\left({}^{\circ}X, {}^{\circ}\in\right)$ is guaranteed by some model theoretic construction done under ZFC. Roughly speaking, let $\left(X_0, \in_0\right) := \left(X, \in\right)$ and after α steps, $\left(X_\alpha, \in_\alpha\right)$ is constructed, we list all $\mathcal{F}$ which are finitely satisfiable in $\left(X_\alpha, \in_\alpha\right)$ and use the Compactness Theorem to show that an elementary extension $\left(X_{\alpha+1}, \in_{\alpha+1}\right)$ exists in which all such $\mathcal{F}$'s are satisfied. So at each step we get more $\mathcal{F}$'s than in the previous step. Moreover, $|X_{\alpha+1}|$ is kept to the minimum, *i.e.* has cardinality no greater than the totality of all the families $\mathcal{F}$. For limit ordinal α, the model $\left(X_\alpha, \in_\alpha\right)$ is obtained by taking the union of models from previous steps. Then a κ-saturated $\left(X_\alpha, \in_\alpha\right)$ appears at some limit ordinal α. Trivially, the κ-saturated elementary extension is a proper extension.

1.2.2 *Superstructure, internal and external sets*

Given a set X in U, we define

$$V_0(X) := X, \quad V_{n+1}(X) := V_n(X) \cup \mathcal{P}(V_n(X)), \quad n < \omega.$$

Then the *superstructure* over X is the union

$$V(X) := \bigcup_{n < \omega} V_n(X).$$

Note that if $X \in U_\alpha$ for some $\alpha \in Ord$, then $V(X) \subset U_{\alpha+\omega}$ and $V(X) \in U_{\alpha+\omega+1}$. Also, $V(\emptyset)$ is just U_ω. One can also think of $V(X)$ as the collection of sets constructed from X by basic set operations in finitely many steps.

Most ordinary mathematics, especially *functional analysis*, takes place in $V(\mathbb{R})$. Most construction of $\mathbb{R}$ puts $\mathbb{R}$ in $U_{\omega+4}$. Except in finite combinatorics, there is very little interest in sets having finite ranks. So, although functional analysis can be done in $U_{\omega+\omega}$, we prefer to choose $V(\mathbb{R})$ as the more convenient framework.

Therefore we will extend and idealize all $V_n(\mathbb{R})$ uniformly. This is done by taking, for some uncountable cardinal κ, a κ-saturated elementary extension:

$$\big(V(\mathbb{R}), \in\big) \npreceq \big(\mathfrak{V}, \varepsilon\big). \tag{1.1}$$

Although $\mathfrak{V}$ is just some set in U and ε needs not agree with $\in$—the true set membership for sets in U (in fact it does not!), a useful portion of certain superstructure is embedded in $\mathfrak{V}$.

First of all, $\mathbb{R} \in V(\mathbb{R}) \subset \mathfrak{V}$. We treat $\mathbb{R}$ as a new constant symbol and form the expanded language $\mathcal{L}_{\{\mathbb{R}\}}$. For $n < \omega$, we let $\phi_n(x)$ be the $\mathcal{L}_{\{\mathbb{R}\}}$-formula expressing $x \in V_n(\mathbb{R})$. So $\phi_0(x)$ is the formula $x \in \mathbb{R}$, where $\in$ is the binary relation symbol and $\mathbb{R}$ the constant symbol, and $\phi_1(x)$ is the formula $(x \in \mathbb{R}) \vee (x \in \mathcal{P}(\mathbb{R}))$ (*i.e.* $(x \in \mathbb{R}) \vee \big(\forall y\,(y \in x \Rightarrow y \in \mathbb{R})\big)$) *etc.* We define

$$^*\mathbb{R} := \big\{ X \in \mathfrak{V} \,\big|\, (\mathfrak{V}, \varepsilon) \models \phi_0(X) \big\}.$$

Equivalently, $^*\mathbb{R}$ consists of the X's from $\mathfrak{V}$ such that $(X, \mathbb{R})$ satisfies the relation ε. That is, $X\varepsilon\mathbb{R}$ holds in $\big(\mathfrak{V}, \varepsilon\big)$. (Keep in mind that ε is not the true set membership relation.)

Observe that $\mathbb{R} \subset {}^*\mathbb{R}$ by (1.1).

$^*\mathbb{R}$ forms the hyperreal field to be discussed in §1.4.

Now define a mapping $\pi : \mathrm{dom}(\pi) \to V(^*\mathbb{R})$, where the domain

$$\mathrm{dom}(\pi) := \{X \in \mathfrak{V} \,|\, (\mathfrak{V}, \varepsilon) \models \phi_n(X) \text{ for some } n < \omega\,\},$$

by induction on $n < \omega$:

$$\pi(A) := \begin{cases} A & \text{if } (\mathfrak{V}, \varepsilon) \models \phi_0(A) \\ \{\pi(X) \,|\, X\varepsilon A\} & \text{if } (\mathfrak{V}, \varepsilon) \models \phi_{n+1}(A) \wedge \neg\phi_n(A). \end{cases}$$

The well-definedness of π follows from the elementary extension in (1.1).

Note that $\pi(\mathbb{R}) = {}^*\mathbb{R}$ and indeed the range of π is included in $V(^*\mathbb{R})$. Moreover, $V(\mathbb{R}) \subset \mathrm{dom}(\pi) \subset \mathfrak{V}$.

On the other hand, keep in mind that trivially

$$V(\mathbb{R}) \subset V(^*\mathbb{R}).$$

For each $X \in V(\mathbb{R})$, we have $X \subset \pi(X)$ by (1.1). Moreover, $(\pi(X), \in)$ is a κ-saturated elementary extension of $(X, \in)$. (Here $\in$ is the true set membership relation.)

Notice that π is injective and the range of π is

$$\bigcup_{n<\omega} \pi(V_n(\mathbb{R})) \subset V(^*\mathbb{R}).$$

We define

$$^*X := \pi(X), \quad \text{where} \quad X \in V(\mathbb{R}).$$

Hence $\pi(V_n(\mathbb{R})) = {}^*V_n(\mathbb{R})$ and

$$* : V(\mathbb{R}) \to \bigcup_{n<\omega} {}^*V_n(\mathbb{R}).$$

So we have sets such as $^*\mathbb{N}$, $^*\mathbb{Q}$, $^*\mathbb{R}$, $^*\mathbb{C}, \ldots$ as well as the $^*V_n(\mathbb{R})$, $n < \omega$. This is the uniform κ-saturated elementary extension of all sets in $V(\mathbb{R})$, hence of all ordinary mathematical objects that we have referred to earlier. The above is summarized by the commutative diagram Fig. 1.1.

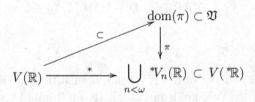

Fig. 1.1   The *-embedding into the nonstandard universe.

There are two important types of sets in $V(^*\mathbb{R})$:

sets in $\displaystyle\bigcup_{n<\omega} {}^*V_n(\mathbb{R})$ are called ***internal sets***,

sets in $V(^*\mathbb{R}) \setminus \displaystyle\bigcup_{n<\omega} {}^*V_n(\mathbb{R})$ are called ***external sets***.

Internal sets include all sets of the form $^*X$, where $X \in V(\mathbb{R})$. As we will see in a moment, these are not the only internal sets. Also all infinite sets in $V(\mathbb{R})$ $(\subset V(^*\mathbb{R}))$ of cardinality $< \kappa$ are external.

The mapping $*$ identifies the idealization of the X's from $V(\mathbb{R})$ as *sets* with the correct set-theoretic relations to each other preserved. But $*$ does not embeds $(V(\mathbb{R}), \in)$ elementarily into $(V(^*\mathbb{R}), \in)$ because of the existence of external sets. For instance,

$$(V(\mathbb{R}), \in) \models \forall X \subset \mathbb{N} \; (\exists \text{ least element in } X \; w.r.t. \; <).$$

But in $V(\,^*\mathbb{R})$ the subset $(\,^*\mathbb{N}\setminus\mathbb{N}) \subset \,^*\mathbb{N}$ does not have a least element.

However, we get a weakened form of elementary embedding by requiring that quantifiers in the formulas we deal with come with a bounded form:

$$\forall x \in V_n(\mathbb{R}), \quad \exists x \in V_n(\mathbb{R}).$$

A formula in which all quantifiers are bounded in this manner is called a **bounded quantifier formula**.

Despite the restrictions, basically all ordinary mathematical properties are expressible by bounded quantifier formulas, because they take place in some $V_n(\mathbb{R})$.

1.2.3 *Two principles*

Modern nonstandard analysis is based on two fundamental principles. Here is the first one:

Transfer Principle:
Let $\phi(x_1,\ldots,x_n)$ be a bounded quantifier formula for the language $\mathcal{L}$ and $A_1,\ldots,A_n \in V(\mathbb{R})$.
Then $\phi(A_1,\ldots,A_n)$ holds in $V(\mathbb{R})$ iff $\phi(\,^*A_1,\ldots,\,^*A_n)$ holds in $V(\,^*\mathbb{R})$.

Consequently $*$ preserves all finitary set Boolean operations. For example, $^*(A \cup B) = \,^*A \cup \,^*B$, $^*(A \cap B) = \,^*A \cap \,^*B$, etc.

The transfer principle reflects the uniform elementary extension of all ordinary mathematical structures. Consider for example the real field structure $(\mathbb{R}, +, \cdot, 0, 1)$. Treat $+, \cdot$ as subsets of $\mathbb{R}^3$, then in $V(\,^*\mathbb{R})$ it has the elementary extension $(\,^*\mathbb{R}, \,^*+, \,^*\cdot, 0, 1)$. This is a true elementary extension for the field theory language (with two binary function symbols and two constant symbols) and properties of fields are inherited from the transfer applied to bounded formulas in the language of set theory. Here those properties are expressible by formulas using bounded quantifiers $\forall x \in \mathbb{R}$ and $\exists x \in \mathbb{R}$. Moreover, $(\,^*\mathbb{C}, \,^*+, \,^*\cdot, 0, 1)$ and $(\,^*\mathbb{R}, \,^*+, \,^*\cdot, \,^*< 0, 1)$ are simultaneously elementary extensions of $(\mathbb{C}, +, \cdot, 0, 1)$ and $(\mathbb{R}, +, \cdot, < 0, 1)$.

For convenience, from now on we fix the cardinal

$$\kappa = |V(\mathbb{R})|.$$

Hence $\kappa > |X|$ for any $X \in V(\mathbb{R})$. *i.e.* κ is greater than the cardinality of any ordinary mathematical object.

(Although in most circumstance $\kappa = \aleph_1$ is sufficient.)

Here is the second fundamental principle:

κ-Saturation Principle:
Let $\mathcal{F} \subset V({}^*\mathbb{R})$ be a collection of internal sets and $|\mathcal{F}| < \kappa$. If $\mathcal{F}$ has the finite intersection property (i.e. $\bigcap \mathcal{F}_0 \neq \emptyset$ for all finite $\mathcal{F}_0 \subset \mathcal{F}$) then $\bigcap \mathcal{F} \neq \emptyset$.

We leave it as an exercise to prove using the κ-saturation of $(\mathfrak{V}, \varepsilon)$ the above two fundamental principles of nonstandard analysis.

In nonstandard analysis, by transfer, we can lift a standard problem, solve it in $V({}^*\mathbb{R})$, and push the solution back to $V(\mathbb{R})$ and make it into a solution of the original problem. In the process, κ-saturation is needed to ensure enough set existence. In for example the nonstandard hull construction in the next chapter, κ-saturation gives the existence of a certain idealized space, then by taking a quotient *w.r.t.* certain properties, one obtains a new standard space. One can see the similarity between saturation and compactness in topology, we will explore this further in §1.6.

As a consequence of the transfer, we have the following useful tool for building internal sets:

Internal Definition Principle:
Let $A, C_1, \ldots, C_n$ be internal sets and $\phi(x_1, \ldots, x_n)$ be a first order formula in language $\mathcal{L}$. Then $\{x \in A, \mid (V({}^*\mathbb{R}), \in) \models \phi(A, C_1, \ldots, C_n)\}$ is internal.
In other words, internal sets satisfy the comprehension axiom.

For example, since ${}^*\mathbb{N}$ is internal and any $n \in {}^*\mathbb{N}$ is internal, so is the set $\{m \in {}^*\mathbb{N} \mid m > n\}$.

Most ordinary mathematical objects are external. (In fact all infinite ones are external.) For example, $\mathbb{N}$ is external. This can be seen by transfer only, but it follows immediately from κ-saturation: for if $\mathbb{N}$ were internal, then sets of the form $\{m \in \mathbb{N} \mid m > n\}$, $n \in \mathbb{N}$, would have been internal and have the finite intersection property. But the intersection of all such sets is obviously empty.

1.2.4 *Internal extensions*

The following simple result will be used frequently when extending an external sequence of internal sets to an internal sequence.

Proposition 1.1. (Internal Extension) *Let A be internal and $I \in V(\mathbb{R})$ with $|I| < \kappa$. Let $f : I \to {}^*\mathcal{P}(A)$, then there is an internal $F : {}^*I \to {}^*\mathcal{P}(A)$ such that $f = F\!\restriction_I$, i.e. the restriction of F on I.*

(Note that f is not necessarily an element of $V(\mathbb{R})$.)

Proof. For each $i \in I$, by the internal definition principle, the following is internal:

$$\mathcal{F}_i = \{\theta \,|\, \theta : {}^*I \to {}^*\mathcal{P}(A) \wedge \theta(i) = f(i)\}.$$

Moreover, by transfer, the $\mathcal{F}_i$'s satisfy the finite intersection property. Now let $F \in \bigcap_{i \in I} \mathcal{F}_i$ by κ-saturation. $\quad\square$

An internal set A is said to be **hyperfinite** if there is $N \in {}^*\mathbb{N}$ and an internal bijection between A and $[0, N] \cap {}^*\mathbb{N}$. We leave it as an exercise to prove the following from κ-saturation.

Lemma 1.1. (Hyperfinite Extension) *Let $\mathcal{F} \in V(\mathbb{R})$ equipped with a function $\theta : \mathcal{F} \to \mathbb{N} \cup \{\infty\}$. Assume that $A \in \mathcal{F}$ with $|A| < \kappa$ and for every finite $A_0 \subset A$ there is $A' \in \mathcal{F}$ satisfying $\theta(A') \in \mathbb{N}$ and $A_0 \subset A' \subset A$.*
Then there is $A' \in {}^*\mathcal{F}$ such that $\theta(A') \in {}^*\mathbb{N}$ and $A \subset A' \subset {}^*A$. $\quad\square$

The above lemma will be useful in situation where $\mathcal{F}$ represents certain class of objects (e.g. Boolean algebras, vector spaces, *etc.* in some $V_n(\mathbb{R})$) and θ represents some characteristic of those objects (e.g. the finite/infinite sizes, dimensions, *etc.*) Then under the given conditions, hyperfinite versions of some of those objects (e.g. hyperfinite Boolean algebras, hyperfinite dimensional vector spaces, *etc.*) can be found so that finite combinatorial properties become applicable.

A **directed set** $(I, \leq)$ is a partial ordering such that

$$\forall i, j \in I \, \exists k \in I \, \big(i \leq k \wedge j \leq k\big).$$

Finally we formulate one more principle which is often used—especially when the directed set is $(\mathbb{N}, \leq)$.

Lemma 1.2. *Let f be an internal function, $(I, \leq)$ a directed set with $|I| < \kappa$ and $S_i \subset \mathrm{Dom}(f)$, $i \in I$, nonempty internal subsets such that $S_i \supset S_j$ whenever $i \leq j$. Then*

$$f\Big[\bigcap_{i \in I} S_i\Big] = \bigcap_{i \in I} f[S_i].$$

Proof. $(\subset)$: trivial.

$(\supset)$: Let $y \in \bigcap_{i \in I} f[S_i]$.

Define for each nonempty finite $J \subset I$ the internal set

$$\mathcal{F}_J := \Big\{x \in \bigcap_{j \in J} S_j \,\big|\, f(x) = y \Big\}.$$

For such J, let $i \in I$ so that $j \leq i$ for all $j \in J$, then $y \in f[S_i] \subset f\left[\bigcap_{j \in J} S_j\right]$ hence $\mathcal{F}_J \neq \emptyset$. Moreover, $\left(\mathcal{F}_{J_1} \cap \mathcal{F}_{J_2}\right) \supset \mathcal{F}_{J_1 \cup J_2} \neq \emptyset$, therefore the $\mathcal{F}_J$'s have the finite intersection property.

By saturation, let $x \in \bigcap\{\mathcal{F}_J \mid \emptyset \neq J \subset I \text{ finite}\}$, then $x \in \bigcap_{i \in I} S_i$ and $y = f(x) \in f\left[\bigcap_{i \in I} S_i\right]$. □

Note that the similar statement $f\left[\bigcup_{i \in I} S_i\right] = \bigcup_{i \in I} f[S_i]$ holds always without any additional requirements.

From now on we fix a κ-saturated extension $(\mathfrak{V}, \varepsilon)$ of $(V(\mathbb{R}), \in)$ and work exclusively in $V(*\mathbb{R})$ with the injection $* : V(\mathbb{R}) \to V(*\mathbb{R})$.

$V(*\mathbb{R})$ is referred to as the **Nonstandard Universe**.

1.2.5 *Notes and exercises*

Nonstandard analysis has its origin in Leibniz's notion of infinitesimals and monads in calculus. The rigorous and general formulation however appeared only almost three hundred years later through A. Robinson's application of model theory for first order logic.

A detailed construction of the nonstandard universe can be found in [Chang and Keisler (1990)]. There are numerous and similar introductory books on nonstandard analysis. For a complete treatment with a wide range of applications, see [Albeverio *et al.* (1986)].

In order to specify a unique nonstandard universe, further conditions in addition to transfer and κ-saturation are necessary. Note that under ZFC there are arbitrarily large cardinals $\lambda > \kappa$ such that $\lambda = |\cup_{\alpha < \lambda} 2^\alpha|$. Let's fix such a λ. By a **special** model of cardinality λ we mean a model formed from the union of a chain of $|\alpha|$-saturated models, $\alpha < \lambda$, with each of them having cardinality $\leq \lambda$. By results in [Chang and Keisler (1990)] Chap. 5.1., there exists a special model of cardinality λ elementarily extending $(V(\mathbb{R}) \in)$. Trivially, such model is κ-saturated. What is more, up to isomorphism, it is the unique elementary extension which is a special model of cardinality λ.

Under the assumption of the existence of a large uncountable cardinal called the *inaccessible cardinal* and let λ be such cardinal, then λ satisfies the above conditions and the above special model is λ-saturated and is up to isomorphism the unique λ-saturated elementary extension of cardinality λ. However, even the relative consistency of the existence of such a cardinal is not provable in ZFC. (See [Jech (2003)] and [Kunen (1983)].)

There are alternative axiomatic approach to nonstandard analysis without the need of constructing nonstandard universe. Most notably is Nelson's *Internal Set Theory* ([Nelson (1977)]) which deals with a conservative extension of ZFC with a unary relation symbol added for standard objects. However, IST has not been developed enough and has not attained the level of strength as the nonstandard universe approach. See also [Kanovei and Reeken (2004)] for other axiomatic systems.

The nonstandard methods used in this book can be formulated in the setting of neometric spaces developed by Fajardo and Keisler ([Fajardo and Keisler (1996)]). Potentially, the neometric setting offers an alternative approach to nonstandard analysis with less demand on logic, however it has not received too much attention so far.

A comment about our terminology: Suppose $\mathcal{N}$ is the name of some $\mathcal{F} \subset V({}^*\mathbb{R})$ of mathematical structures specified using a finite language.

- If the structures with the name $\mathcal{N}$ are regarded as consisting of only one sort of elements, then internal sets from $\bigcup_{n \in \omega} {}^*(\mathcal{F} \cap V_n(\mathbb{R}))$ are still called $\mathcal{N}$. For example, groups, rings, fields belongs to such category. So $({}^*\mathbb{R}, {}^*+, 0)$ is still called a group.
- On the other hand, if the structures with the name $\mathcal{N}$ consist of more than one sort of elements in a natural way, internal sets from $\bigcup_{n \in \omega} {}^*(\mathcal{F} \cap V_n(\mathbb{R}))$ are called *internal* $\mathcal{N}$ or ${}^*\mathcal{N}$. For example, vector spaces consist of two sorts of elements: vectors and scalars from a field, and falls into this category. Hence, if $X \in V(\mathbb{R})$ is a vector space over $\mathbb{C}$, then ${}^*X$ is an internal vector space (or a *vector space) over the field ${}^*\mathbb{C}$. However, if restricted to scalar multiplication by scalars in $\mathbb{C}$, the resulted structure on ${}^*X$ is simply called a vector space instead and it would be an external structure in $V({}^*\mathbb{R})$ only. Other examples include normed linear spaces, Banach algebras, measure spaces, *etc.*

EXERCISES

(1) Show that $\mathfrak{V}$ contains sets on which ε is not well-founded, i.e. there are $X_n \in \mathfrak{V}$, $n < \omega$, such that $X_{n+1} \varepsilon X_n$. In particular, the whole $\mathfrak{V}$ cannot be identified with a superstructure, nor given a hierarchy similar to U. (See [Kanovei and Reeken (2004)] for an axiomatization of nonstandard analysis based on non-well-founded sets.)

(2) Investigate what other kind of portions of a superstructure is embedded into $\mathfrak{V}$.

(3) Let $X \in V(\mathbb{R})$ be infinite. Show that X is necessarily an external set.
(4) Verify the transfer and κ-saturation principles from the κ-saturated
$$(\mathfrak{V}, \varepsilon) \succ (V(\mathbb{R}), \in)$$
(5) Verify the internal definition principle and show that the set of internal subsets of a fixed internal set is closed under union, intersection and complement. *i.e.* it forms a Boolean algebra.
(6) Consider a cardinal $\lambda < \kappa$ and let C be internal. Show that if $A_\alpha \subset C$, $\alpha < \lambda$, are pairwise disjoint nonempty internal sets, then $\bigcup_{\alpha<\lambda} A_\alpha$ is external. Moreover, if A is internal and $A \subset \bigcup_{\alpha<\lambda} A_\alpha$, then there is $n \in \mathbb{N}$ and $\alpha_m < \lambda$, $m < n$, such that $A \subset \bigcup_{m<n} A_{\alpha_m}$
(7) Show that every internal subset of $^*\mathbb{N}$ has a least element.
(8) Prove Lem. 1.1.

1.3   The Ultraproduct Construction

Ultraproduct provides a direct but less poetic construction of the nonstandard universe.

A **filter** over a nonempty set I is some family $F \subset \mathcal{P}(I) \setminus \{\emptyset\}$ which is closed upward and closed under intersection. *i.e.*

- $\forall X \subset I\big((\exists Y \in F\,(Y \subset X)) \Rightarrow X \in F\big)$;
- $\forall X, Y \in F\,(X \cap Y) \in F$.

It is an **ultrafilter** if in addition

- $\forall S \subset I\,\big((S \in F) \Leftrightarrow ((I \setminus S) \notin F)\big)$.

Of course, each $n \in I$ generates an ultrafilter given by $\{S \subset I \mid n \in S\}$, called a principal ultrafilter.

A filter F over I is called a **nonprincipal** filter if

- $\forall n \in I\,(\{n\} \notin F)$.

One regards a filter as a notion of being a large subset and an ultrafilter leaves no room between being large and small.

Clearly an ultrafilter over a finite set is necessarily a principal ultrafilter. By the Fréchet filter over I we mean the filter

$$\{S \subset I \mid (I \setminus S) \text{ is finite}\}.$$

Given any $S \subset I$, one can easily construct a filter extending the Fréchet filter so that either S or $I \setminus S$ is an element. Using AC, one can even extend the Fréchet filter to an ultrafilter. Note the result is always a nonprincipal ultrafilter.

Viewed as a characteristic function on $\mathcal{P}(A)$, a nonprincipal ultrafilter over I corresponds to a finitely additive $\{0, 1\}$-measure over $\mathcal{P}(I)$, and vice versa.

Given an ultrafilter U over a set I and a family of sets X_i indexed by $i \in I$, we define for $f, g \in \prod_{i \in I} X_i$ that

$$f \sim_U g \quad \text{iff} \quad \{i \in I \mid f_i = g_i\} \in U.$$

We leave it as an exercise to show that $\sim_U$ forms an equivalence relation. Then define the **ultraproduct**

$$\prod_U X_i := \prod_{i \in I} X_i / \sim_U$$

as the quotient of $\prod_{i \in I} X_i$ w.r.t. the equivalence relation $\sim_U$. The equivalence classes in $\prod_U X_i$ are denoted by $[f]_U$.

When all $X_i = X$, for some X, we call $\prod_U X$ the **ultrapower** of X w.r.t. U. Furthermore, we identify $X \subset \prod_U X$ via the constant functions $f_i \equiv x$, for $x \in X$.

Trivially, if U is principal, then $\prod_U X_i$ is naturally identified with some X_i. So from now on we only work with nonprincipal ultrafilters U.

As models in the language of set theory, the ultraproduct $\prod_U (X_i, \in)$ is a model $\left(\prod_U X_i, \in_U \right)$, where $\in_U \subset \prod_U X_i \times \prod_U X_i$ is given by

$$[f]_U \in_U [g]_U \quad \text{iff} \quad \{i \in I \mid f_i \in g_i\} \in U.$$

It is not hard to check that $\in_U$ is well-defined and moreover, in the case of an ultrapower, we have:

Theorem 1.4. (Łoś' Theorem) $(X, \in) \prec \left(\prod_U X, \in_U \right)$. $\square$

Take for example a nonprincipal ultrafilter U over ω (any extension of the Fréchet filter to a ultrafilter will do). We take for the time being $\left(\prod_U V(\mathbb{R}), \in_U \right)$ as the $(\mathfrak{V}, \varepsilon)$ in the nonstandard universe construction. Then the transfer principle is satisfied, by the Łoś' Theorem.

One verifies that

$$^*\mathbb{R} = \left\{ [f]_U \in \prod_U V(\mathbb{R}) \,\middle|\, [f]_U \in_U \mathbb{R} \right\} = \left\{ [f]_U \,\middle|\, f \in \mathbb{R}^\omega \wedge \{n \mid f_n \in \mathbb{R}\} \in U \right\}$$

i.e. $^*\mathbb{R} = \prod_U \mathbb{R}$. In general, we have

$$^*X = \prod_U X, \quad \text{where } X \in V(\mathbb{R}).$$

Now let A_n, $n \in \mathbb{N}$, be countably many internal sets having the finite intersection property. Replacing the A_n by $\bigcap_{m \leq n} A_m$, we can assume that $A_n \supset A_{n+1}$. Let $A_0 \in {}^*V_p(\mathbb{R}) = \prod_U V_p(\mathbb{R})$ for some $p < \omega$. Represent each A_n as $[f_n]_U$ for some $f_n \in V_p(\mathbb{R})^\omega$. Define $I_0 := \omega$ and for $0 < n \in \mathbb{N}$,

$$I_n := \left\{ k \in \omega \,\middle|\, k > \min I_{n-1} \wedge \bigwedge_{m < n} \left(f_{m+1}(k) \subset f_m(k) \right) \right\}.$$

Then all $I_n \in U$, $I_n \supsetneq I_{n+1}$ and $\bigcap_{n \in \mathbb{N}} I_n = \emptyset$.

Let $f \in V_p(\mathbb{R})^\omega$ be given by $f(k) := f_n(k)$, where $k \in I_n \setminus I_{n+1}$. Then for each $n \in \mathbb{N}$, we have

$$\{k \mid f(k) \subset f_n(k)\} \supset I_n,$$

hence $[f]_U \subset A_n$. In particular, $\bigcap_{n \in \mathbb{N}} A_n \neq \emptyset$.

Therefore $\left(\prod_U V(\mathbb{R}), \in_U \right)$ satisfies the $\aleph_1$-saturation principle.

For κ-saturation, where $\kappa = |V(\mathbb{R})|$ as fixed in the previous section, we need some more sophisticated ultrafilter.

We say that an ultrafilter over I is **countably incomplete**, if there is a partition $\bigcup_{n \in \mathbb{N}} A_n = I$ such that $A_n \notin U$ for all $n \in \mathbb{N}$. Note that any nonprincipal ultrafilter over a countable set is countably incomplete. This property has been used in the above definition of f.

Let $\mathcal{F}$ denote the set of finite subsets of κ. We say that an ultrafilter U over κ is **good** if for any $\theta : \mathcal{F} \to U$ which is monotonic (*i.e.* $\theta(X) \subset \theta(Y)$ whenever $Y \subset X$) then there is $\phi : \mathcal{F} \to U$ which is additive (*i.e.* $\phi(X \cup Y) = \phi(X) \cap \phi(Y)$ always holds) such that $\phi(X) \subset \theta(X)$ for all $X \in U$.

It has been proved in ZFC the existence of a countably incomplete good ultrafilter over κ, and the ultrapower $\left(\prod_U V(\mathbb{R}), \in_U \right)$ is κ-saturated. This is a result due to Keisler, see [Chang and Keisler (1990)] Chap. 6.1.

Roughly speaking, given a family of fewer than κ internal sets, the finite intersection property is coded in $\mathcal{F}$ and the additive function is used to construct an element in the intersection.

Therefore the ultrapower method provides an alternative construction of the nonstandard universe satisfying both the transfer and κ-saturation principles.

We will see the use of ultrapower again in the next chapter.

1.3.1   *Notes and exercises*

Under ZF, the requirement that every filter extends to some ultrafilter is strictly weaker than AC ([Moore (1982)]).

Construction of a non-Archimedean ordered field similar to the ultrapower of $\mathbb{R}$ dates back as early as more than a half century ago in [Hewitt (1948)].

The measure corresponds to a nonprincipal ultrafilter over a countable set is never σ-additive. But over an uncountable set, the σ-additivity of such measure is equivalent to the existence of a large cardinal called the *measurable cardinal*, which is much larger than inaccessible cardinals. In particular the relative consistency with ZFC of such existence is not provable in ZFC. (See [Jech (2003)] and [Chang and Keisler (1990)].)

The ultrapower construction of the nonstandard universe is a special construction that avoids the direct use of model theoretic results such as the Completeness Theorem. However, there is a generalization of such construction called the *bounded limit ultrapower* and it can be proved ([Chang and Keisler (1990)] Thm. 6.4.17) that any nonstandard universe can be constructed by that method.

EXERCISES

(1) Use AC to show that every filter extends to an ultrafilter.
(2) Show that the measure corresponds to a nonprincipal ultrafilter over a countable set is never σ-additive.
(3) Let U be an ultrafilter over I, show that $\sim_U$ is an equivalence relation on $\prod_{i \in I} X_i$ and check that $\in_U$ is well-defined.
(4) Prove Łoś' Theorem $\left(X, \in \right) \prec \left(\prod_U X, \in_U \right)$ by induction on the complexity of the formulas.
(5) Suppose $(\mathfrak{V}, \varepsilon)$ is taken to be the ultrapower $\left(\prod_U V(\mathbb{R}), \in_U \right)$. Then show that $^*X = \prod_U X$ for all $X \in V(\mathbb{R})$.
(6) Identify the equivalence class of an infinite element in $\prod_U \mathbb{R}$.
(7) Given ultrafilters U_1 and U_2, is $\prod_{U_1} \left(\prod_{U_2} X \right)$ identifiable with $\prod_U X$ for some ultrafilter U?
(8) Consider a construction analogous to the ultraproduct where the ultrafilter is replaced by a filter. Investigate what sort of transfer is possible.

1.4   Application: Elementary Calculus

Analytic properties of the real number system are physically coded inside hyperreal numbers.

By the **hyperreal number system** we mean the nonstandard extension $^*\mathbb{R}$. For the ease of reading, common operation and relation symbols on $^*\mathbb{R}$, such as $^*+$, $^*\cdot$, $^*<$, ..., are simply written as $+$, $\cdot$, $<$, ... Note that, by transfer, $\left({}^*\mathbb{R}, +, \cdot, <, 0, 1 \right)$ forms an ordered field whose theory is expressed either as axioms in the language of an ordered field or as set-theoretic axioms about the given sets $+$, $\cdot$, $<$, 0, 1.

1.4.1   *Infinite, infinitesimals and the standard part*

Let $r \in {}^*\mathbb{R}$. Then we say r is

- **finite**, if $|r| < n$ for some $n \in \mathbb{N}$; (notation: $|r| < \infty$ or $r \in \mathrm{Fin}({}^*\mathbb{R})$);
- **infinite**, otherwise; (notation: $r \approx \pm\infty$ depending on $\mathrm{sgn}(r)$);
- **infinitesimal**, if $|r| < 1/n$ for all $0 \neq n \in \mathbb{N}$; (notation: $r \approx 0$.)

(Note the $|r|$ is written instead of $^*|r|$.) For $0 \neq r \in {}^*\mathbb{R}$ it follows from the transfer that $r \approx 0$ iff $|1/r| = \infty$ and so the only infinitesimal in $\mathbb{R}$ is 0.

Given $r, s \in {}^*\mathbb{R}$, we say that

- r and s are **infinitely close**, if $|r - s| \approx 0$; in notation: $r \approx s$.

Clearly, $\approx$ is an equivalence relation on $^*\mathbb{R}$, with equivalence classes

- $\mu(r) := \{ s \in {}^*\mathbb{R} \mid s \approx r \}$, the **monad** of r.

Proposition 1.2. *Let* $r \in \mathrm{Fin}({}^*\mathbb{R})$. *Then there is a unique* $s \in \mathbb{R}$ *such that* $r \approx s$.

Proof.   Define $s := \sup\{ u \in \mathbb{R} \mid u \leq r \}$.

By r being finite and the Monotone Convergence Theorem, $s \in \mathbb{R}$. If $s \not\approx r$, let $\epsilon \in \mathbb{R}$ such that $0 < \epsilon < |r - s|$. Then either $s + \epsilon < r$ or $s - \epsilon > r$, both cases contradict to the definition of s.

Now let $s_1, s_2 \in \mathbb{R}$ and $s_1 \approx r \approx s_2$, then $s_1 \approx s_2$, so $s_1 - s_2 = 0$, the only infinitesimal in $\mathbb{R}$. $\square$

By the proposition, the following is well-defined:

$^\circ : \mathrm{Fin}({}^*\mathbb{R}) \to \mathbb{R}$, where $^\circ r \in \mathbb{R}$ and $r \approx {}^\circ r$, the **standard part** of r.

Everything mentioned so far generalizes to any finite dimensional Euclidean space $\mathbb{R}^n$.

Observe that if $r \in {}^*\mathbb{R} \setminus \mathbb{R}$, then $(r - {}^\circ r)^{-1}$ is infinite. Therefore the hyperreals ${}^*\mathbb{R}$ contains infinite elements forming a **non-Archimedean field**.

Each $r \in \mathrm{Fin}({}^*\mathbb{R})$ has unique representation as $s + \epsilon$ for some $s \in \mathbb{R}$ and $\epsilon \approx 0$. The main features of ${}^*\mathbb{R}$ are summarized in Fig. 1.2.

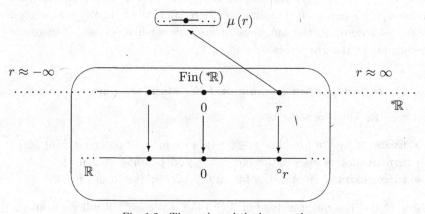

Fig. 1.2   The reals and the hyperreals.

Internal subsets of ${}^*\mathbb{R}$ behave like subsets of $\mathbb{R}$. Here is an example.

Proposition 1.3. *Let $\emptyset \neq A \subset {}^*\mathbb{R}$ be an internal set and has an upper bound in ${}^*\mathbb{R}$. Then A has a least upper bound.*

Proof.   Apply transfer to the formula

$$\forall X \subset \mathbb{R}\left[\left(\exists r \in \mathbb{R}\,(X \le r)\right) \Rightarrow \left(\exists s \in \mathbb{R}\,(X \le s \wedge \forall x < s\; X \not\le x)\right)\right],$$

where $X \le r$ abbreviates $\forall x \in X\,(x \le r)$. $\square$

1.4.2   *Overspill, underspill and limits*

The usage of the transfer often takes place in the following form.

Proposition 1.4. *Let $A \subset {}^*\mathbb{R}$ be an internal subset.*

(i) *(Overspill) Suppose A contains arbitrarily large positive finite numbers. Then A contains an infinite number.*

(ii) *(Underspill) Suppose A contains arbitrarily small positive non-infinitesimal numbers. Then A contains a positive infinitesimal.*

Proof. (i): Assume without loss of generality that A has an upper bound, so by Prop. 1.3 it has a least upper bound, say $r \in {}^*\mathbb{R}$. Then $r \approx \infty$. As r is least among such bounds and as there are infinite numbers less than r, A contains an infinite number.

(ii) follows from (i) by considering the internal set

$$\{r^{-1} \,|\, r \neq 0 \wedge r \in A\}. \qquad \square$$

Note that $\mu(0)$ is external as ${}^*\mathbb{R} \setminus \mu(0)$ is external by the underspill— since it contains arbitrarily small positive non-infinitesimal numbers but no infinitesimals. Translating by r, it is clear that $\mu(r)$ is also external for every $r \in {}^*\mathbb{R}$.

Now consider sequences. A *sequence* $\{a_n\}_{n\in\mathbb{N}}$ in $\mathbb{R}$ is just a function $a : \mathbb{N} \to \mathbb{R}$. So it extends to ${}^*a : {}^*\mathbb{N} \to {}^*\mathbb{R}$ *i.e.* the internal sequence $\{{}^*a(n)\}_{n\in {}^*\mathbb{N}}$. The latter is often written

either as $\{{}^*a_n\}_{n\in {}^*\mathbb{N}}$ or as $\{a_n\}_{n\in {}^*\mathbb{N}}$

depending on the emphasis or clarity in presentation.

Proposition 1.5. *Given a sequence $\{a_n\}_{n\in\mathbb{N}} \subset \mathbb{R}$, $\lim_{n\to\infty} a_n = a$ iff $a_N \approx a$ for all $N \in {}^*\mathbb{N} \setminus \mathbb{N}$.*

Proof. ($\Rightarrow$) : Suppose $\lim_{n\to\infty} a_n = a$ then there are $m_k \in \mathbb{N}$, $k \in \mathbb{N}$ such that $\forall n \in \mathbb{N} \left(n \geq m_k \Rightarrow |a_n - a| < 1/k\right)$.

Transfer this for each $k \in \mathbb{N}$, then for each $N \approx \infty$ $|a_N - a| < 1/k$, for all $k \in \mathbb{N}$; that is, $a_N \approx a$.

($\Leftarrow$) : Suppose $\lim_{n\to\infty} a_n \neq a$, then there is $\epsilon \in \mathbb{R}^+$ and increasing sequence $n_k \in \mathbb{N}$ so that $|a_{n_k} - a| > \epsilon$. So $\forall m \in \mathbb{N} \left(\exists n \geq m \; |a_n - a| > \epsilon\right)$. By transfer, it follows that for some $N \approx \infty$, $|a_N - a| > \epsilon$. $\qquad \square$

There is a similar statement for an internal sequence not necessarily of the form $\{{}^*a(n)\}_{n\in {}^*\mathbb{N}}$. The proof is similar.

Proposition 1.6. *Let $\{a_n\}_{n\in {}^*\mathbb{N}} \subset {}^*\mathbb{R}$ be internal.*
Then $\lim_{n\to\infty} {}^\circ a_n = a$ for some $a \in \mathbb{R}$ iff $a_N \approx a$ for all small $N \in$ ${}^*\mathbb{N} \setminus \mathbb{N}$. $\qquad \square$

(By "for all small $N \in {}^*\mathbb{N} \setminus \mathbb{N}$" we mean "there is some $M \in {}^*\mathbb{N} \setminus \mathbb{N}$, for all $N \in ({}^*\mathbb{N} \setminus \mathbb{N}) \cap [0, M]$.")

Intuitively, limit points of a sequence are elements having infinite indices.

Theorem 1.5. (Bolzano-Weierstrass Theorem) *Every bounded sequence in $\mathbb{R}$ has a convergent subsequence.*

Proof.   Let $\{a_n\}_{n\in\mathbb{N}} \subset \mathbb{R}$ be bounded. Therefore $\{a_n\}_{n\in {}^*\mathbb{N}} \subset \mathrm{Fin}({}^*\mathbb{R})$. Take any $N \in {}^*\mathbb{N} \setminus \mathbb{N}$. Let $a = {}^\circ a_N$. For each $m \in \mathbb{N}$, an application of transfer shows that $\exists n \in \mathbb{N}\left(|a_n - a| < m^{-1}\right)$. Fix any such n as n_m. Then the subsequence $\{a_{n_m}\}_{m\in\mathbb{N}}$ converges to a. $\square$

1.4.3   Infinitesimals and continuity

Let's turn our attention to functions $f : \mathbb{R} \to \mathbb{R}$. Similar to Prop. 1.5, we have

Proposition 1.7. *Let $r \in \mathbb{R}$. Then*
$\lim_{x\to r} f(x) = a$ *iff* $\forall x \in {}^*\mathbb{R}\left(x \approx r \Rightarrow {}^*f(x) \approx f(r)\right)$. $\square$

The set of *positive reals* is denoted by $\mathbb{R}^+$, *i.e.* $(0,\infty)$.

Proposition 1.8. *Let $r \in {}^*\mathbb{R}$. Then*

f *is continuous at* r *iff* $\forall s \in {}^*\mathbb{R}\left(s \approx r \Rightarrow {}^*f(r) \approx {}^*f(s)\right)$.

That is, ${}^*f\left[\mu(r)\right] \subset \mu({}^*f(r))$.

Proof. $(\Rightarrow)$: Let $s \approx r$. Let $\epsilon \in \mathbb{R}^+$. By continuity, there is $\delta \in \mathbb{R}^+$

$$\forall u \in \mathbb{R}\left(|u - r| < \delta \Rightarrow |f(u) - f(r)| < \epsilon\right).$$

But $|s - r| < \delta$, so by transferring the above, we have $|{}^*f(s) - {}^*f(r)| < \epsilon$. Because this holds for all $\epsilon \in \mathbb{R}^+$, it follows that ${}^*f(s) \approx {}^*f(r)$.

$(\Leftarrow)$: If f is not continuous at r, then for some $\epsilon \in \mathbb{R}^+$, for any $n \in \mathbb{N}$, there are $r_n \in \mathbb{R}$ such that $|r_n - r| < n^{-1}$ and $|f(r_n) - f(r)| > \epsilon$. So the following internal sets

$$\left\{u \in {}^*\mathbb{R} \mid |u - r| < n^{-1} \wedge |{}^*f(u) - {}^*f(r)| > \epsilon\right\}, \quad n \in \mathbb{N},$$

satisfy the finite intersection property. By κ-saturation (actually it suffices to use $\aleph_1$-saturation), let s be in the intersection, then $s \approx r$ and $|{}^*f(s) - {}^*f(r)| > \epsilon$, *i.e.* ${}^*f(s) \not\approx {}^*f(r)$. $\square$

In fact the proof of the above proposition requires the transfer principle only, we leave it as an exercise to demonstrate this.

The differential calculus was originally developed by Leibniz and Newton using infinitesimals. Now this is given rigorous meaning in nonstandard analysis.

Proposition 1.9. *Let $f : \mathbb{R} \to \mathbb{R}$ and $r \in \mathbb{R}$. Then f is differentiable at r iff there is $a \in \mathbb{R}$ such that*

$$\forall s \in {}^*\mathbb{R} \left[(s \approx r \wedge s \neq r) \Rightarrow \frac{{}^*f(s) - {}^*f(r)}{s - r} \approx a \right].$$

Proof. $(\Rightarrow)$: Suppose $f'(r) = a$. Let $s \approx r$ but $s \neq r$. As $|r - s| < \delta$ for each $\delta \in \mathbb{R}^+$, it follows by transferring the definition of differentiability that

$$\left| \frac{{}^*f(s) - {}^*f(r)}{s - r} - a \right| < \epsilon,$$

for each $\epsilon \in \mathbb{R}^+$. *i.e.* $\dfrac{{}^*f(s) - {}^*f(r)}{s - r} \approx a$.

$(\Leftarrow)$: If f is not differentiable at r, then for each $a \in \mathbb{R}$, there are $\epsilon > 0$ and $r_n \in \mathbb{R}$, $n \in \mathbb{N}$, such that $|r - r_n| < 1/n$ and

$$\left| \frac{f(r_n) - f(r)}{r_n - r} - a \right| \geq \epsilon.$$

The conclusion follows from either the saturation or the transfer, similar to the proof of Prop. 1.8. $\square$

In §1.6 topological notions such as openness, closeness and compactness will be given simple and intuitive definitions using nonstandard analysis. Here we consider these notions for $\mathbb{R}$.

Intuitively, a set is open if every point is cushioned inside the set—in fact by its monad.

Proposition 1.10. $S \subset \mathbb{R}$ *is open iff* $\mu(r) \subset {}^*S$ *for all* $r \in S$.

Proof. $(\Rightarrow)$: Let S be open and $r \in S$, so S includes an open interval containing r, *i.e* for some $\epsilon \in \mathbb{R}^+$, the interval $(r - \epsilon, r + \epsilon) \subset S$. Then by transfer, $\forall u \in {}^*\mathbb{R} \left(|u - r| < \epsilon \Rightarrow u \in {}^*S \right)$. In particular $\forall u \approx r \left(u \in {}^*S \right)$. That is, $\mu(r) \subset {}^*S$.

$(\Leftarrow)$: If S is not open, for some $r \in S$, there are $\epsilon_n \in \mathbb{R}^+$, ϵ_n decreasing to 0, so that $r + \epsilon_n \notin S$. Then the internal sets

$$\{x \in {}^*\mathbb{R} \mid x \notin {}^*S \wedge |x - r| < 1/n\}, \quad n \in \mathbb{N},$$

satisfy the finite intersection property. Let u be in the intersection, then $u \notin {}^*S$ but $u \approx r$, *i.e.* $\mu(r) \not\subset {}^*S$. $\square$

A subset is closed iff the complement is open, so the following dual is an easy consequence of the above proposition.

Corollary 1.1. $S \subset \mathbb{R}$ *is closed iff* ${}^\circ r \in S$ *for all* $r \in {}^*S \cap \mathrm{Fin}({}^*\mathbb{R})$. $\square$

Intuitively one thinks of compactness as being tight enough so that no true expansion is possible. This is made precise by the following characterization:

Corollary 1.2. *Let $C \subset \mathbb{R}$. Then C is compact iff $\forall r \in {}^*C \left({}^\circ r \in C \right)$.*

Proof.   Assuming the Heine-Borel Theorem (see Thm. 1.22), $C \subset \mathbb{R}$ is compact iff it is closed and bounded.

($\Rightarrow$) : Since ${}^*C \subset \mathrm{Fin}({}^*\mathbb{R})$, ${}^\circ r$ is defined for each $r \in {}^*C$. Therefore $\forall r \in {}^*C \left({}^\circ r \in C \right)$ by Cor. 1.1.

($\Leftarrow$) : If ${}^\circ r$ is defined for all r from the internal set ${}^*C$, then, by overspill, ${}^*C \subset \mathrm{Fin}({}^*\mathbb{R})$. Hence C is closed by Cor. 1.1 again.   □

Proposition 1.11. *If $f : \mathbb{R} \to \mathbb{R}$ is continuous and $C \subset \mathbb{R}$ is compact, then $f(C)$ is compact.*

Proof.   Let $r \in {}^*(f(C))$. Then $r = {}^*f(c)$ for some $c \in {}^*C$. Let $d = {}^\circ c$, then by Cor. 1.2, $d \in C$. Moreover, since $d \approx c$, ${}^*f(c) \approx {}^*f(d) = f(d)$, by Prop. 1.8. *i.e.* ${}^\circ r = {}^\circ ({}^*f(c)) = f(d) \in f(C)$, so $f(C)$ is compact by Cor. 1.2.   □

For internal sets A and B, we say that they have the same ***internal cardinality*** if there is an internal bijection $\theta : A \to B$. (Note that automatically $|A| = |B|$ as a possibly external cardinal number.)

Recall that an internal set A is ***hyperfinite***, if it has the same internal cardinality as $\{0, 1, \cdots, N\}$ for some $N \in {}^*\mathbb{N}$. Although finite sets are hyperfinite, for most of the time hyperfinite refers to *infinite but hyperfinite*.

Hyperfinite sets are quite useful for the discretization of a continuous object. Moreover, by transfer, finite combinatorial techniques are applicable in hyperfinite settings.

Let $N \in {}^*\mathbb{N} \setminus \mathbb{N}$. Write $\Delta t = N^{-1}$. Then define

$$\mathbb{T} := \left\{ n\Delta t \mid n = 0, 1, \ldots, N \right\}.$$

So $\mathbb{T}$ is a hyperfinite subset of ${}^*[0, 1]$. We also think of $\mathbb{T}$ as the unit discrete timeline consists of discrete time points 0, Δt, $2\Delta t$, $\ldots$, $N\Delta t = 1$. It is also called the ***hyperfinite timeline***.

Note that since $\mathbb{T} \subset {}^*\mathbb{Q}$ it follows from transfer that $\mathbb{T} \not\supseteq [0, 1]$. However, it is possible to choose $N \in {}^*\mathbb{N} \setminus \mathbb{N}$ so that $\mathbb{T}$ contains all rational numbers in $[0, 1]$.

In the following we fix an arbitrary $N \in {}^*\mathbb{N} \setminus \mathbb{N}$ and hence also the $\mathbb{T}$.

We leave it as an exercise to prove the following representation of an integral as a hyperfinite Riemann sum. As one can see, this is more in the spirit of the infinitesimal calculus of Leibniz.

Proposition 1.12. *Let $f : [0,1] \to \mathbb{R}$ be continuous. Then*

$$\int_0^1 f(x)dx = {}^{\circ} \sum_{n=0}^{N} {}^*f(n\Delta t)\,\Delta t.$$

$\square$

Now we combine the nonstandard characterization of continuity and the hyperfinite timeline in the following results.

Theorem 1.6. (Intermediate Value Theorem) *Let $f : [0,1] \to \mathbb{R}$ be continuous. Suppose that $f(0) \le 0 \le f(1)$, then $\exists r \in [0,1]\,\big(f(r) = 0\big)$.*

Proof. Consider the hyperfinite set $\{\,{}^*f(t) : t \in \mathbb{T}\}$. By transfer of properties of finiteness, there is a least $t \in \mathbb{T}$ so that ${}^*f(t) \le 0 \le {}^*f(t + \Delta t)$.
By continuity, we get ${}^*f(t) \approx {}^*f(t + \Delta t)$, so both have value ≈ 0.
Let $r = {}^{\circ}t$. By continuity again, $f(r) \approx {}^*f(t) \approx 0$, *i.e.* $f(r) = 0$. $\square$

Theorem 1.7. (Extreme Value Theorem) *Let $f : [0,1] \to \mathbb{R}$ be continuous, then f attains a maximum on $[0,1]$.*

Proof. Let $m = \max\{\,{}^*f(t) : t \in \mathbb{T}\}$. Then since $\mathbb{T}$ is hyperfinite, we get by transfer $m = {}^*f(t)$ for some $t \in \mathbb{T}$. Let $r = {}^{\circ}t$. By continuity, $f(r) = {}^{\circ}m$ and we claim that it is the maximum attained by f on $[0,1]$:
For each $s \in [0,1]$, take $u \in \mathbb{T}$ such that $u \approx s$. Then $m \ge {}^*f(u)$. By continuity, ${}^*f(u) \approx f(s)$, so ${}^{\circ}m \ge f(s)$. $\square$

We end this section with the following existence result. In contrast to the classical proofs, the following is direct and no reference to the Ascoli's Lemma is necessary.

Theorem 1.8. (Peano's Existence Theorem) *Let $f : \mathbb{R} \times [0,1] \to \mathbb{R}$ be bounded and continuous. Then for any given $y(0) = y_0 \in \mathbb{R}$, the differential equation*

$$\frac{dy}{dt} = f\big(y(t), t\big)$$

has a solution.

Proof.   We work with $\mathbb{T}$ instead of $[0,1]$. Define $Y : \mathbb{T} \to {}^*\mathbb{R}$ by hyperfinite iteration: (Replace dt by Δt.)

$$Y(k\Delta t) = y_0 + \sum_{i=0}^{k-1} {}^*f\big(Y(i\Delta t), i\Delta t\big)\Delta t.$$

Since f is bounded, Y has a finite bound.

By f being bounded, whenever $t_1 \approx t_2$ in $\mathbb{T}$, $Y(t_1) \approx Y(t_2)$.

Now we define $y : [0,1] \to \mathbb{R}$ by $y(t) = {}^\circ Y(\bar{t})$, where $\bar{t}$ is the point in $\mathbb{T}$ to the immediate left of $t \in [0,1]$. So y is a continuous function. By Prop. 1.12, we have

$$\int_0^t f\big(y(s), s\big)ds \approx \sum_{i=0}^{\bar{t}N} {}^*f\big({}^*y(i\Delta t), i\Delta t\big)\Delta t$$

$$\approx \sum_{i=0}^{\bar{t}N} {}^*f\big(Y(i\Delta t), i\Delta t\big)\Delta t \quad \text{(by continuity of } f\text{)}$$

$$\approx y(t).$$

So y is a solution. $\square$

However, in general, the above solution fails to be unique.

1.4.4   *Notes and exercises*

Elementary calculus was developed in the 17th century by Leibniz and Newton based on the notion of infinitesimals. However, Leibniz' theory of infinitesimals was not rigorous enough and so by the 19th century the use of infinitesimals was replaced by $\epsilon - \delta$-style rigorous treatments given by Cauchy and Weierstrass. Due to Robinson's effort in the 1960's, infinitesimals re-emerged in the rigorous theory of the nonstandard analysis. By 1970's attempts have been made to re-introduce infinitesimals in teaching elementary calculus, with the belief that educators and students may be attracted to its intuitive approach. See [Keisler (1986)].

In applied mathematics such as numerical analysis, mathematical physics, mathematical finance, it is often more useful to model using the hyperfinite timeline rather than using a real interval. The main reason is that a lot of phenomena are of discrete nature and yet it is not realistic to pre-fix a real increment. But a hyperfinite timeline does satisfy both requirements.

EXERCISES

(1) Show that $\mu(0)$ forms a ring, i.e. it is closed under $+$, $-$, $\cdot$.

(2) Construct $(\mathbb{R}, +, \cdot, 0, 1)$ as a quotient field from $({}^*\mathbb{Q}, +, \cdot, 0, 1)$.

(3) Show that ${}^*\mathbb{Q}$ is almost a real closed field in the sense that for any odd $n \in \mathbb{N}$ and integers $a_0, \cdots, a_n$, $a_n \neq 0$, there is $r \in {}^*\mathbb{Q}$ such that $a_n r^n + \cdots a_1 r + a_0 \approx 0$.

(4) If ${}^*\mathbb{R}$ is constructed by an ultrapower using a nonprincipal ultrafilter over ω, write down explicitly an infinite number and a nonzero infinitesimal number.

(5) Given an example of a polynomial of the form $p(x) = \sum_{n=1}^{N} a_n x^n$, where $a_1 \approx \infty$, $a_n \in {}^*\mathbb{R}$, $N \in {}^*\mathbb{N}$ such that $p(\epsilon) \approx 0$ for all large enough $\epsilon \approx 0$.

(6) Formulate and prove the corresponding overspill and underspill for internal subsets of ${}^*\mathbb{N}$.

(7) Show that $\left(1 + rN^{-1}\right)^N \approx \left(1 + rM^{-1}\right)^M \in \mathrm{Fin}({}^*\mathbb{R})$ for any $N, M \in {}^*\mathbb{N} \setminus \mathbb{N}$ and $r \in \mathrm{Fin}({}^*\mathbb{R})$. Moreover, let $f : \mathbb{R} \to \mathbb{R}$ be given by $f(x) := {}^{\circ}\left(1 + rN^{-1}\right)^N$, $N \in {}^*\mathbb{N} \setminus \mathbb{N}$, use the Binomial Theorem to show that $f(x + y) = f(x)f(y)$ for all $x, y \in \mathbb{R}$. (Of course, $f(x) = e^x$.)

(8) Let $N, M \in {}^*\mathbb{N} \setminus \mathbb{N}$ with $M^2/N \in \mathrm{Fin}({}^*\mathbb{R})$. Show that

$$\left(\frac{N^2}{N^2 - M^2}\right)^N \left(\frac{N - M}{N + M}\right)^M \approx e^{-\frac{M^2}{N}}.$$

(9) Let $\{a_n\}_{n \in \mathbb{N}} \subset \mathbb{R}$. Show that $a_N \approx a$ for $N \approx \infty$ iff $\{a_n\}_{n \in \mathbb{N}}$ contains a subsequence converging to a.

(10) Is there a chain of vector spaces $\{V_n \mid n < N\}$, for some $N \in {}^*\mathbb{N}$, such that $V_0 \neq V_N$ but $\forall n < N \left(V_n = V_{n+1}\right)$?

(11) Show that

$$\lim_{m \to \infty} \lim_{n \to \infty} a_{mn} = a \text{ iff } \forall M \approx \infty \exists K \approx \infty \forall N > K \left(a_{MN} \approx a\right).$$

Find a similar characterization for

$$\lim_{m \to \infty} \left(\lim_{n \to \infty} a_{mn}\right) = \lim_{n \to \infty} \left(\lim_{m \to \infty} a_{mn}\right).$$

(12) Prove the ($\Leftarrow$) direction in Prop. 1.8 using the overspill only.

(13) Show that $f : \mathbb{R} \to \mathbb{R}$ is uniformly continuous iff

$$\forall s, r \in {}^*\mathbb{R} \left(s \approx r \Rightarrow {}^*f(s) \approx {}^*f(r)\right).$$

(14) Find nonstandard characterizations of continuity and uniform continuity for functions f with open domain $\mathrm{Dom}(f) \subset \mathbb{R}$.

(15) Use nonstandard characterizations to prove that if $f : \mathbb{R} \to \mathbb{R}$ is differentiable at $r \in \mathbb{R}$ then it is continuous at r.

(16) Use nonstandard characterizations to prove that if f is continuous on a compact set C, then it is uniformly continuous on C.

(17) Find $N \in {}^*\mathbb{N}$ so that $\mathbb{T} \supset \mathbb{Q} \cap [0, 1]$, *i.e.* $\mathbb{T}$ contains all rational numbers in the unit interval.

(18) Use infinitesimals to prove the chain rule of differentiation.

(19) Prove Prop. 1.12. Use this to prove the Fundamental Theorem of Calculus.

(20) Let st : ${}^*\mathbb{R} \to \mathbb{R} \cup \{\pm\infty\}$ be given by $\mathrm{st}(r) = {}^\circ r$, if $r \in \mathrm{Fin}({}^*\mathbb{R})$, and $\mathrm{st}(r) = \mathrm{sgn}(r) \cdot \infty$ otherwise. Show that $\mathrm{st}[A]$ is closed for any internal $A \subset {}^*\mathbb{R}$.

1.5   Application: Measure Theory

Any nonstandard measure is convertible to an ordinary measure.
Measure algebras are given nonstandard recognition.

A measure is a function defined on some collection of sets and its definition
originated from the usual length function which assigns a total length to a
finite union of intervals from $\mathbb{R}$. Measure theory is not only an important
collection of tools applicable in functional analysis, it is also a major source
of examples investigated by functional analysts.

We first begin with some necessary precise definitions and terminologies.
For properties and definitions of the topological notions mentioned in this
section, confer § 1.6 if needed.

1.5.1   *Classical measures*

Given a set Ω, an *algebra* over Ω is some collection $\mathcal{B} \subset \mathcal{P}(\Omega)$ which is
closed under $\cup$, $\cap$ and complement in Ω. In particular, $\emptyset, \Omega \in \mathcal{B}$. So an
algebra $\mathcal{B}$ over a set forms a Boolean algebra *w.r.t.* the set operations. We
sometimes call such $\mathcal{B}$ a *set algebra*.

If $\mathcal{B}$ is also closed under countable union (equivalently, under countable
intersection), then $\mathcal{B}$ is called a σ-*algebra*.

If $\mathcal{C} \subset \mathcal{P}(\Omega)$, the algebra generated by $\mathcal{C}$ is obtained by iterating arbitrarily finitely many times the operations of finite intersection, finite union
and complement. The result is the intersection of all subalgebras of $\mathcal{P}(\omega)$
that include $\mathcal{C}$. Likewise, the σ-algebra generated by $\mathcal{C}$ is obtained by iterating arbitrarily finitely many times the operations of countable intersection,
countable union and complement. It coincides with the intersection of all
σ-subalgebras of $\mathcal{P}(\omega)$ that include $\mathcal{C}$.

The σ-algebra generated by $\mathcal{C}$ is denoted by $\sigma\mathcal{C}$.

Given either an algebra or a σ-algebra over Ω, the pair $(\Omega, \mathcal{B})$ is called
a *measurable space*.

With attention to a given measurable space $(\Omega, \mathcal{B})$, elements of $\mathcal{B}$ are
called *measurable subsets* of Ω.

Given measurable spaces $(\Omega_1, \mathcal{B}_1)$ and $(\Omega_2, \mathcal{B}_2)$, a function $f : \Omega_1 \to \Omega_2$
is called a *measurable function* if $f^{-1}[\mathcal{B}_2] \subset \mathcal{B}_1$, *i.e.* $f^{-1}[X] \in \mathcal{B}_1$ for
every $X \in \mathcal{B}_2$. We also say that f is $\mathcal{B}_1$-*measurable* for emphasis.

Let $\mathbb{F}$ temporarily denote any one of the following sets:

$$\{0,1\}, \quad [0,1], \quad [0,\infty), \quad [0,\infty], \quad \mathbb{R}, \quad \mathbb{R} \cup \{\infty\}, \quad \mathbb{C},$$

with the usual meaning attached to 0 and $+$.

Given a measurable space $(\Omega, \mathcal{B})$, a **finitely additive measure** on $\mathcal{B}$ is a function $\mu : \mathcal{B} \to \mathbb{F}$ such that

- $\mu(\emptyset) = 0$;
- $\forall X, Y \in \mathcal{B}\left((X \cap Y = \emptyset) \Rightarrow (\mu(X \cup Y) = \mu(X) + \mu(Y))\right)$.

We also say μ is a finitely additive measure on Ω if the reference to $\mathcal{B}$ is implicitly understood.

Note that if $\mathbb{F}$ does not include ∞, the first condition is redundant. Also, additivity holds for any finitely many disjoint sets from $\mathcal{B}$ by iterating the second condition.

That is, if $X_1, \ldots, X_n$ is a list of finitely many pairwise disjoint elements from $\mathcal{B}$, then

$$\mu\left(X_1 \cup \cdots \cup X_n\right) = \mu(X_1) + \cdots + \mu(X_n).$$

In the case that $\mathcal{B}$ is a σ-algebra, the term σ-**additive measure** is used for a finitely additive measure μ on $\mathcal{B}$ that satisfies the following countable additivity condition:

- $\mu\left(\bigcup_{n \in \mathbb{N}} A_n\right) = \sum_{n \in \mathbb{N}} \mu(X_n)$ for any pairwise disjoint $\{X_n\}_{n \in \mathbb{N}} \subset \mathcal{B}$.

Here it is required that $\sum_{n \in \mathbb{N}} \mu(X_n)$ converges absolutely if $\mu\left(\bigcup_{n \in \mathbb{N}} X_n\right)$ is finite. In the case $\mathbb{F} \subset \left(\mathbb{R} \cup \{\infty\}\right)$ and $\mu\left(\bigcup_{n \in \mathbb{N}} X_n\right) = \infty$, the countable additivity condition is only required when $\mu(X_n) \geq 0$ for all $n \in \mathbb{N}$.

In particular, the limit given by $\sum_{n \in \mathbb{N}} \mu(X_n)$ remains unchanged for any permutations of the X_n's.

By a **measure** on $\mathcal{B}$ we mean either a finitely additive measure or a σ-additive measure when the context makes it clear which is meant.

In either case, the triple $(\Omega, \mathcal{B}, \mu)$ is then called a **measure space**. Note the difference between a measurable space and a measure space. For emphasis, we may specifically mention either a **finitely additive** or σ-**additive measure space**.

Elements of $\mathcal{B}$ in a measure space $(\Omega, \mathcal{B}, \mu)$ are celled μ-**measurable**.

Let $(\Omega_1, \mathcal{B}_1, \mu_1)$ and $(\Omega_2, \mathcal{B}_2, \mu_2)$ be measure spaces. Then we call a function $f : \Omega_1 \to \Omega_2$ a **measure preserving function** if f is surjective and $f^{-1}[X] \in \mathcal{B}_1$ with $\mu_1\left(f^{-1}[X]\right) = \mu_2(X)$ for every $X \in \mathcal{B}_2$.

Let μ be a measure on Ω. The following is a list of common types of measures.

- If $\mathbb{F} = \{0, 1\}$, then μ is called a $\{0, 1\}$-*measure*, *i.e.* a *zero-one-measure*.
- If $\mathbb{F} = [0, 1]$, then μ is called a **probability measure**, or simply a **probability**, and $(\Omega, \mathcal{B}, \mu)$ is called a **probability space**.
- If $\mu(\Omega)$ is finite, then μ is called a **finite measure**, or a **bounded measure**. Otherwise, it is an **infinite measure** or an **unbounded measure**.
- If there are $\Omega_n \in \mathcal{B}$ such that $\mu(\Omega_n)$ is finite for all $n \in \mathbb{N}$ and $\lim_{n \to \infty} \mu(\Omega_n) = \mu(\Omega)$, then μ is called a σ-**finite measure**.
- If $\mathbb{F} \subset [0, \infty]$, then μ is called a **positive measure**.
- If $\mathbb{F} \subset \mathbb{R}$, then μ is called a **real-valued measure**, or a $\mathbb{R}$-**valued measure**.
- If $\mathbb{F} = \mathbb{C}$, then μ is called a **complex measure** or a $\mathbb{C}$-**valued measure**.
- Similarly, $\mathbb{F}$-**valued measures** refer to measures $\mu : \Omega \to \mathbb{F}$.
- If $\mathbb{F} \subset \mathbb{R} \cup \{\infty\}$, then μ is called a **signed measure**.

Given a σ-additive complex measure space $(\Omega, \mathcal{B}, \mu)$, the **total variation** of μ is the function $|\mu| : \mathcal{B} \to [0, \infty]$ given by

$$|\mu|(X) := \sup \sum_{n=1}^{\infty} |\mu(X_n)|, \quad X \in \mathcal{B},$$

where the supremum is over all **partitions** of X, *i.e.* disjoint sets $\{X_n\}_{n \in \mathbb{N}} \subset \mathcal{B}$ such that $X = \cup_{n \in \mathbb{N}} X_n$.

The resulted $|\mu|$ remains unchanged if only finite partitions are used in the definition.

It can be shown that the total variation of a σ-additive complex measure is a bounded positive measure. (See [Rudin (1987)].)

Given a σ-additive signed measure $\mu : \mathcal{B} \to (\mathbb{R} \cup \{\infty\})$ a classical result known as the **Hahn-Jordan Decomposition** shows that there are unique σ-additive measures $\mu_1 : \mathcal{B} \to [0, \infty]$ and $\mu_2 : \mathcal{B} \to [0, \infty)$ so that $\mu = \mu_1 - \mu_2$. (See [Rudin (1987)].)

For a σ-additive real measure space $(\Omega, \mathcal{B}, \mu)$, by regarding μ as complex-valued, it is easy to see that the Hahn-Jordan decomposition is given by $\mu_1 = \frac{1}{2}(|\mu| + \mu)$ and $\mu_2 = \frac{1}{2}(|\mu| - \mu)$.

A σ-additive complex measure μ naturally decomposed as $\mu = \mu_1 + i\mu_2$, where μ_1, μ_2 are the real and imaginary part of μ respectively. Clearly μ_1, μ_2 are real-valued measures, hence it follows from the Hahn-Jordan Decomposition that $\mu = (\mu_{11} - \mu_{12}) + i(\mu_{21} - \mu_{22})$ for some finite real-valued σ-additive measures $\mu_{11}, \mu_{12}, \mu_{21}, \mu_{22}$.

A σ-additive measure space $(\Omega, \mathcal{B}, \mu)$ is said to be **complete**, if whenever $Y \subset X \in \mathcal{B}$ is such that $\mu(X) = 0$, then $Y \in \mathcal{B}$. Of course, $\mu(Y) = 0$ in such case.

For $Y \subset \Omega$, if there is $X \in \mathcal{B}$ such that $Y \subset X$ and $\mu(X) = 0$, we say that Y is a **null set** or a μ-**null set**. Hence $(\Omega, \mathcal{B}, \mu)$ is complete if $\mathcal{B}$ include all null sets.

Given a σ-additive measure space $(\Omega, \mathcal{B}, \mu)$, where μ is a positive measure, let $\mathcal{N} := \{Y \subset \Omega \mid \exists X \in \mathcal{B}\,(\mu(X) = 0 \wedge Y \subset X)\}$ and let $\overline{\mathcal{B}}$ be the σ-algebra generated by $\mathcal{B} \cup \mathcal{N}$. Define

$$\bar{\mu} : \overline{\mathcal{B}} \to [0, \infty] \quad \text{by} \quad \bar{\mu}(X) := \inf\left\{\mu(Z) \mid \exists Z \in \mathcal{B}\,(Z \supset X)\right\}.$$

Then it is not hard to see that $\bar{\mu}$ is the unique σ-additive measure on $\overline{\mathcal{B}}$ extending μ and $(\Omega, \overline{\mathcal{B}}, \bar{\mu})$ is a complete σ-additive measure space.

By the above-mentioned Hahn-Jordan Decomposition results, any σ-additive measure space $(\Omega, \mathcal{B}, \mu)$, where μ is either a σ-additive signed measure or a complex measure, extends uniquely to a complete σ-additive measure space $(\Omega, \overline{\mathcal{B}}, \bar{\mu})$, called the **completion** of $(\Omega, \mathcal{B}, \mu)$. We also say that $\overline{\mathcal{B}}$ is the **completion** of $\mathcal{B}$ w.r.t. $\bar{\mu}$.

In many occasions, measures considered in functional analysis have a topological origin.

If Ω is a topological space, the σ-algebra generated by the class of open sets is called a **Borel algebra** over Ω, with its elements called **Borel subsets** of Ω. A σ-additive measure μ on the Borel algebra $\mathcal{B}$ over Ω, is called a **Borel measure** on Ω and $(\Omega, \mathcal{B}, \mu)$ a **Borel measure space**.

Let $(\Omega_1, \mathcal{B}_1, \mu_1), (\Omega_2, \mathcal{B}_2, \mu_2)$ be Borel measure spaces. Then any measurable function $f : \Omega_1 \to \Omega_2$ is called a **Borel-measurable function**. That is to say, $f^{-1}[X]$ is Borel whenever X is a Borel subset of Ω_2. Note that this is equivalent to $f^{-1}[U]$ being Borel for every open subset of Ω_2.

However, if $(\Omega, \mathcal{B}, \mu)$ is a measure space and X is a topological space, a function $f : \Omega \to X$ such that $f^{-1}[U] \in \mathcal{B}$ for every open $U \subset X$ is just called a μ-**measurable function**.

Note that in general a Borel measure algebra is not complete.

Let $(\Omega, \mathcal{B}, \mu)$ be a Borel measure space.

- μ is called **inner-regular** if

$$\mu(X) = \sup\{\mu(C) \mid C \subset X \text{ and } C \text{ is closed}\}, \quad X \in \mathcal{B}.$$

- μ is called **outer-regular** if

$$\mu(X) = \inf\{\mu(U) \mid U \supset X \text{ and } U \text{ is open}\}, \quad X \in \mathcal{B}.$$

- μ is called **regular** if it is both inner-regular and outer-regular.
- μ is called a **Radon measure** if it is a completion of a Borel measure and for any μ-measurable X,

$$\mu(X) = \sup\{\mu(C) \,|\, C \subset X \text{ and } C \text{ is compact}\}$$
$$= \inf\{\mu(U) \,|\, U \supset X \text{ and } U \text{ is open}\}.$$

Observe that if a Borel measure is finite, then inner-regularity is equivalent to outer-regularity.

If Ω is a locally compact Hausdorff space and is a countable union of compact sets, then every Borel measure on Ω which is finite on compact sets is regular. Many of the spaces that we will study do satisfy this property and is also metrizable.

For each $n \in \mathbb{N}$, the Lebesgue measure on $\mathbb{R}^n$ is an example of a Radon measure which is the completion of a Borel measure.

1.5.2 *Internal measures and Loeb measures*

First recall the remarks on p.17) concerning the use of $*$.

By a ***finitely additive measure*** , we mean an internal function

$$\mu : \mathcal{B} \to {}^*\mathbb{F}$$

where $\mathbb{F} = \{0,1\}$, $[0,1]$, $[0,\infty)$, $[0,\infty]$, $\mathbb{R}$, $\mathbb{R}\cup\{\infty\}$ or $\mathbb{C}$, and where for some internal Ω, the triple $(\Omega, \mathcal{B}, \mu)$ forms an $*$measure space, *i.e.* an ***internal measure space***.

By transfer, μ is **hyperfinitely additive**. This means that for any hyperfinite internal sequence $\{A_n\}_{0 \le n < N} \subset \mathbb{B}$, where $N \in {}^*\mathbb{N}$,

$$\Big(\forall n < m < N \,(A_n \cap A_m = \emptyset)\Big) \Rightarrow \Big(\mu\big(\bigcup_{n<N} A_n\big) = \sum_{n<N} \mu(A_n)\Big).$$

Observe that the latter is a hyperfinite sum.

- In an internal measure space $(\Omega, \mathcal{B}, \mu)$, μ is called a ***finite internal measure*** or a ***bounded internal measure*** if $|\mu(\Omega)| < \infty$.

Given a hyperfinite nonempty internal set Ω, the probability measure μ on Ω that assigns measure $|\Omega|^{-1}$ to each singleton $\{\omega\}, \omega \in \Omega$, is called the ***normalized counting measure***, or the ***counting probability*** on Ω. (Recall that $|\Omega|$ denotes the internal cardinality of Ω.)

It follows then that

$$\mu(X) := \frac{|X|}{|\Omega|} \quad \text{for every } X \in {}^*\mathcal{P}(\Omega).$$

Nonstandard Methods in Functional Analysis

Consider the hyperfinite timeline $\mathbb{T}$ from §1.4. For convenience, as $|\mathbb{T}| = N + 1$, we deviate from the above definition in a non-essential way by referring to the counting probability on $\mathbb{T}$ as the probability measure μ such that $\mu(\{0\}) = 0$ and is the normalized counting measure on $\mathbb{T} \setminus \{0\}$.

In Prop. 1.12, the representation of an integral as a hyperfinite sum, suppose we take f to be the **indicator function** χ_S where $S \subset [0,1]$ is a Lebesgue measurable subset, then

$$\text{Leb}(S) = \int_0^1 \chi_S(x)dx = \overset{\circ}{\sum_{n=0}^{N}} {}^*\chi_S(n\Delta t)\,\Delta t \approx \mu({}^*S \cap \mathbb{T}),$$

where μ is the counting probability on $\mathbb{T}$. Of course, χ_S need not be continuous, but we shall make this connection precise.

Let $(\Omega, \mathcal{B}, \mu)$ be an internal probability space. Define

$$\,^\circ\mu : \mathcal{B} \to [0,1],$$

where, for any $A \in \mathcal{B}$, set $\,^\circ\mu(A) := \,^\circ\big(\mu(A)\big)$, the standard part of the hyperreal $\mu(A)$, then $(\Omega, \mathcal{B}, \,^\circ\mu)$ is clearly a finitely additive probability space. In applications, one often encounters sets in $\sigma\mathcal{B}$, so an extension of $\,^\circ\mu$ to a σ-additive probability on $\sigma\mathcal{B}$ is desired. (Except in trivial cases, $\mathcal{B}$ is never a σ-algebra, see exercises in §1.2.)

Before we construct the σ-additive extension, we first define the **inner** and **outer measure** of $\,^\circ\mu$: let $A \subset \Omega$ be *any* subset, then

$$\underline{\mu}(A) := \sup\{\,^\circ\mu(X) : X \in \mathcal{B} \wedge X \subset A\} \text{ and}$$
$$\overline{\mu}(A) := \inf\{\,^\circ\mu(X) : X \in \mathcal{B} \wedge X \supset A\}.$$

It is easily seen that $\underline{\mu}(A) \leq \overline{\mu}(A)$ holds for any $A \subset \Omega$.

Define $L(\mathcal{B}) := \{A \subset \Omega \,|\, \underline{\mu}(A) = \overline{\mu}(A)\}$. We also define

$$L(\mu) : L(\mathcal{B}) \to [0,1] \quad \text{by} \quad L(\mu)(A) := \underline{\mu}(A) \quad \big(= \overline{\mu}(A)\big).$$

For sets A and B, we write

$$A \triangle B \quad \text{for} \quad (A \setminus B) \cup (B \setminus A),$$

the **symmetric difference**. (Here $A \setminus B$ denotes $\{x \in A \,|\, x \notin B\}$.)

Theorem 1.9. *Let $(\Omega, \mathcal{B}, \mu)$ be an internal probability space with the corresponding $(\Omega, L(\mathcal{B}), L(\mu))$ given as above. Then*

(i) *$L(\mathcal{B})$ is a σ-algebra extending $\mathcal{B}$.*
(ii) *$(\Omega, L(\mathcal{B}), L(\mu))$ forms a σ-additive probability space.*
(iii) *$L(\mathcal{B})$ is the completion of $\sigma\mathcal{B}$ w.r.t $L(\mu)$.*

(iv) (Internal Approximation) $\forall A \in L(\mathcal{B}) \exists B \in \mathcal{B} \left(L(\mu)(A \triangle B) = 0\right)$.

Proof. (i): It is clear that $\mathcal{B} \subset L(\mathcal{B})$.

It is also clear that $A \in L(\mathcal{B})$ iff $(\Omega \setminus A) \in L(\mathcal{B})$. To show that $L(\mathcal{B})$ forms a σ-algebra, it suffices to check that it is a monotone class. (See [Ash (2000)].) Let $A_n \in L(\mathcal{B})$ be increasing and $A = \bigcup_{n \in \mathbb{N}} A_n$. Let $\epsilon \in \mathbb{R}^+$, then there are $C_n, D_n \in \mathcal{B}$ with

$$C_n \subset A_n \subset D_n \quad \text{and} \quad \mu\big(D_n \setminus C_n\big) < \frac{\epsilon}{2^n}, \quad n \in \mathbb{N}.$$

We can assume that the C_n's are increasing.

Let $r := \sup_{n \in \mathbb{N}} {}^{\circ}\mu(C_n)$. Then $\underline{\mu}(A) \geq r$. We now show that $r \geq \overline{\mu}(A)$.

By internal extension Prop. 1.1 and by Prop. 1.6, let $N \in {}^*\mathbb{N} \setminus \mathbb{N}$ such that $r \approx \mu(C_N)$. Then it holds for *any* $m \in \mathbb{N}$ that

$$\mu\Big(\big(\bigcup_{n<m} D_n\big) \setminus C_N \Big) \leq \mu\big(\bigcup_{n<m} (D_n \setminus C_n)\big) \leq \sum_{n=0}^{m} \frac{\epsilon}{2^n} \leq 2\epsilon.$$

By the overspill and the $\bigcup_{n<m} D_n$, C_N being internal,

$$\mu\Big(\big(\bigcup_{n<M} D_n\big) \setminus C_N \Big) \leq 2\epsilon \quad \text{for some } M \in {}^*\mathbb{N}.$$

Hence ${}^*\mu(\bigcup_{n<M} D_n) \leq r + 2\epsilon$. Since $\epsilon \in \mathbb{R}^+$ is arbitrary and $A \subset \bigcup_{n<M} D_n \in \mathcal{B}$, we have $\overline{\mu}(A) \leq r$.

Therefore $\underline{\mu}(A) = r = \overline{\mu}(A)$, *i.e.* $A \in L(\mathcal{B})$. So we conclude that $L(\mathcal{B})$ forms a σ-algebra.

(iv): Let $A \in L(\mathcal{B})$. So there are $C_n, D_n \in \mathcal{B}$ such that $C_n \subset A \subset D_n$ with $\mu(D_n \setminus C_n) \approx 0$. Let $C = \bigcup_{n \in \mathbb{N}} C_n$ and $D = \bigcap_{n \in \mathbb{N}} D_n$. By saturation, there is $B \in \mathcal{B}$ such that $C \subset B \subset D$.

In particular, $L(\mu)(A \triangle B) = 0$, *i.e.* A is approximated by some $B \in \mathcal{B}$ under $L(\mu)$.

(ii): Clearly $L(\mu)$ extends μ. So it suffices to show that $L(\mu)$ is σ-additive.

First let $A_n \in L(\mathcal{B})$ with $L(\mu)(A_n) = 0$, $n \in \mathbb{N}$. Let $\epsilon \in \mathbb{R}^+$ and $D_n \in \mathcal{B}$ so that $A_n \subset D_n$ and $\mu(D_n) \leq 2^{-n}\epsilon$. Extend the D_n to an internal sequence by Prop. 1.1, then for small $N \in {}^*\mathbb{N} \setminus \mathbb{N}$, we have

$$\bigcup_{n \in \mathbb{N}} A_n \subset \bigcup_{n<N} D_n \quad \text{and} \quad \mu(\bigcup_{n<N} D_n) \leq 2\epsilon.$$

Therefore $L(\mu)\big(\bigcup_{n \in \mathbb{N}} A_n \big) = 0$. *i.e.* the countable union of null sets is again a null set under $L(\mu)$.

40   *Nonstandard Methods in Functional Analysis*

Now consider disjoint $A_n \in L(\mathcal{B})$, $n \in \mathbb{N}$. Write $A = \bigcup_{n \in \mathbb{N}} A_n$. By (iv), there are B, $B_n \in \mathcal{B}$ so that

$$L(\mu)(A \triangle B) = 0 \quad \text{and} \quad L(\mu)(A_n \triangle B_n) = 0, \ n \in \mathbb{N}.$$

Replacing B_n by $(B_n \cap B) \setminus \bigcup_{m < n} B_m$, we can assume that the B_n's are disjoint subsets of B.

Let $\epsilon \in \mathbb{R}^+$, then by saturation there is $D \in \mathcal{B}$ such that

$$\bigcup_{n \in \mathbb{N}} B_n \subset D \subset B \quad \text{and} \quad \mu(D) \leq \epsilon + \sum_{n \in \mathbb{N}} {}^\circ\mu(B_n).$$

Since this holds for any $\epsilon \in \mathbb{R}^+$,

$$L(\mu)\Big(\bigcup_{n \in \mathbb{N}} B_n \Big) \leq \sum_{n \in \mathbb{N}} {}^\circ\mu(B_n) \leq L(\mu)(B) = L(\mu)\Big(\bigcup_{n \in \mathbb{N}} A_n \Big). \tag{1.2}$$

But ${}^\circ\mu(B_n) = L(\mu)(A_n)$ and $L(\mu)\Big(\bigcup_{n \in \mathbb{N}} B_n \triangle \bigcup_{n \in \mathbb{N}} A_n \Big) = 0$, so by what was just proved about countable union of null sets, we obtain from (1.2)

$$L(\mu)\Big(\bigcup_{n \in \mathbb{N}} A_n \Big) \leq \sum_{n \in \mathbb{N}} L(\mu)(A_n) \leq L(\mu)\Big(\bigcup_{n \in \mathbb{N}} A_n \Big).$$

Therefore $L(\mu)$ is σ-additive.

(iii): Let $A \subset \Omega$ be such that $A \subset D$ and $L(\mu)(D) = 0$ for some $D \in L(\mathcal{B})$. But $L(\mu)(D) = \overline{\mu}(D)$, hence $\overline{\mu}(A) = 0$, implying $\underline{\mu}(A) = \overline{\mu}(A) = 0$. Therefore $A \in L(\mathcal{B})$.

Moreover, by (iv), each element in $L(\mathcal{B})$ is approximated within measure zero by an element of $\mathcal{B}$, so $L(\mathcal{B})$ is the least algebra extending $\mathcal{B}$ that contains all null sets, *i.e.* $L(\mathcal{B})$ is the completion of $\mathcal{B}$ w.r.t. $L(\mu)$. $\square$

The measure constructed in Thm. 1.9 was first given by P. Loeb in [Loeb (1975)], hence $L(\mathcal{B})$ is called the **Loeb algebra** of $\mathcal{B}$, $L(\mu)$ the **Loeb measure** of μ and $\big(\Omega, L(\mathcal{B}), L(\mu)\big)$ the **Loeb space**.

There is an alternative Loeb construction utilizing the fact that $(\Omega, \mathcal{B}, {}^\circ\mu)$ satisfies the Carathéodory's criteria, *i.e.* if $\{A_n \,|\, n \in \mathbb{N}\} \subset \mathcal{B}$ is decreasing to $\emptyset$, then $\lim_{n \to \infty} {}^\circ\mu(A_n) = 0$. This is the case because the A_n's are internal so

$$\Big[\forall n \in \mathbb{N}\, (A_n \supset A_{n+1}) \wedge \bigcap_{n \in \mathbb{N}} A_n = \emptyset \Big] \Rightarrow \Big[\exists n \in \mathbb{N}\, (A_n = \emptyset) \Big],$$

by saturation. Therefore, by the Carathéodory's Extension Theorem (see [Ash (2000)]), ${}^\circ\mu$ has a unique σ-additive extension $L_0(\mu) : \sigma\mathcal{B} \to [0, 1]$. Let $\mathcal{N}$ be the collection of null sets, *i.e.*

$$\mathcal{N} := \big\{ X \subset \Omega \,\big|\, \exists Y \in \sigma\mathcal{B}\, \big((L_0(\mu)(Y) = 0) \wedge (X \subset Y) \big) \big\}.$$

Then $L(\mathcal{B})$ is just $\sigma(\mathcal{B} \cup \mathcal{N})$, the σ-algebra generated by $\mathcal{B} \cup \mathcal{N}$, and $L(\mu)$ is just the unique extension of $L_0(\mu)$ on $L(\mathcal{B})$.

In either constructions, we obtain:

Theorem 1.10. Let $(\Omega, \mathcal{B}, \mu)$ be an internal probability space, then $^\circ\mu$ extends uniquely to a σ-additive probability measure on $\sigma\mathcal{B}$ with its completion given by the Loeb space $(\Omega, L(\mathcal{B}), L(\mu))$. □

In [Loeb (1975)] the Carathéodory's Extension Theorem for an unbounded internal positive measure μ was used to produce a σ-additive measure. In [Henson (1979)](*Cf.* also [Živaljević (1992)]) this extension was shown to be unique on the σ-algebra generated by the internal algebra on which μ is defined.

Given two internal measures μ, ν on the same internal algebra $\mathcal{B}$, we write $\mu \approx \nu$ if $\forall X \in \mathcal{B} \left(\mu(X) \approx \nu(X) \right)$.

We use Re and Im to denote the **real part** and **imaginary part** of a complex number. Hence given a complex measure μ, $\mathrm{Re}(\mu)$ and $\mathrm{Im}(\mu)$ denote the measures corresponding to restricting the values of μ to its real part and imaginary part respectively.

The following is an internal version of the Hahn-Jordan Decomposition.

Proposition 1.13. Let $(\Omega, \mathcal{B}, \mu)$ be a finite internal complex measure space. Then there are finite internal positive measures $\mu_{11}, \mu_{12}, \mu_{21}, \mu_{22}$ such that $\mu \approx (\mu_{11} - \mu_{12}) + i(\mu_{21} - \mu_{22})$.

Moreover, the decomposition

$$^\circ\mu = (\,^\circ\mu_{11} - \,^\circ\mu_{12}) + i(\,^\circ\mu_{21} - \,^\circ\mu_{22})$$

is unique.

Proof. Let $\mu_1 := \mathrm{Re}(\mu)$ and $\mu_2 := \mathrm{Im}(\mu)$, so μ_1, μ_2 are finite internal real-valued measures.

Since μ_1 is finite, for some $r \in \mathrm{Fin}(\,{}^*\mathbb{R})$, $\mu_1 : \mathcal{B} \to [-r, r]$.

By saturation, there is $A \in \mathcal{B}$ such that $\mu_1(A) \approx \inf_{X \in \mathcal{B}} \mu_1(X) \leq 0$.

Now define finite internal positive measures $\mu_{11}, \mu_{12} : \Omega \to [0, r]$ by

$$\mu_{11}(X) = \mu_1(X \setminus A) \quad \text{and} \quad \mu_{22}(X) = -\mu_1(X \cap A), \quad X \in \mathcal{B}.$$

Then $\mu_1 \approx \mu_{11} - \mu_{12}$.

Note that if μ'_{11}, μ'_{12} are similarly defined, then $(\mu_{11} - \mu_{12}) \approx (\mu'_{11} - \mu'_{12})$, hence the decomposition $^\circ\mu_1 = \,^\circ\mu_{11} - \,^\circ\mu_{12}$ is unique.

The decomposition of μ_2 is similar. □

Clearly the conclusion in Thm. 1.10 also holds for a finite internal positive measure space $(\Omega, \mathcal{B}, \mu)$ by scaling μ by a positive finite factor if necessary. Meanwhile, for a finite internal complex measure μ, Prop. 1.13 provides the unique decomposition $°\mu = (°\mu_{11} - °\mu_{12}) + i(°\mu_{21} - °\mu_{22})$. Therefore Thm. 1.10 generalizes to the case of finite internal complex measures. Together with the earlier remark on unbounded internal positive measures, the following generalization holds.

Theorem 1.11. *Let* $(\Omega, \mathcal{B}, \mu)$ *be an internal measure space, where* μ *is either a finite internal complex measure or an internal* $\mathbb{F}$*-valued measure, where* $\mathbb{F} = [r, \infty]$ *for some* $r \in \mathrm{Fin}(^*\mathbb{R})$. *Then* $°\mu$ *extends uniquely to a* σ*-additive measure on* $\sigma\mathcal{B}$. $\qquad\square$

In the $\mathbb{F}$-valued measure case above, if μ takes an infinite positive hyperreal values, we denote the resulted values of $°\mu$ by ∞. It is convenient to denote the completion of the above measure space as $(\Omega, L(\mathcal{B}), L(\mu))$.

1.5.3 *Lebesgue measure, probability and liftings*

Consider the internal counting probability $\mu : {}^*\mathcal{P}(\mathbb{T}) \to {}^*[0, 1]$. Then there is measure preserving mapping between $(\mathbb{T}, {}^*\mathcal{P}(\mathbb{T}), L(\mu))$ and the Lebesgue measure space on $[0, 1]$, as the following shows.

Theorem 1.12. *Let* $\mathrm{st} : \mathbb{T} \to [0, 1]$ *be the standard part mapping* $t \mapsto °t$. *Let* μ *be the internal counting probability on* $\mathbb{T}$.
Then for any Lebesgue measurable $A \subset [0, 1]$, $\mathrm{Leb}(A) = L(\mu)\big(\mathrm{st}^{-1}[A]\big)$.

Proof. First let $A \subset [0, 1]$ be a Borel subset.
  If $A = [r, s]$ for some $r, s \in \mathbb{Q} \cap [0, 1]$, then
$$\mathrm{st}^{-1}[A] = \mathrm{st}^{-1}[r, s] = \bigcap_{n \in \mathbb{N}} \big([r - n^{-1}, s + n^{-1}] \cap \mathbb{T}\big),$$
hence
$$L(\mu)\big(\mathrm{st}^{-1}[A]\big) = \lim_{n \to \infty} L(\mu)\Big([r - n^{-1}, s + n^{-1}] \cap \mathbb{T}\Big)$$
$$= \lim_{n \to \infty} (s - r + 2n^{-1}) = s - r = \mathrm{Leb}(A).$$
Now suppose $L(\mu)\big(\mathrm{st}^{-1}[A_n]\big) = \mathrm{Leb}(A_n)$, where $A_n \subset [0, 1]$, $n \in \mathbb{N}$, are Borel. Then
$$L(\mu)\Big(\mathrm{st}^{-1}\Big[\bigcap_{n \in \mathbb{N}} A_n\Big]\Big) = L(\mu)\Big(\bigcap_{n \in \mathbb{N}} \mathrm{st}^{-1}[A_n]\Big) = \lim_{n \to \infty} L(\mu)\big(\mathrm{st}^{-1}[A_n]\big)$$
$$= \lim_{n \to \infty} \mathrm{Leb}(A_n) = \mathrm{Leb}\Big(\bigcap_{n \in \mathbb{N}} A_n\Big).$$

Similarly, we have $L(\mu)\left(\mathrm{st}^{-1}\left[\bigcup_{n\in\mathbb{N}} A_n\right]\right) = \mathrm{Leb}\left(\bigcup_{n\in\mathbb{N}} A_n\right)$, therefore we have proved (using the Monotone Class Theorem or by transfinite induction on the complexity) that $L(\mu)(\mathrm{st}^{-1}[A]) = \mathrm{Leb}(A)$ for all Borel $A \subset [0,1]$.

Finally, let $A \subset [0,1]$ and $\mathrm{Leb}(A) = 0$. So for each $\epsilon \in \mathbb{R}^+$ there is Borel $A' \subset [0,1]$, such that $A \subset A'$ and $\mathrm{Leb}(A') \leq \epsilon$, then the outer measure $\overline{\mu}(A) \leq L(\mu)(A') = \mathrm{Leb}(A') \leq \epsilon$. Since $\epsilon \in \mathbb{R}^+$ is arbitrary, $L(\mu)(A) = 0$.

Since the algebra of Lebesgue measurable subsets of $[0,1]$ is the completion of the Borel subsets $w.r.t.$ the Lebesgue measure, we conclude that $L(\mu)(\mathrm{st}^{-1}[A]) = \mathrm{Leb}(A)$ for all Lebesgue measurable $A \subset [0,1]$. $\square$

Combing with Thm. 1.9(iv), we obtain the following.

Corollary 1.3. *Let $A \subset [0,1]$ be Lebesgue measurable.*

Then there is an internal $S \subset \mathbb{T}$ such that $L(\mu)\left(S \triangle \mathrm{st}^{-1}[A]\right) = 0$ and $\mathrm{Leb}(A) = L(\mu)\left(S\right) \approx \mu(S)$. $\square$

Given a measure space $(\Omega, \mathcal{B}, \mu)$, two μ-measurable functions f_1 and f_2 are said to be equal *almost everywhere* ($w.r.t.$ μ), if

$$\mu\left(\{x \in \Omega \mid f_1(x) \neq f_2(x)\}\right) = 0.$$

In notation: $f_1 = f_2$ a.e. μ.

We also write "$\cdots$ *a.e.* μ" in other similar circumstances.

Let $(\Omega, \mathcal{B}, \mu)$, be an internal probability space, then by Thm. 1.9 (iv), for every $S \in L(\mathcal{B})$, the indicator function χ_S is $L(\mu)$-measurable and there is an internal $F : \Omega \to {}^*\mathbb{R}$ which is μ-measurable and ${}^\circ F = \chi_S$ a.e. $L(\mu)$. (For an internal function $F : \Omega \to {}^*\mathbb{R}$ we let ${}^\circ F$ be the function $x \to {}^\circ F(x)$, if $F(x) \in \mathrm{Fin}({}^*\mathbb{R})$, and ${}^\circ F(x) = \mathrm{sgn}(F(x)) \cdot \infty$ otherwise.)

More generally, for an internal measure space $(\Omega, \mathcal{B}, \mu)$ of the types in Thm. 1.11, we say that an internal μ-measurable function $F : \Omega \to {}^*\mathbb{F}$ is a *lifting* of a $L(\mu)$-measurable $f : \Omega \to \mathbb{F}$ if $f = {}^\circ F$ a.e. $L(\mu)$.

(Notice that $\{x \in \Omega \mid f(x) \approx F(x)\} \in L(\mathcal{B})$.)

The following characterizes Loeb measurable functions. See [Stroyan and Bayod (1986)] for its original version.

Theorem 1.13. (Anderson's Theorem) *Let $(\Omega, \mathcal{B}, \mu)$ be a finite internal complex measure space. Let X be a Hausdorff space.*

(i) *Let $F : \Omega \to {}^*X$ be a μ-measurable internal function that lifts $f : \Omega \to X$. Then f is $L(\mu)$-measurable.*

(ii) *Suppose X has a countable basis. If $f : \Omega \to X$ is $L(\mu)$-measurable, then there is a μ-measurable internal function $F : \Omega \to {}^*X$ that lifts f.*

Proof.   First of all, by Prop. 1.13, we only need to deal with the case that μ is a finite internal positive measure. Scaling by a finite factor if needed, we can further assume that μ is an internal probability.

In the following, a nonstandard characterization of open sets is used. See Prop. 1.14 in § 1.6.1 if necessary, or consider only the case $X \subset [-\infty, \infty]^n$ for some $n \in \mathbb{N}$ and replace open sets by hypercubes having rational vertices.

(i): Let $F : \Omega \to {}^*X$ be a μ-measurable lifting of $f : \Omega \to X$. Define $\Omega_0 := \{x \in \Omega \mid f(x) \approx F(x)\}$. Then $L(\mu)(\Omega_0) = 1$.

Let $U \subset X$ be any open subset. Then

$$\forall x \in \Omega_0 \left(\big(f(x) \in U\big) \Leftrightarrow \big(F(x) \in {}^*U\big)\right).$$

Hence $\big(f^{-1}[U] \cap \Omega_0\big) = \big(F^{-1}[{}^*U] \cap \Omega_0\big) \in L(\mathcal{B})$.

By $L(\mu)(\Omega_0) = 1$, $f^{-1}[U] \in L(\mathcal{B})$. Therefore f is $L(\mu)$-measurable.

(ii): Let $\{U_n\}_{n \in \mathbb{N}}$ be a countable base of open sets generating the topology on X. For a $L(\mu)$-measurable $f : \Omega \to X$, we have $f^{-1}[U_n] \in L(\mathcal{B})$ for all $n \in \mathbb{N}$. By the Loeb construction, there are $A_{nm} \in \mathcal{B}$, $n, m \in \mathbb{N}$ such that

$$A_{nm} \subset f^{-1}[U_n] \quad \text{and} \quad L(\mu)\big(f^{-1}[U_n] \setminus A_{nm}\big) < m^{-1}.$$

Clearly, for any finitely many n's and m's, there is an internal $F : \Omega \to {}^*X$ which is μ-measurable and satisfies $F[A_{nm}] \subset {}^*U_n$ for all n, m from the finite list.

Therefore, by saturation, there is an internal μ-measurable $F : \Omega \to {}^*X$ such that $F[A_{nm}] \subset {}^*U_n$ for all $n, m \in \mathbb{N}$.

Then

$$\{x \in \Omega \mid f(x) \not\approx F(x)\} = \bigcup_{n \in \mathbb{N}} \{x \in \Omega \mid (f(x) \in U_n) \wedge (F(x) \notin {}^*U_n)\}$$

$$\subset \bigcup_{n \in \mathbb{N}} \left(f^{-1}[U_n] \setminus \big(\bigcup_{m \in \mathbb{N}} A_{nm} \big)\right).$$

By σ-additivity, the set in the last term is an $L(\mu)$-null set, so we conclude that $L(\mu)\big(\{x \in \Omega \mid f(x) \not\approx F(x)\}\big) = 0$, *i.e.* F lifts f. $\square$

Observe how the second-countability was used in the above proof to show that certain set has measure zero.

We mostly use Thm. 1.13 when X is separable and metrizable (hence second-countable) such as $X = \mathbb{R}, [-\infty, \infty], \mathbb{C}$ or $[-\infty, \infty]^n$, $n \in \mathbb{N}$.

As a consequence of both Thm. 1.12 and Thm. 1.13,

Corollary 1.4. *Let μ be the internal counting probability on $\mathbb{T}$. Then a function $f : [0,1] \to \mathbb{R}$ is Lebesgue measurable iff $f \circ \mathrm{st} : \mathbb{T} \to \mathbb{R}$ is $L(\mu)$- measurable iff $f \circ \mathrm{st}$ has a μ-measurable lifting.* $\square$

Likewise, Lebesgue measurable functions on $\mathbb{R}$ can be lifted to internal measurable functions *w.r.t.* some counting probability.

Recall that for a measure space $(\Omega, \mathcal{B}, \mu)$, $f : \Omega \to \mathbb{C}$ is a **simple measurable function** if it is of the form $\sum_{m=0}^{n} \alpha_m \chi_{A_m}$ where $n \in \mathbb{N}$ and $\alpha_m \in \mathbb{C}$, $A_m \in \mathcal{B}$ with $|\mu(A_m)| < \infty$. (The latter condition is imposed to avoid the situation of $\infty - \infty$ when μ takes infinite value.)

Then we define the integral as

$$\int f \, d\mu := \sum_{m=0}^{n} \alpha_m \mu(A_m).$$

For a signed measure μ and a μ-measurable $f : \Omega \to [0, \infty)$, we define the **Lebesgue integral** of f *w.r.t.* μ as

$$\int f \, d\mu := \sup \int \theta \, d\mu \in \mathbb{R} \cup \{\infty\},$$

where sup is taken over all positive simple functions θ dominated by f, *i.e.* over simple functions $0 \le \theta \le f$. For any μ-measurable $f : \Omega \to \mathbb{R}$, we say that f is **Lebesgue integrable** *w.r.t.* μ (or simply as μ-**integrable**) if $\int |f| \, d\mu < \infty$. In that case, we let

$$\int f \, d\mu := \int f^+ \, d\mu - \int (-f^-) \, d\mu,$$

where $f^+ = \max\{f, 0\}$ and $f^- = \min\{f, 0\}$.

If $f : \Omega \to \mathbb{C}$, we say that f is μ-**integrable** iff both the real part $\mathrm{Re}(f)$ and imaginary part $\mathrm{Im}(f)$ of f are μ-integrable. In this case we define

$$\int f \, d\mu := \int \mathrm{Re}(f) \, d\mu + \int \mathrm{Im}(f) \, d\mu.$$

This naturally generalizes to the case when μ is a σ-additive complex measure by using the Hahn-Jordan Decomposition.

Now consider an internal probability space $(\Omega, \mathcal{B}, \mu)$. Then for a $L(\mu)$- integrable $f : \Omega \to \mathbb{C}$, unless f is bounded, the lifting $F : \Omega \to {}^*\mathbb{C}$ given by Thm. 1.13 is not necessary μ-integrable. Even if it is, it is not necessary true that $\int f \, dL(\mu) \approx \int F \, d\mu$. The correct notion requires an additional condition.

Given a finite internal complex measure space $(\Omega, \mathcal{B}, \mu)$, a μ-measurable function $F : \Omega \to {}^*\mathbb{C}$ is called S-**integrable** if

$$\int |F| \, d\mu \in \mathrm{Fin}({}^*\mathbb{C}) \quad \text{and} \quad \forall A \in \mathcal{B} \left(\mu(A) \approx 0 \Rightarrow \int_A |F| \, d\mu \approx 0 \right).$$

(Here $\int_A f \, d\mu$ is defined as $\int f \cdot \chi_A \, d\mu$.)

The proof of the following is not hard and is left as an exercise.

Lemma 1.3. *Let $(\Omega, \mathcal{B}, \mu)$ be be a finite internal complex measure space. Suppose $F : \Omega \to {}^*\mathbb{C}$ is bounded (i.e. $\sup_{x \in \Omega} |F(x)| < \infty$) and lifts some $f : \Omega \to \mathbb{C}$. Then F is S-integrable and $\int F \, d\mu \approx \int f \, dL(\mu)$.* $\qquad\square$

Theorem 1.14. *Let $(\Omega, \mathcal{B}, \mu)$ be a finite complex measure space. Consider μ-measurable $f : \Omega \to \mathbb{C}$. Then*

(i) *f is $L(\mu)$-integrable iff f has an S-integrable lifting w.r.t μ.*
(ii) *For any S-integrable lifting F of f, $\int f \, dL(\mu) \approx \int F \, d\mu$.*

Proof. As in the proof of Thm. 1.13, we assume that $(\Omega, \mathcal{B}, \mu)$ is a probability space. By dealing with $\mathrm{Re}(f)$ and $\mathrm{Im}(f)$ separately, we further assume that $f : \Omega \to \mathbb{R}$.

(i): By Thm. 1.13, let $F : \Omega \to {}^*\mathbb{R}$ be μ-measurable and lift f. Clearly F^+, F^- are respectively μ-measurable liftings of f^+, f^-, so we assume without loss of generality that $f \geq 0$ and $F \geq 0$.

Write $f_n = \max\{f, n\}$, $n \in \mathbb{N}$, and $F_n = \max\{F, n\}$, $n \in {}^*\mathbb{N}$. Then by Lem. 1.3, each F_n, $n \in \mathbb{N}$, is an S-integrable lifting of f_n and

$$\forall n \in \mathbb{N} \left(\int F_n d\mu \approx \int f_n dL(\mu) \right).$$

Moreover, because F lifts f, for any $N \in {}^*\mathbb{N} \setminus \mathbb{N}$, F_N is a μ-measurable lifting of f.

($\Rightarrow$) : Assume that f is $L(\mu)$-integrable. Since simple functions are bounded, it follows then $\lim_{n \to \infty} \int |f_n - f| \, dL(\mu) = 0$. Hence

$$\lim_{n,m \to \infty} \int |f_n - f_m| dL(\mu) = 0 \quad \text{and} \quad \lim_{n,m \to \infty} {}^\circ\!\int |F_n - F_m| d\mu = 0.$$

Then for any small $N \in {}^*\mathbb{N} \setminus \mathbb{N}$, $\lim_{n \to \infty} {}^\circ \int (F_N - F_n) d\mu = 0$.

In particular, for such N,

$$\int F_N \, d\mu \approx \lim_{n \to \infty} {}^\circ\!\int F_n \, d\mu = \lim_{n \to \infty} \int f_n \, dL(\mu) = \int f \, dL(\mu) < \infty.$$

Now let $A \in L(\mathcal{B})$ with $\mu(A) \approx 0$. Since

$$\int_A F_N \, d\mu \leq \int_A \left(F_N - F_n \right) d\mu + \int_A F_n \, d\mu, \quad n \in \mathbb{N},$$

from $\lim_{n\to\infty} {}^\circ \int (F_N - F_n)\, d\mu = 0$ and, by $F_n \cdot \chi_A$ being an S-integrable lifting of $f_n \cdot \chi_A$, we have $\int_A F_n\, d\mu \approx \int_A f_n\, dL(\mu) = 0$, consequently $\int_A F_N\, d\mu \approx 0$.

Therefore F_N is an S-integrable lifting of f.

($\Leftarrow$) : Suppose the lifting F of f is S-integrable. (Still assuming that $F \geq 0$, because F is S-integrable iff both F^+ and F^- are.)

Since ${}^\circ F = f$ a.e. $L(\mu)$,

$$L(\mu)\big(\{x \in \Omega \mid F(x) > N \}\big) = 0 \quad \text{for all} \quad N \in {}^*\mathbb{N}\setminus\mathbb{N},$$

hence

$$\int |F - F_N|\, d\mu \leq \int_{\{F \geq N\}} |F|\, d\mu \approx 0 \quad \text{for all} \quad N \in {}^*\mathbb{N}\setminus\mathbb{N},$$

by the S-integrability of F.

So $\lim_{n\to\infty} {}^\circ \int (F - F_n)d\mu = 0$ and $\lim_{n\to\infty} {}^\circ \int F_n d\mu = {}^\circ \int F\, d\mu$.

By approximating the f_n's by simple functions, we have $\int f\, dL(\mu) = \lim_{n\to\infty} \int f_n\, dL(\mu)$.

Therefore

$$\int f\, dL(\mu) = \lim_{n\to\infty} {}^\circ\!\int F_n\, d\mu = {}^\circ\!\int F\, d\mu < \infty,$$

showing that f is $L(\mu)$-integrable.

(ii): Let F be any S-integrable lifting of f, then by (i), f is $L(\mu)$-integrable and the proof shows that $\int f\, dL(\mu) = {}^\circ\!\int F\, d\mu$.

Moreover, if G is another S-integrable lifting of f, then

$$\int G\, d\mu = \int_{\{F=G\}} F\, d\mu + \int_{\{F \neq G\}} G\, d\mu \approx \int F\, d\mu,$$

because $\mu(\{F \neq G\}) \approx 0$, and F and G are S-integrable. $\qquad\square$

Combining Thm. 1.12 and Thm. 1.14, one obtains

Corollary 1.5. *Let μ be the internal counting probability on $\mathbb{T}$. Then a function $f : [0,1] \to \mathbb{R}$ is Lebesgue integrable iff $f \circ$ st $: \mathbb{T} \to \mathbb{R}$ is $L(\mu)$-integrable iff $f \circ$ st has a S-integrable lifting w.r.t. μ.* $\qquad\square$

As a consequence, Prop. 1.12 is generalized: Lebesgue integrals over the $[0,1]$ are represented by hyperfinite sums. (Likewise for integrals over $\mathbb{C}$.)

For $p \in \mathbb{R}^+$, we say that $F : \Omega \to {}^*\mathbb{C}$ is SL^p if F^p is S-integrable. These are the internal counterparts of L^p-functions, and we will mention them again in § 2.6.2.

1.5.4 *Measure algebras and Kelley's Theorem*

We end this section with a characterization of measure algebras.

Recall that $\mathcal{B} = (\mathcal{B}, 0, 1, \wedge, \vee, \smallsetminus)$, is a **Boolean algebra** if the binary functions $\wedge, \vee, \smallsetminus$ satisfies rules analogous to $\cap, \cup, \backslash$ on subsets of some set Ω, with $0, 1$ playing the rôles of $\Omega, \emptyset$. In fact, by Stone's Theorem, any Boolean algebra is isomorphic to some set algebra over some set Ω.

Measures on a Boolean algebra $\mathcal{B}$ are defined in the same way as those on a set algebra. So a finitely additive probability on $\mathcal{B}$ is some $\mu : \mathcal{B} \to [0, 1]$ such that $\mu(1) = 1$ and

$$\forall a, b \in \mathcal{B} \left((a \wedge b = 0) \Rightarrow (\mu(a \vee b) = \mu(a) + \mu(b)) \right).$$

A Boolean algebra $\mathcal{B}$ is called a **measure algebra** if there is a finitely additive probability μ on it such that $\forall a \in \mathcal{B} \left(\mu(a) = 0 \Leftrightarrow a = 0 \right)$.

So given a probability space $(\Omega, \mathcal{B}, \mu)$, if we define an equivalence relation on $\mathcal{B}$ by $A \approx B$ if $\mu(A \triangle B) = 0$, then the quotient algebra $\mathcal{B}/\approx$ is a measure algebra. (It is easy to check that $\mathcal{B}/\approx$ forms a Boolean algebra, where on the equivalence classes, we define $[A] \wedge [B] := [A \cap B]$, $[A] \vee [B] := [A \cup B]$, etc.)

Clearly not all Boolean algebras $\mathcal{B}$ are measure algebras, so we need to find ways to identify them.

For $\mathcal{A} \subset \mathcal{B}$, we write

$$\mathcal{A}^n := \left\{ \sigma \mid \sigma : \{0, \ldots, n\} \to \mathcal{A} \right\}, \, n \in \mathbb{N},$$

and $\mathcal{A}^{<\omega} := \bigcup_{n \in \mathbb{N}} \mathcal{A}^n$. Then define

$$\widehat{\alpha}(\sigma) := \max \left\{ \frac{|I|}{n+1} \mid I \subset \{0, \ldots, n\}, \bigwedge_{i \in I} \sigma_i \neq 0 \right\}, \quad \text{where } \sigma \in \mathcal{B}^n, \, n \in \mathbb{N}.$$

Note that $\widehat{\alpha}(\sigma)$ is a characteristic of σ, as a listing with possible repetition, not just as a subset, and is invariant under permutations of σ.

Given $\mathcal{A} \subset \mathcal{B}$, the **intersection number** of $\mathcal{A}$ is defined as

$$\alpha(\mathcal{A}) := \inf \left\{ \widehat{\alpha}(\sigma) \mid \sigma \in \mathcal{A}^{<\omega} \right\}.$$

and the **measure number** of $\mathcal{A}$ as

$$\beta(\mathcal{A}) := \sup \left\{ r \in [0, 1] \mid \text{for some finitely additive probability } \mu \text{ on the} \right.$$
$$\left. \text{subalgebra of } \mathcal{B} \text{ generated by } \mathcal{A}, \, \forall a \in \mathcal{A} \left(\mu(a) \geq r \right) \right\}.$$

Note that $\beta(\mathcal{A}) = 0$ whenever $0 \in \mathcal{A}$.

Let $\mathcal{A} \subset \mathcal{B}$ be finite. Then the subalgebra of $\mathcal{B}$ generated by $\mathcal{A}$ can be identified with the algebra $\mathcal{P}(\{p_0, \ldots, p_k\})$, with $p_0, \ldots, p_k$ enumerating some finite set.

Let $\mathcal{A}$ be enumerated as $a_0, \ldots, a_n$ and form the following matrix

$$M := \left[m_{ij}\right]_{\substack{i=0,\ldots,k \\ j=0,\ldots,n}}, \quad \text{where } m_{ij} = \chi_{a_j}(p_i).$$

For $\sigma \in \mathcal{A}^{<\omega}$ of the form

$$\sigma = \left(\underbrace{a_0 \cdots a_0}_{h_0} \underbrace{a_1 \cdots a_1}_{h_1} \cdots \cdots \underbrace{a_n \cdots a_n}_{h_n} \right),$$

observe then

$$\widehat{\alpha}(\sigma) = \max_{0 \le i \le k} \sum_{j=0}^{n} \frac{h_j}{h} m_{ij}, \quad \text{where } h = \sum_{j=0}^{n} h_j.$$

Now let $\alpha \in \mathbb{R}$ be minimal such that

$$M \cdot \begin{bmatrix} x_0 \\ \vdots \\ x_n \end{bmatrix} \le \begin{bmatrix} \alpha \\ \vdots \\ \alpha \end{bmatrix}, \tag{1.3}$$

where $x_0, \ldots, x_n \in [0, \infty)$ with $x_0 + \cdots + x_n \ge 1$.

From linear programming (see [Gass (2003)]), we know that the solution is given by some extreme point $(\alpha_0, \ldots, \alpha_n, \alpha) \in [0, \infty)^{n+2}$ determined by a subset of the following hyperplanes:

$$x_0 + \cdots + x_n = 1,$$
$$m_{i0} + \cdots + m_{in} x_n = x_{n+1}, \quad i = 0, \ldots, k.$$

Since the coefficients $m_{ij} = 0, 1$, we further conclude that the extreme point has rational coordinates of the form

$$(\alpha_0, \ldots, \alpha_n, \alpha) = (h_0/h, \ldots, h_n/h, \alpha), \quad \text{for some } h_j \in \mathbb{N}, \sum_{j=0}^{n} h_j = h.$$

Now let $\sigma = \left(\underbrace{a_0 \cdots a_0}_{h_0} \underbrace{a_1 \cdots a_1}_{h_1} \cdots \cdots \underbrace{a_n \cdots a_n}_{h_n} \right)$, then $\widehat{\alpha}(\sigma) = \alpha$ and $\widehat{\alpha}(\tau) \ge \alpha$ for all $\tau \in \mathcal{A}^{<\omega}$.

That is, $\alpha(\mathcal{A}) = \alpha$ and the infimum value is attained.

Next we let $\beta \in \mathbb{R}$ be maximal such that

$$M^{\mathrm{T}} \cdot \begin{bmatrix} y_0 \\ \vdots \\ y_k \end{bmatrix} \ge \begin{bmatrix} \beta \\ \vdots \\ \beta \end{bmatrix}, \tag{1.4}$$

where $y_0, \ldots, y_k \in [0, \infty)$ with $y_0 + \cdots + y_k \leq 1$.

Similar to the above, the solution β corresponds to some extreme point $(\beta_0, \ldots, \beta_k, \beta) \in [0, \infty)^{k+2}$ having rational coordinates. Moreover, it represents a probability measure assigning weights $\beta_0, \ldots, \beta_k$ to the points $p_0, \ldots, p_k$. Conversely, any probability measure on the algebra generated by $\mathcal{A}$ with weights $\beta_0, \ldots, \beta_k$ assigned to the points $p_0, \ldots, p_k$ satisfies (1.4) for some $\beta \in \mathbb{R}$. Therefore $\beta(\mathcal{A}) = \beta$ and the supremum value is attained.

The problems (1.3) and (1.4) are dual to each other, so, by a theorem from linear programming, we conclude that $\alpha = \beta$. Hence we have proved the following:

Lemma 1.4. *Let $\mathcal{A} \subset \mathcal{B}$ be a finite subset. Then $\alpha(\mathcal{A}) = \beta(\mathcal{A})$ and both the infimum and supremum are attained.* $\qquad\square$

Then a combinatorial criteria for measure algebra is given as follows.

Theorem 1.15. (Kelley's Theorem) *A Boolean algebra $\mathcal{B}$ is a measure algebra iff*

$$\exists \{\mathcal{A}_n\}_{n \in \mathbb{N}} \subset \mathcal{P}(\mathcal{B}) \left((\forall n \in \mathbb{N} \, (\alpha(\mathcal{A}_n) > 0)) \wedge \mathcal{B} = \bigcup_{n \in \mathbb{N}} \mathcal{A}_n \cup \{0\} \right).$$

Proof. $(\Rightarrow)$: Let $\mu : \mathcal{B} \to [0, 1]$ be a probability measure so that $\mu(a) = 0$ iff $a = 0$. Let $\mathcal{A}_n = \{a \in \mathcal{B} \mid \mu(a) \geq n^{-1}\}$, $n \in \mathbb{N}$. Then for each $n \in \mathbb{N}$, $\beta(\mathcal{A}_n) > 0$, and, by Lem. 1.4,

$$\alpha(\mathcal{A}_n) = \inf \{\alpha(\mathcal{C}) \mid \mathcal{C} \subset \mathcal{A}_n \text{ is finite}\}$$
$$= \inf \{\beta(\mathcal{C}) \mid \mathcal{C} \subset \mathcal{A}_n \text{ is finite}\} \geq \beta(\mathcal{A}_n) > 0,$$

therefore the conclusion follows.

$(\Leftarrow)$: Suppose $\mathcal{B} = \bigcup_{n \in \mathbb{N}} \mathcal{A}_n \cup \{0\}$, with $\alpha(\mathcal{A}_n) > 0$. By saturation (more precisely, Lem. 1.1), there is a hyperfinite Boolean algebra $\mathcal{B}'$ such that $\mathcal{B} \subset \mathcal{B}' \subset {}^*\mathcal{B}$.

Let $n \in \mathbb{N}$ and $\mathcal{A}_n' := {}^*\mathcal{A}_n \cap \mathcal{B}'$. Then

$${}^*\beta(\mathcal{A}_n') = {}^*\alpha(\mathcal{A}_n') \approx \alpha(\mathcal{A}_n) \gtrapprox 0, \quad n \in \mathbb{N},$$

by transferring Lem. 1.4 and $\mathcal{A}_n \subset \mathcal{A}_n' \subset {}^*\mathcal{A}_n$.

Consequently, there are internal hyperfinitely additive probabilities γ_n on the algebra generated by $\mathcal{A}_n'$ such that $\gamma_n(a) \geq {}^*\beta(\mathcal{A}_n') \gtrapprox 0$, for each $a \in \mathcal{A}_n'$. Extend γ_n to a hyperfinitely additive probability ρ_n on $\mathcal{B}'$.

For each $n \in \mathbb{N}$, define

$$\mu_n := \sum_{m \in {}^*\mathbb{N}} 2^{-m-1} \nu_m, \quad \text{where } \nu_m = \rho_{\min\{m,n\}},$$

then μ_n is hyperfinitely additive probability on $\mathcal{B}'$ so that

$$\mu_n(a) \geq 2^{-m-1} \, {}^*\beta(\mathcal{A}'_m) \gtrapprox 0,$$

for each $a \in \mathcal{A}'_m$, $m \leq n$.

Extend $\{\mu_n \mid n \in \mathbb{N}\}$ to an internal sequence by saturation (Prop. 1.1) and fix $N \in {}^*\mathbb{N} \setminus \mathbb{N}$. Let $\mu : \mathcal{B} \to [0,1]$ be given by

$$\mu(a) := {}^\circ\mu_N(a), \quad a \in \mathcal{B},$$

then μ is a finitely additive probability with $\mu(a) = 0$ iff $a = 0$. $\square$

A σ-algebra admitting a σ-additive probability μ on it so that $\mu(a) = 0$ iff $a = 0$ is called a σ-*measure algebra*.

Every Boolean algebra has a partial ordering given by $a \leq b$ iff $a - b = 0$. A σ-algebra $\mathcal{B}$ is *weakly ω-distributive* if for any $\{a_{i,j} \mid i, j < \omega\} \subset \mathcal{B}$ such that $a_{i,j+1} \leq a_{i,j}$, there are $\theta_n : \omega \to \omega$, $n \in \mathbb{N}$ such that $\theta_n \leq \theta_{n+1}$ and

$$\bigvee_{0 \leq i < \omega} \bigwedge_{0 \leq j < \omega} a_{i,j} = \bigwedge_{n \in \mathbb{N}} \bigvee_{0 \leq i < \omega} a_{i, \theta_n(i)}.$$

We leave as an exercise to prove the following:

Theorem 1.16. (Kelley's Theorem—σ-additive version) *A σ-algebra $\mathcal{B}$ is a σ-measure algebra iff $\mathcal{B}$ is weakly ω-distributive and satisfies the condition in Thm. 1.15.* $\square$

1.5.5 *Notes and exercises*

During the early development of nonstandard analysis, conversion of an internal measure to an ordinary measure was met with obstacles, until the need of the saturation principle was realized and the landmark achievement of the Loeb measure and Loeb integration theory in [Loeb (1975)]. Soon afterward, Anderson gave a nonstandard construction of Brownian motion in [Anderson (1976)], leading to very fruitful applications of nonstandard methods in probability and measure theory. More details can be found in [Albeverio et al. (1986)].

Thm. 1.12 is a special case of a general result proved in [Render (1993)] that any Radon measure is obtainable as the image of a measure preserving function from a Loeb measure on a hyperfinite set.

There is some work on extending the Loeb measure and Loeb integration to vector measures, dealing with Banach space-valued measures and Banach space-valued integrable functions. See § 4.1 and the references mentioned within.

Cor. 1.4 is a representation of the Lebesgue measure on $[0,1]$ by the Loeb measure of an internal probability on a hyperfinite set. In [Anderson (1982)], this was generalized for all Radon measures.

The Kelley's Theorems, Thm. 1.15 and Thm. 1.16, were originally proved in [Kelley (1959)]. The proof here using hyperfinite algebra and linear programming is reproduced from [Ng (1991)].

EXERCISES

(1) Let $A \subset \mathbb{T}$ be $L(\mu)$-measurable, where μ is the internal counting probability on $\mathbb{T}$. Show that $st(A)$ is Lebesgue measurable.

(2) Give the Loeb measure construction for the counting probability on $\bigcup_{n<N^2} n\mathbb{T}$, *i.e.* the internal set $\{n\Delta t \mid n < N^2\}$. Then generalize Thm. 1.12 for the Lebesgue measure on $\mathbb{R}$.

(3) Prove Thm. 1.13.

(4) Prove Lem. 1.3.

(5) Let $(\Omega, \mathcal{B}, \mu)$ be an internal probability space and $F : \Omega \to {}^*\mathbb{R}$ a lifting of $f : \Omega \to \mathbb{R}$. Suppose $\theta : \mathbb{R} \to \mathbb{R}$ is bounded continuous, show that ${}^*\theta(F)$ is an S-integrable lifting of $\theta(f)$.

(6) Let $(\Omega, \mathcal{B}, \mu)$ be an internal probability space and $F : \Omega \to {}^*\mathbb{R}$ be μ-measurable. Show that F is S-integrable iff

$$\forall r \in {}^*\mathbb{R}^+ \setminus \mathbb{R}^+ \left(\int_{\{|F|>r\}} |F| \, d\mu \approx 0 \right).$$

(7) Apply Cor. 1.5 to prove the Dominated Convergence Theorem for Lebesgue measurable functions on the unit interval: Suppose

$$f_n, f : [0,1] \to \mathbb{R}, \ n \in \mathbb{N},$$

where the f_n's are Lebesgue measurable and f is Lebesgue integrable such that $\forall x \in [0,1] \left(\lim_{n\to\infty} f_n(x) \in \mathbb{R} \right)$ and $|f_n| \leq f$, then

$$\lim_{n\to\infty} \int_0^1 f_n(x) \, dx = \int_0^1 \left(\lim_{n\to\infty} f_n(x) \right) dx.$$

(8) Let μ be the internal counting probability on $\mathbb{T}$. Let $\mu \otimes \mu$ be the counting probability on $\mathbb{T}^2$. Suppose $f : \mathbb{T}^2 \to \mathbb{R}$ is $L(\mu \otimes \mu)$-integrable. Prove that

$$L(\mu)\Big(\big\{t \in \mathbb{T} \mid f(t, \cdot) \ L(\mu)\text{-integrable}\big\}\Big) = 1$$

and that

$$\int f \, dL(\mu \otimes \mu) = \int \left(\int f(x,y) \, dL(\mu)(x) \right) dL(\mu)(y).$$

(This is a special case of the *Keisler's Fubini's Theorem*. See [Albeverio *et al.* (1986)].

NONSTANDARD ANALYSIS   53

(9) Given an example of a Boolean algebra which is not a measure algebra.
(10) Apply the hyperfinite Boolean algebra method to prove Thm. 1.16.
(11) Let $X \in V(\mathbb{R})$, does $\sigma\big({}^*\mathcal{P}(X)\big)$ satisfy weakly ω-distributivity?

1.6   Application: Topology

Topological properties are simply properties about monads.
Compactness means no room for further idealization.

A **topological space** X corresponds to a pair $(X, \mathcal{T}_X)$, or simply $(X, \mathcal{T})$, consisting of a *set* X together with a **topology**—some $\mathcal{T} \subset \mathcal{P}(X)$ containing $\emptyset$ and is closed under arbitrary $\cup$ and finite $\cap$. Elements in $\mathcal{T}$ are called **open** subsets of the space X. A subset $C \subset X$ is called **closed** if $(X \setminus C)$ is open. Hence closed sets are closed under arbitrary $\cap$ and finite $\cup$. For $A \subset X$,

$$\overline{A} := \bigcap \{ C \mid A \subset C \subset X, \ C \text{ is closed} \},$$

the **closure** of A.

Given two topologies $\mathcal{T}_1$ and $\mathcal{T}_2$ on X, if $\mathcal{T}_1 \subset \mathcal{T}_2$ we say that $\mathcal{T}_2$ is **finer** (or **stronger**) than $\mathcal{T}_1$ and likewise $\mathcal{T}_1$ is **coarser** (or **weaker**) than $\mathcal{T}_2$.

The finest topology is the **discrete topology** given by $\mathcal{P}(X)$, *i.e.* every singleton $\{x\}$, $x \in X$, is open. Such X is called a **discrete space**.

Given a topological space $(X, \mathcal{T})$, and $Y \subset X$, the topological space $(Y, \{U \cap Y, \mid U \in \mathcal{T}\})$ is called a **subspace** of X.

1.6.1   *Monads and topologies*

We consider only topological spaces $X \in V(\mathbb{R})$, although results here hold also for topological spaces X in any $V(S)$.

Given $x \in X$, the **monad** of x is denoted and defined as

$$\mu(x) = \mu_{\mathcal{T}}(x) = \mu_X(x) := \bigcap \{ {}^*U \mid x \in {}^*U, U \in \mathcal{T} \}.$$

The **nearstandard** part of ${}^*X$ is defined as:

$$\mathrm{ns}({}^*X) := \bigcup_{x \in X} \mu(x).$$

The notion of open sets and closed sets can be characterized in terms of monads.

Proposition 1.14. *Let $A \subset X$, then*

(i) *A is open iff $\forall x \in A \left(\mu(x) \subset {}^*A \right)$;*

(ii) *A is closed iff $\forall x \in X \left((\mu(x) \cap {}^*A) \neq \emptyset \Rightarrow x \in A \right)$.*

(iii) *For $x \in X$, $x \in \overline{A}$ iff $\mu(x) \cap {}^*A \neq \emptyset$.*

Proof. (i) ($\Rightarrow$) : follows trivially from the definition.

For ($\Leftarrow$), fix any $x \in A$. If $\forall U \in \mathcal{T}\left(x \in U \Rightarrow \exists y \in {}^*U \setminus {}^*A\right)$, then by saturation, $\exists y \in \left(\mu(x) \setminus {}^*A\right)$, a contradiction.

So, by transfer $\forall x \in A\,\exists U \in \mathcal{T}\left(x \in U \subset A\right)$, and therefore

$$A = \bigcup\{U \in \mathcal{T}\,|\,U \subset A\}.$$

In particular, A is open.

(ii) follows from a dual statement.

(iii) follows from the saturation. □

For general spaces, the monads $\{\mu(x)\,|\,x \in X\}$ need not be pairwise disjoint. We will consider in the next subsection a large class of spaces for which the monads are indeed pairwise disjoint.

If the monads are disjoint, they form a partition of ns($^*X$), producing equivalence classes on ns($^*X$) whose equivalence relation is denoted by $\approx_{\mathcal{T}}$, $\approx_X$ or simply as $\approx$ (**infinitely close** to each other *w.r.t.* $\mathcal{T}$), *i.e.*

$$\forall x, y \in \text{ns}({}^*X)\left(x \approx y \Leftrightarrow \left(\exists z \in X\,(x, y \in \mu(z))\right)\right).$$

So, roughly, $x \approx y$ if there is no separation of x, y originated from $(X, \mathcal{T})$ and this signifies the intuitive idea about monads. In this setting, the definition of a monad extends to all nearstandard elements in a natural way: let $x \in \text{ns}({}^*X)$, we define $\mu(x)$ as $\{y \in {}^*X\,|\,y \approx x\}$. So $x \approx y$ iff $\mu(x) = \mu(y)$, for $x, y \in \text{ns}({}^*X)$. Note that it would not be a good idea here to define for $x \in \left(\text{ns}({}^*X) \setminus X\right)$ its monad as the intersection $\bigcap_{x \in U \in \mathcal{T}} {}^*U$. For example, let $\epsilon \in {}^*\mathbb{R}$, be a nonzero positive infinitesimal, then $\epsilon \in {}^*(0, 1)$ and $0 \notin {}^*(0, 1)$, but we like to keep 0 and ϵ in the same monad.

Still assuming that the monads are disjoint. Let $A \subset X$. We write $x \approx A$ to abbreviate $\exists y \in A\,(y \approx x)$. Then from Prop. 1.14, the openness of A means $\forall x \in {}^*X\left(x \approx A \Rightarrow x \in {}^*A\right)$, which expresses the intuitive meaning that, ideally, an open set includes all nearby points from $^*X$. Likewise, A is closed iff $\forall x \in X\left(x \approx {}^*A \Rightarrow x \in A\right)$, *i.e.* no new points from X are admitted into the set through idealization.

A **pseudometric** on a set X is a function $d : X^2 \to [0, \infty)$ with the property that $\forall x, y, z \in X$

$$\left(d(x, x) = 0\right) \wedge \left(d(x, y) = d(y, x)\right) \wedge \left(d(x, y) \leq d(x, z) + d(y, z)\right),$$

i.e. a symmetric binary function vanishing on the diagonal and satisfying the **Triangle Inequality**.

For $x \in X$ and $r \in \mathbb{R}^+$, let $B(x, r) := \{y \in X\,|\,d(x, y) < r\}$ and $S(x, r) := \{y \in X\,|\,d(x, y) = r\}$, the **open ball** and the **sphere**. Then a repeated application of the union and finite intersection operations to these $B(x,r)$ generates a topology T. The resulted (X, T) is called a ***pseudo-metric space***. We also write (X, d) for such (X, T).

The above function $d : X^2 \rightarrow [0, \infty)$ is called a ***metric*** if we strengthen the condition $\forall x \in X \left(d(x,x) = 0 \right)$ to

$$\forall x, y \in X \left(d(x,y) = 0 \Leftrightarrow x = y \right).$$

Then under the topology generated, the space is called a ***metric space***.

It is possible to have two different pseudometrics (or metrics) d_1 and d_2 generating the same topology T on X. In which case we say that d_1 and d_2 are ***equivalent pseudometrics*** (or ***equivalent metrics***).

Let $\mathcal{D}$ be a family of pseudometrics on X. For $d \in \mathcal{D}$, $x \in X$, $r \in \mathbb{R}^+$, we define

$$B_d(x,r) := \{ y \in X \mid d(x,y) < r \}.$$

Let T be the topology generated by these $B_d(x,r)$, then the space (X, T) is called a ***uniform space***.

For a uniform space such as the above we extend the definition of ***monad*** to any points $x \in {}^*X$ by

$$\mu(x) := \bigcap_{d \in \mathcal{D}, n \in \mathbb{N}} {}^*B_d(x, n^{-1}).$$

Note that here, for $x \in {}^*X \setminus \mathrm{ns}({}^*X)$, the set $\bigcap \{ {}^*U \mid x \in {}^*U, U \in T \}$ would not be taken as a useful definition of a monad, for it coincides with ${}^*X$.

Furthermore, if (X, d) is an internal metric space, the monad of any $x \in X$ is defined as

$$\mu(x) := \bigcap_{n \in \mathbb{N}} {}^*B(x, n^{-1}).$$

For these extended definition of monads, we note that $y \in \mu(x)$ implies that $\mu(y) = \mu(x)$, by the symmetry of the pseudometrics. In particular, $\{ \mu(x) \mid s \in {}^*X \}$ forms a partition of ${}^*X$. We then write $x \approx y$ for the equivalence relation given by the monads and call x, y infinitely close to each other as before. Clearly, this equivalence relation extends that given before on $\mathrm{ns}({}^*X)$ (where the monads of elements from X are pairwise disjoint), so the use of the same symbol $\approx$ is justified.

The monads of a hyperreal given in §1.4 are of course a special case of the above. Moreover, under the usual metric topology, $\mathrm{ns}({}^*\mathbb{R}) = \mathrm{Fin}({}^*\mathbb{R})$.

1.6.2 *Monads and separation axioms*

A topological space $(X, \mathcal{T})$ is called **Hausdorff** iff

$$\forall x, y \in X \left(x \neq y \Rightarrow \exists U, V \in \mathcal{T}((U \cap V = \emptyset) \wedge (x \in U \wedge y \in V)) \right).$$

Note that in a Hausdorff space, singletons are closed. The class of Hausdorff spaces includes clearly all metric spaces as well as most spaces that we will deal with in this book.

Hausdorff spaces are characterized by distinct points having distinct monads.

Proposition 1.15. *A topological space X is Hausdorff iff*

$$\forall x, y \in X \left(x \neq y \Leftrightarrow \mu(x) \cap \mu(y) = \emptyset \right).$$

Proof. Let $a, b \in X$. By the definition of monads and by saturation,

$$\mu(a) \cap \mu(b) = \emptyset \Leftrightarrow \bigcap_{a \in U \in \mathcal{T}} {}^*U \cap \bigcap_{b \in V \in \mathcal{T}} {}^*V = \emptyset$$

$$\Leftrightarrow \exists U, V \in \mathcal{T} \left({}^*U \cap {}^*V = \emptyset \wedge a \in {}^*U \wedge b \in {}^*V \right),$$

and so, by transfer,

$$\mu(a) \cap \mu(b) = \emptyset \Leftrightarrow \exists U, V \in \mathcal{T} \left(U \cap V = \emptyset \wedge a \in U \wedge b \in V \right). \qquad \square$$

We remark that in a Hausdorff space, monads form a partition of the nearstandard part, so we freely use the definition of monads of nearstandard elements and the relation $\approx$.

Being Hausdorff means certain separation property is satisfied. Here are more separation axioms: a topological space X is called

- **regular**, if $\forall x \in X \ \forall$ closed $C \subset X$

$$\left(x \notin C \Rightarrow \exists U, V \in \mathcal{T}((U \cap V = \emptyset) \wedge (x \in U \wedge C \subset V)) \right);$$

- **completely regular**, if $\forall x \in X \ \forall$ closed $C \subset X$

$$\left(x \notin C \Rightarrow \exists \text{ continuous } f : X \to \mathbb{R} \left((f(x) = 1) \wedge (f[C] = \{0\})\right) \right);$$

- **Tychonoff**, if X is completely regular and Hausdorff;
- **normal**, if $\forall$ closed $C, D \subset X$

$$\left(C \cap D = \emptyset \Rightarrow \exists U, V \in \mathcal{T}((U \cap V = \emptyset) \wedge (C \subset U \wedge D \subset V)) \right).$$

So in a completely regular space, any closed subset is separated from a point outside in a continuous manner. Here the continuity of $f : X \to \mathbb{R}$ means that $f^{-1}[U]$ is an open subset of X whenever $U \subset \mathbb{R}$ is open.

It is not hard to see that completely regular spaces are regular.

An equivalent characterization of regularity is this: X is regular iff

$$\forall x \in X \left(\forall U \in \mathcal{T} \left((x \in U) \Rightarrow \exists V \in \mathcal{T} \left(x \in V \subset \overline{V} \subset U \right) \right) \right).$$

Likewise, X is normal iff $\forall$ closed $C \subset X$

$$\forall U \in \mathcal{T} \left((C \subset U) \Rightarrow \exists V \in \mathcal{T} \left(C \subset V \subset \overline{V} \subset U \right) \right).$$

Similar to Prop. 1.15, we have the following:

Proposition 1.16.

(i) X *is regular iff* $\forall x \in X \; \forall$ *closed* $C \subset X \; (x \notin C) \Rightarrow$

$$\mu(x) \cap \bigcap_{C \subset U \in \mathcal{T}} {}^*U = \emptyset.$$

(ii) X *is normal iff* $\forall$ *closed* $C, D \subset X \; (C \cap D = \emptyset) \Rightarrow$

$$\bigcap_{C \subset U \in \mathcal{T}} {}^*U \cap \bigcap_{D \subset U \in \mathcal{T}} {}^*U = \emptyset.$$

$\square$

Because of Prop. 1.15, Hausdorff spaces admit a useful notion of monads. Moreover this class includes most spaces we will be interest in. Therefore we assume form now on that:

> Unless otherwise stated, all topological spaces $(X, \mathcal{T})$ are Hausdorff.

1.6.3   *Standard part and continuity*

Now given a topological space X, we define the injection

$$\mathrm{st} : \mathrm{ns}({}^*X) \to X$$

by taking $\mathrm{st}(x)$ to be the unique $y \in X$ such that $x \approx y$. Here $\mathrm{st}(x)$ is uniquely defined since X is Hausdorff. We also write $^\circ x$ for $\mathrm{st}(x)$.

The function st is referred to as the **standard part mapping** and $\mathrm{st}(x)$ is called the **standard part** of x.

Consequently, Prop. 1.14 can be rewritten as: $A \subset X$ is open iff $\text{st}^{-1}[A] \subset {}^*A$ and A is closed iff $\text{st}^{-1}[A] \supset \left({}^*A \cap \text{ns}({}^*X)\right)$, since $\text{ns}({}^*X)$ is the disjoint union of $\text{st}^{-1}[A]$ and $\text{st}^{-1}[X \setminus A]$.

For $x \in X$, note that $\mu(x) = \text{st}^{-1}[\{x\}]$.

The image of internal sets under the standard part mapping turns out to be quite simple. The following is a special example.

Proposition 1.17. *Let $A \subset {}^*X$ be the intersection of fewer than κ many internal sets. Then $\text{st}[A \cap \text{ns}({}^*X)]$ is closed.*

Proof. Write $C = \text{st}[A \cap \text{ns}({}^*X)]$. To avoid triviality, we assume $C \neq \emptyset$. Fix $a \in X$ such that $\mu(a) \cap {}^*C \neq \emptyset$.

Define $\mathcal{F} := \{\, {}^*U \mid a \in U \in \mathcal{T} \,\}$, so $\mu(a) = \bigcap \mathcal{F}$.

For any ${}^*U \in \mathcal{F}$, ${}^*U \cap {}^*C \neq \emptyset$, hence $U \cap C \neq \emptyset$, by transfer. So $A \cap {}^*U \neq \emptyset$. Note that $\mathcal{F}$ is closed under finite intersection, so for any finite $\mathcal{F}_0 \subset \mathcal{F}$, $A \cap \bigcap \mathcal{F}_0 \neq \emptyset$. Then it follows from saturation that $A \cap \bigcap \mathcal{F}$, an intersection of fewer that κ internal sets having the finite intersection property, is nonempty. *i.e.* $A \cap \mu(a) \neq \emptyset$, *i.e.* $a \in C$.

As this holds for all $a \in X$, therefore, by Prop. 1.14(ii), C is closed. $\square$

Let $f : X_1 \to X_2$, where $(X_1, \mathcal{T}_1)$ and $(X_2, \mathcal{T}_2)$ are topological spaces. Generalizing the real-valued function case, f is said to be **continuous** at $x \in X_1$ if

$$\forall V \in \mathcal{T}_2 \left(f(x) \in V \Rightarrow \exists U \in \mathcal{T}_1 \left(x \in U \wedge f[U] \subset V\right)\right)$$

and f is continuous if it is continuous at every $x \in X_1$.

For simplicity of notation, μ stands for monads in either ${}^*X_1$ or ${}^*X_2$. The following generalizes Prop. 1.8.

Proposition 1.18. $f : X_1 \to X_2$ *is continuous at $x \in X_1$ iff*

$${}^*f[\mu(x)] \subset \mu(f(x)), \quad \text{i.e. } \forall y \in {}^*X_1 \left(y \approx x \Rightarrow {}^*f(y) \approx f(x)\right).$$

Proof. $(\Rightarrow)$: Note that

$${}^*f[\mu(x)] = {}^*f\Big[\bigcap_{x \in U \in \mathcal{T}_1} {}^*U\Big] \subset \Big(\bigcap_{x \in U \in \mathcal{T}_1} {}^*f[U]\Big) \subset \Big(\bigcap_{f(x) \in V \in \mathcal{T}_2} {}^*V\Big) = \mu(f(x)),$$

where the second inclusion follows from the definition of continuity and transfer.

$(\Leftarrow)$: If $f(x) \in V \in \mathcal{T}_2$ are such that $\forall U \in \mathcal{T}_1 \left(x \in U \Rightarrow f[U] \setminus V \neq \emptyset\right)$, then in particular, for any $n \in \mathbb{N}$,

$$\forall U_0, \ldots, U_n \in \mathcal{T}_1 \left(x \in \cap_{i \leq n} U_i \Rightarrow f[[\cap_{i \leq n} U_i] \setminus V \neq \emptyset\right),$$

consequently, by saturation, $^*f[\mu(x)] \setminus {}^*V \neq \emptyset$.

Therefore $^*f[\mu(x)] \not\subset \mu(f(x))$.   □

Let $x \in \mathrm{ns}(^*X)$, so $x \approx y$ for some $y \in X$. If f is continuous at y, then $^*f[\mu(x)] = {}^*f[\mu(y)] \subset \mu(^*f(y))$, so $^*f(x) \approx f(y)$ and $^*f[\mu(x)] \subset \mu(^*f(x))$. Hence we have:

Corollary 1.6. $f : X_1 \to X_2$ *is continuous iff*

$$\forall x \in X_1 \left({}^*f[\mu(x)] \subset \mu(f(x)) \right) \quad \textit{iff} \quad \forall x \in \mathrm{ns}(^*X_1) \left({}^*f[\mu(x)] \subset \mu(^*f(x)) \right).$$

                                                                        □

Corollary 1.7. $f : X_1 \to X_2$ *is continuous iff* $\forall V \in \mathcal{T}_2 \left(f^{-1}[V] \in \mathcal{T}_1 \right)$.

Proof.   By Prop. 1.14(i),

$$\left(\forall V \in \mathcal{T}_2 \left(f^{-1}[V] \in \mathcal{T}_1 \right) \right) \Leftrightarrow \left(\forall V \in \mathcal{T}_2 \, \forall x \in f^{-1}[V] \left(\mu(x) \subset {}^*f^{-1}[V] \right) \right).$$

It is not hard to check that the latter is equivalent to

$$\forall x \in X_1 \left({}^*f[\mu(x)] \subset \mu(f(x)) \right).$$

(Saturation is needed in one direction.)   □

If $\mathcal{T}, \mathcal{T}'$ are topologies on X_1 with $\mathcal{T}'$ finer than $\mathcal{T}$ and let μ and μ' be the corresponding monads, then $\forall x \in X \left(\mu'(x) \subset \mu(x) \right)$. Moreover, by Cor. 1.7, any continuous $f : X_1 \to X_2$ w.r.t. $\mathcal{T}$ is continuous w.r.t. $\mathcal{T}'$.

Other than those $^*f$ for some continuous f, there exist other internal functions that capture continuity using conditions in Cor. 1.6.

Proposition 1.19. Let $f : {}^*X_1 \to {}^*X_2$ be an internal function satisfying conditions $f[\mathrm{ns}(^*X_1)] \subset \mathrm{ns}(^*X_2)]$ and $\forall x \in \mathrm{ns}(^*X_1) \left(f[\mu(x)] \subset \mu(f(x)) \right)$. (i.e. $\forall x, y \in \mathrm{ns}(^*X_1) \left(x \approx y \Rightarrow f(x) \approx f(y) \right)$.)
Define $^\circ f : X_1 \to X_2$ by $x \mapsto {}^\circ\left(f(x) \right)$. Then $^\circ f$ is continuous.

Proof.   Suppose $^\circ f$ is discontinuous at some $a \in X_1$. Then for some $V \in \mathcal{T}_2$ with $^\circ f(a) \in V$, we have for each U, where $a \in U \in \mathcal{T}_1$, there is some $b \in U$ such that $^\circ f(b) \notin V$, hence $f(b) \notin {}^*V$.

Therefore the sets

$$\{ x \in {}^*U \mid f(x) \in {}^*X \setminus {}^*V \}, \quad \text{where } a \in U \in \mathcal{T}_1,$$

satisfy the finite intersection property. Now, by saturation, let b belong to the intersection of all these sets, then $b \approx a$, but $f(b) \notin {}^*V \ni f(a)$, contradicting the assumptions on f.   □

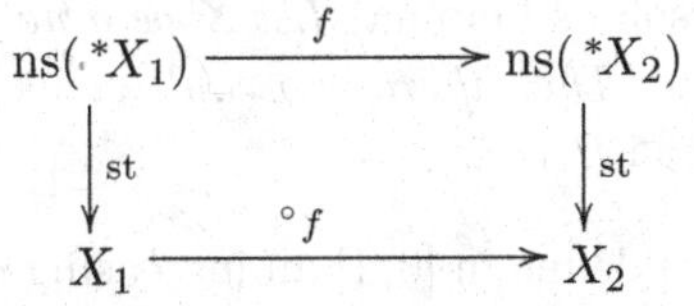

$$\text{ns}(^*X_1) \xrightarrow{\quad f \quad} \text{ns}(^*X_2)$$

$$\downarrow \text{st} \qquad\qquad \downarrow \text{st}$$

$$X_1 \xrightarrow{\quad {}^\circ f \quad} X_2$$

Fig. 1.3   The commutative diagram for Prop. 1.19.

In Prop. 1.19, we also say that $^\circ f$ is the **standard part** of the function f. Note that $^\circ x \mapsto {}^\circ (f(x))$ is well-defined and is the same as $^\circ f$. Therefore the diagram in Fig. 1.3 commutes.

A mapping $f : X_1 \to X_2$ is said to be **open** if $\forall U \in T_1 \left(f[U] \in T_2 \right)$. The following is an analog of Cor. 1.6

Proposition 1.20. *Let $f : X_1 \to X_2$. Then f is open iff*

$$\forall x \in X_1 \left({}^*f[\mu(x)] \supset \mu(f(x)) \right).$$

Proof. $(\Rightarrow)$: Let $x \in X$ and $\mathcal{F} = \{ {}^*U \mid x \in U \in T_1 \}$.

Consider $z \in \bigcap_{U \in \mathcal{F}} {}^*f[U]$. Since $\mathcal{F}$ is closed under finite intersection, the following sets have the finite intersection property:

$$\left\{ y \in \bigcap \mathcal{F}_0 \mid {}^*f(y) = z \right\}, \quad \text{with } \mathcal{F}_0 \text{ ranging over finite subsets of } \mathcal{F}.$$

Therefore, by saturation, for some $y \in \bigcap \mathcal{F}$, $z = {}^*f(y)$. *i.e.*

$$\bigcap_{U \in \mathcal{F}} {}^*f[U] \subset {}^*f\left[\bigcap \mathcal{F} \right] = {}^*f[\mu(x)].$$

The inclusion in the opposite direction way is trivial, so actually $\bigcap_{U \in \mathcal{F}} {}^*f[U] = {}^*f[\mu(x)]$. Then

$$^*f[\mu(x)] = \left(\bigcap_{x \in U \in T_1} {}^*f[U] \right) \supset \left(\bigcap_{f(x) \in V \in T_2} {}^*V \right) = \mu(f(x)),$$

where we use the fact that $f[U] \in T_2$ for every $U \in T_1$.

$(\Leftarrow)$: Let $U \in T_1$ and $y \in f[U]$. So $y = f(x)$ for some $x \in U$. Then

$$\mu(y) = \mu(f(x)) \subset {}^*f[\mu(x)] \subset {}^*f[U],$$

and therefore $f[U] \in T_2$ by Prop. 1.14(i). $\square$

A bijective continuous open mapping $f : X_1 \to X_2$ is called a **homeomorphism**. Note in such case that T_2 is generated by $\{ f[U] \mid U \in T_1 \}$. When such mapping exists, X_1 and X_2 are said to be **homeomorphic**. If f is a homeomorphism, then f^{-1} is also a homeomorphism, making homeomorphism an equivalence relation between topological spaces.

Theorem 1.17. (Urysohn's Lemma) *Let X be a normal space, $A, B \subset X$ be disjoint closed subsets. Then there is a continuous $f : X \to [0,1]$ so that $f^{-1}(0) \supset A$ and $f^{-1}(1) \supset B$.*

Proof. For $n \in {}^*\mathbb{N}$, we define in $[0,1]$ an increasing sequence of finite sets $\mathbb{T}_n := \{ m\, 2^{-n} \mid m = 0, \ldots, 2^n \}$.

We first define inductively for $n \in \mathbb{N}$ two functions

$$U_n : \{-1, 0, \ldots, 2^n\} \to \mathcal{T} \quad \text{and} \quad f_n : X \to \mathbb{T}_n,$$

so that, for $m = -1, \ldots (2^n - 1)$, we have

$$\overline{U_n(m)} \subset U_n(m+1) \quad \text{and} \quad \overline{U_n(2^n - 1)} \cap B = \emptyset$$

and where f_n is defined to be

$$f_n := \sum_{m=-1}^{2^n - 1} (m+1) 2^{-n}\, \chi_{U_n(m+1) \setminus U_n(m)}.$$

So it is only necessary to define the U_n.

For all $n \in \mathbb{N}$, we set $U_n(-1) := \emptyset$ and $U_n(2^n) := X$.

For $n = 0$, we use normality to chose $U_0(0)$ to be any $U \in \mathcal{T}$ such that $A \subset U$ and $\overline{U} \cap B = \emptyset$.

Suppose U_n is defined. Let U_{n+1} agree with U_n at even numbers, *i.e.*

$$U_{n+1}(2m) := U_n(m), \quad \text{where } m = 0, \ldots, 2^n.$$

For the odd numbers, apply the inductive hypothesis and normality: for $m = 0, \ldots, (2^n - 2)$, let $U_{n+1}(2m+1)$ be any $U \in \mathcal{T}$ such that

$$U_n(m) = U_{n+1}(2m) \subset U \subset \overline{U} \subset U_{n+1}(2m+2) = U_n(m+1),$$

and $U_{n+1}(2^{n+1} - 1)$ is any $U \in \mathcal{T}$ such that

$$\overline{U_n(2^n - 1)} = \overline{U_{n+1}(2^{n+1} - 2)} \subset U \quad \text{and} \quad \overline{U} \cap B = \emptyset.$$

Finally, consider ${}^*\{f_n \mid n \in \mathbb{N}\}$ *i.e.* $\{ {}^*f_n \mid n \in {}^*\mathbb{N}\}$.

Fix any $N \in {}^*\mathbb{N} \setminus \mathbb{N}$, consider ${}^*f_N : {}^*X \to {}^*[0,1]$. Take $x, y \in \text{ns}({}^*X)$ such that $x \approx y$. Then $x \in U_n(m)$ iff $y \in U_n(m)$ for any $n, m \in \mathbb{N}$ with $m \leq 2^n$, therefore ${}^*f_N(x) \approx {}^*f_N(y)$. Hence

$$f : X \to [0,1] \quad \text{given by} \quad x \mapsto {}^\circ {}^*f_N(x)$$

is well defined and is continuous by Prop. 1.19.

Moreover, as $U_N(0) \supset {}^*A$ and $\left(U_N(2^N) \setminus U_N(2^N - 1)\right) \supset {}^*B$, we have $f^{-1}(0) \supset A$ and $f^{-1}(1) \supset B$. $\qquad\square$

As an immediate consequence, we have:

Corollary 1.8. *Normal spaces are completely regular.*   □

In a pseudometric space (X, d), a sequence $\{a_n\}_{n \in \mathbb{N}}$ is called **Cauchy**, if $\lim_{n,m \to \infty} d(a_n, a_m) = 0$, *i.e.*

$$\forall \epsilon \in \mathbb{R}^+ \, \exists N \in \mathbb{N} \left(\forall n, m > N \left(d(a_n, a_m) < \epsilon \right) \right).$$

Similar to Prop. 1.5, by using the internal extension $\{a_n\}_{n \in \, {}^*\mathbb{N}}$ it is easy to see the following.

Proposition 1.21. *In a pseudometric space (X, d), a sequence $\{a_n\}_{n \in \mathbb{N}}$ is Cauchy iff $\forall M, N \in ({}^*\mathbb{N} \setminus \mathbb{N}) \left({}^*a_M \approx {}^*a_N \right)$.*
Moreover, if $\lim_{n \to \infty} a_n \in X$, then $\forall N \in ({}^*\mathbb{N} \setminus \mathbb{N}) \left(\lim_{n \to \infty} a_n \approx {}^*a_N \right)$.
     □

A metric space (X, d), is called **complete** if every Cauchy sequence $\{a_n\}_{n \in \mathbb{N}} \subset X$ converges in X. That is,

$$\exists a \in X \lim_{n \to \infty} a_n = a; \quad \text{equivalently,} \quad \exists a \in X \left(\forall N \in {}^*\mathbb{N} \setminus \mathbb{N} \left({}^*a_N \approx a \right) \right).$$

If we are given a double sequence $\{a_{nm}\}_{n,m \in \mathbb{N}}$ in a metric space, it is easy to see that in general

$$\lim_{n \to \infty} \lim_{m \to \infty} a_{nm} \neq \lim_{m \to \infty} \lim_{n \to \infty} a_{nm}.$$

Similar to the remark above, one can easily show the following:

Proposition 1.22. *Let $\{a_{nm}\}_{n,m \in \mathbb{N}}$ be a double sequence in a complete metric space (X, d). Then*

$$\lim_{n,m \to \infty} a_{nm} \quad \text{exists iff} \quad \forall M_1, M_2, N_1, N_2 \in ({}^*\mathbb{N} \setminus \mathbb{N}) \left({}^*a_{N_1 M_2} \approx {}^*a_{N_2 M_2} \right).$$

In such case, $\lim_{n,m \to \infty} a_{nm} \approx {}^*a_{NM}$ for any $N, M \in ({}^*\mathbb{N} \setminus \mathbb{N})$.   □

Now we consider a useful uniform boundedness condition that ensures path-independence of the limit.

Theorem 1.18. *Let (X, d) be a complete metric space and $\{a_{nm}\}_{n,m \in \mathbb{N}} \subset X$ such that, for each $n \in \mathbb{N}$, $\{a_{nm}\}_{m \in \mathbb{N}}$ is Cauchy.*
Suppose the the following uniform boundedness condition is satisfied:

$$\forall \epsilon \in \mathbb{R}^+ \, \exists n \in \mathbb{N} \, \forall n_1, n_2 \in \mathbb{N} \left((n_1, n_2 \geq n) \Rightarrow \left(\forall m \in \mathbb{N} \left(d(a_{n_1 m}, a_{n_2 m}) < \epsilon \right) \right) \right).$$

$$(1.5)$$

Then the following hold:

(i) $\forall N_1, N_2, M_1, M_2 \in ({}^*\mathbb{N} \setminus \mathbb{N}) \left({}^*a_{N_1 M_1} \approx {}^*a_{N_2 M_2} \right)$;

(ii) $\forall N, M \in {}^*\mathbb{N} \left({}^*a_{NM} \in \mathrm{ns}({}^*X) \right)$;

(iii) $\forall N, M \in ({}^*\mathbb{N} \setminus \mathbb{N}) \left(\lim_{n,m \to \infty} a_{nm} = {}^\circ({}^*a_{NM}) \right)$.

Proof.   (i): In (1.5), for each $\epsilon \in \mathbb{R}^+$, choose $n_\epsilon \in \mathbb{N}$ such that whenever $n_\epsilon \le n_1, n_2 \in \mathbb{N}$, we have $\forall m \in \mathbb{N} \left(d(a_{n_1 m}, a_{n_2 m}) < \epsilon \right)$. Then by transfer,

$$\forall N, L \in {}^*\mathbb{N} \left((N, L \ge n_\epsilon) \Rightarrow \left(\forall M \in {}^*\mathbb{N} \left({}^*d({}^*a_{NM}, {}^*a_{LM}) < \epsilon \right) \right) \right), \quad (1.6)$$

where $\epsilon \in \mathbb{R}^+$.

Now, by the Triangle Inequality, for each $\epsilon \in \mathbb{R}^+$,

$$
{}^*d({}^*a_{N_1 M_1}, {}^*a_{N_2 M_2}) \le \overbrace{{}^*d({}^*a_{N_1 M_1}, {}^*a_{N_2 M_1})}^{(A)} + \overbrace{{}^*d({}^*a_{N_2 M_1}, {}^*a_{n_\epsilon M_1})}^{(B)}
$$
$$
+ \underbrace{{}^*d({}^*a_{n_\epsilon M_1}, {}^*a_{n_\epsilon M_2})}_{(C)} + \underbrace{{}^*d({}^*a_{n_\epsilon M_2}, {}^*a_{N_2 M_2})}_{(D)}.
$$

Apply (1.6) to (A) for all $\epsilon \in \mathbb{R}^+$, we see that (A)$\approx 0$. By (1.6), we have both (B), (D)$< \epsilon$. Since $\{a_{n_\epsilon m}\}_{m \in \mathbb{N}}$ is Cauchy, we have (C)≈ 0. Therefore

$${}^*d({}^*a_{N_1 M_1}, {}^*a_{N_2 M_2}) \lesssim 2\epsilon \quad \text{for any } \epsilon \in \mathbb{R}^+,$$

hence ${}^*a_{N_1 M_1} \approx {}^*a_{N_2 M_2}$.

(ii): Clearly we only need to show that ${}^*a_{NM} \in \mathrm{ns}({}^*X)$ in the case when at least one of N, M is infinite.

Let $n \in \mathbb{N}$ and $M \in ({}^*\mathbb{N} \setminus \mathbb{N})$. Since $\{a_{nm}\}_{m \in \mathbb{N}}$ is Cauchy, we have ${}^*a_{nM} \in \mathrm{ns}({}^*X)$ and $\lim_{m \to \infty} a_{nm} \approx {}^\circ({}^*a_{nM})$.

Let $N \in ({}^*\mathbb{N} \setminus \mathbb{N})$ and $m \in \mathbb{N}$. By (1.5), $\{a_{nm}\}_{n \in \mathbb{N}}$ is Cauchy. An application of (1.6) shows that

$$\lim_{n \to \infty} {}^\circ({}^*a_{nm}) \approx {}^*a_{Nm}.$$

Hence ${}^*a_{Nm} \in \mathrm{ns}({}^*X)$.

Finally, let $N, M \in ({}^*\mathbb{N} \setminus \mathbb{N})$. Then as above, we have $\{ {}^*a_{nM} \}_{n \in \mathbb{N}} \subset \mathrm{ns}({}^*X)$ and $\lim_{m \to \infty} a_{nm} \approx {}^\circ({}^*a_{nM})$ for each $n \in \mathbb{N}$.

Again, by (1.6), we have $\{ {}^\circ({}^*a_{nM}) \}_{n \in \mathbb{N}}$ is Cauchy and

$$\lim_{n \to \infty} {}^\circ({}^*a_{nM}) \approx {}^*a_{NM},$$

hence ${}^*a_{NM} \in \mathrm{ns}({}^*X)$.

(iii): Let $N, M \in ({}^*\mathbb{N} \setminus \mathbb{N})$. The computations in the proof of (ii) shows that $\lim_{n \to \infty} \lim_{m \to \infty} a_{nm} = {}^\circ({}^*a_{NM})$. So, by (i) and Prop. 1.22, we have $\lim_{n,m \to \infty} a_{nm} = {}^\circ({}^*a_{NM})$. $\square$

A mapping $f : X \to X$ on a metric space (X, d) is called **Lipschitz** if

$$\exists r \in [0, \infty) \forall x, y \in X \left(d(f(x), f(y)) \le r\, d(x, y) \right).$$

When this condition is satisfied, we say that f has **Lipschitz constant** r. Clearly, by Prop. 1.18, Lipschitz functions are continuous.

If f has Lipschitz constant $r \in [0, 1)$, then f is said to be a **contraction**.

The following basic result is often used in proving fixed point theorems, *i.e.* results about a function f having a point a so that $f(a) = a$.

Theorem 1.19. (Banach Contraction Principle) *Let (X, d) be a complete metric space and $f : X \to X$ be a contraction. Then $\exists x \in X \left(f(x) = x \right).$*

Proof. Let f have Lipschitz constant $r \in [0, 1)$. Fix any $a \in X$.

Then by iterating the Lipschitz condition, we have

$$\forall n \in \mathbb{N} \left(d(f^{n+1}(a), f^n(a)) \le r^n\, d(f(a), a) \right). \tag{1.7}$$

Then for any $m < n$ in $\mathbb{N}$, by the Triangle Inequality and (1.7),

$$d(f^n(a), f^m(a)) \le \sum_{0 \le k < n-m} d(f^{m+k+1}(a), f^{m+k}(a)) \le \frac{r^m}{1 - r} d(f(a), a),$$

which $\to 0$ as $m \to \infty$.

Hence $\{f^n(a)\}_{n \in \mathbb{N}} \subset X$ is Cauchy and $\lim_{n \to \infty} f^n(a) = c$ for some $c \in X$. By Prop. 1.21, $c \approx {}^*f^N(a)$ for any $N \in ({}^*\mathbb{N} \setminus \mathbb{N})$.

Fix any $N \in ({}^*\mathbb{N} \setminus \mathbb{N})$. By transferring (1.7), we have

$${}^*d\left({}^*f({}^*f^N({}^*a)), {}^*f^N({}^*a) \right) \le r^N\, d(f(a), a) \approx 0.$$

This, together with the continuity of f implies that

$$f(c) \approx {}^*f({}^*f^N({}^*a)) \approx {}^*f^N({}^*a) \approx c,$$

hence $f(c) = c$, since both $c, f(c) \in X$. □

The most fundamental fixed point theorem is perhaps the following classical result. See for example [Munkres (2000)] for a proof.

Theorem 1.20. (Brouwer's Fixed Point Theorem) *Let $\|\cdot\|$ be the Euclidean norm on $\mathbb{R}^n$, $n \in \mathbb{N}$. Let $\bar{B} := \{x \in \mathbb{R}^n \mid \|x\| \le 1\}$ be the closed unit ball.*

Suppose $f : \bar{B} \to \bar{B}$ is continuous. Then $\exists x \in \bar{B} \left(f(x) = x \right).$ □

1.6.4   *Robinson's characterization of compactness*

A subset $K \subset X$ is said to be **compact**, if

$$\forall \mathcal{F} \subset \mathcal{T} \left(K \subset \bigcup \mathcal{F} \Rightarrow \exists \mathcal{F}_0 \subset \mathcal{F} \left(|\mathcal{F}_0| < \aleph_0 \wedge K \subset \bigcup \mathcal{F}_0 \right) \right).$$

That is, every open cover of K contains a finite subcover of K.

When X is compact, the space X is called a **compact space**.

The following intuitive and elegant re-formulation of compactness, due to Robinson, will be quoted frequently throughout.

Theorem 1.21.   (Robinson's Characterization of Compactness) *In a topological space X, a subset $K \subset X$ is compact iff*

$$\forall x \in {}^*K \left(\exists y \in K \, (x \approx y) \right).$$

(Equivalently, ${}^*K \subset \mathrm{ns}({}^*X) \wedge \mathrm{st}[{}^*K] = K.)$

Proof. $(\Rightarrow)$: If there is $c \in {}^*K \setminus \mathrm{st}^{-1}[K]$, then for any $a \in K$ there is $U_a \in \mathcal{T}$ such that $a \in U_a$ and $c \notin {}^*U_a$. Now $\{U_a \,|\, a \in K\}$ is an open covering of K and for any finite $K_0 \subset K$, $K \not\subset \bigcup_{a \in K_0} U_a$, for otherwise the transfer implies $c \in {}^*K \subset \bigcup_{a \in K_0} {}^*U_a$, a contradiction.

$(\Leftarrow)$: Suppose ${}^*K \subset \mathrm{st}^{-1}[K]$ and $\mathcal{F} \subset \mathcal{T}$ such that $K \subset \bigcup \mathcal{F}$. If for any finite $\mathcal{F}_0 \subset \mathcal{F}$, $K \not\subset \bigcup \mathcal{F}_0$, then the following internal sets have the finite intersection property:

$$ {}^*K \setminus \bigcup_{U \in \mathcal{F}_0} {}^*U, \quad \text{where } \mathcal{F}_0 \subset \mathcal{F} \text{ is finite.}$$

Therefore, by saturation, there is $c \in {}^*K \setminus \bigcup_{U \in \mathcal{F}} {}^*U$, and hence there is $c \in {}^*K \setminus \mathrm{st}^{-1}[K]$. $\square$

The condition $\mathrm{st}[{}^*K] = K$ intuitively captures the fact that the idealization of K always stays close to K. Note that $\mathrm{st}[{}^*K] \supset K$ always holds trivially for any $K \subset X$.

Since $\mathrm{st}[{}^*K] = K$ implies $\forall x \in X \left((\mu(x) \cap {}^*K \neq \emptyset) \Rightarrow (x \in K) \right)$, compact sets are closed, by Prop. 1.14(ii).

The intersection of a decreasing sequence of nonempty compact sets is nonempty: Let $K_n \subset X$, $n \in \mathbb{N}$, be compact such that $K_n \supset K_{n+1} \neq \emptyset$. Let $a_n \in K_n$, $n \in \mathbb{N}$. Extend it to an internal sequence, let $a = a_N$ for some $N \in {}^*\mathbb{N} \setminus \mathbb{N}$. Then $a \in \cap_{n \in \mathbb{N}} {}^*K_n$, so ${}^\circ a \in \cap_{n \in \mathbb{N}} \mathrm{st}[{}^*K_n] = \cap_{n \in \mathbb{N}} K_n$.

Let $C \subset X$ be closed with ${}^*C \subset \mathrm{ns}({}^*X)$. Then by Prop. 1.14(ii), $\mathrm{st}[{}^*C] = C$, hence C is compact. In particular, closed subsets of a compact set are compact.

If $\mathcal{T}$, $\mathcal{T}'$ are topologies on X with $\mathcal{T}'$ finer than $\mathcal{T}$, then, as before, from $\forall x \in X \left(\mu'(x) \subset \mu(x)\right)$, if $K \subset X$ is compact $w.r.t.$ $\mathcal{T}'$, then it is compact $w.r.t.$ $\mathcal{T}$.

One also compares Prop. 1.17 with the following.

Proposition 1.23. *Let* $K \subset \mathrm{ns}(\,^*X)$ *be internal. Then* $\mathrm{st}[K]$ *is compact.*

Proof. Write $C = \mathrm{st}[K]$. We first present a purely nonstandard proof.

Let $a \in \,^*X$. Suppose $a \not\approx C$. So for each $c \in C$. there is $U_c \in \mathcal{T}$ such that $c \in U_c$ and $a \notin \,^*U_c$.

But $C \subset \bigcup_{c \in C} U_c$ and $C = \mathrm{st}[K]$, so $K \subset \bigcup_{c \in C} \,^*U_c$. Then by saturation, for some finite $C_0 \subset C$, we have $K \subset \bigcup_{c \in C_0} \,^*U_c$. By $C = \mathrm{st}[K]$ again, $C \subset \bigcup_{c \in C_0} U_c$.

Apply transfer, then $\,^*C \subset \bigcup_{c \in C_0} \,^*U_c$. In particular, we have $a \notin \,^*C$.

Therefore $\forall x \in \,^*C \left(\exists c \in C\, (x \approx c)\right)$, and we conclude that $\,^*C \subset \mathrm{ns}(\,^*X)$ and $\mathrm{st}[\,^*C] \subset C$, so C is compact by Thm. 1.21.

Alternatively, one shows that every open cover of C contains a finite subcover. Let $\mathcal{F} \subset \mathcal{T}$ such that $C \subset \bigcup \mathcal{F}$. Then $K \subset \bigcup_{U \in \mathcal{F}} \,^*U$. So, by saturation, $K \subset \bigcup_{U \in \mathcal{F}_0} \,^*U$, for some finite $\mathcal{F}_0 \subset \mathcal{F}$. Hence $C \subset \bigcup \mathcal{F}_0$. $\square$

Now let $K_i \subset X$ be compact, $i \in I$. If I is finite, then

$$\mathrm{st}\left[\,^*(\bigcup_{i \in I} K_i)\right] = \mathrm{st}\left[\bigcup_{i \in I} \,^*K_i\right] = \bigcup_{i \in I} \mathrm{st}[\,^*K_i] = \bigcup_{i \in I} K_i,$$

i.e. a finite union of compact sets is compact. For finite or infinite I,

$$\mathrm{st}\left[\,^*(\bigcap_{i \in I} K_i)\right] \subset \mathrm{st}\left[\bigcap_{i \in I} \,^*K_i\right] \subset \bigcap_{i \in I} \mathrm{st}[\,^*K_i] = \bigcap_{i \in I} K_i,$$

giving $\mathrm{st}\left[\,^*(\bigcap_{i \in I} K_i)\right] = \bigcap_{i \in I} K_i$, *i.e.* the intersection of arbitrarily many compact sets is compact.

Another useful property is the following.

Proposition 1.24. *Let* $(X, \mathcal{T})$ *be a compact space,* $C \subset X$ *be closed and* $a \in \,^*X$. *Then* $^\circ a \in C$ *iff* $a \in \,^*U$ *for every open* $U \supset C$.

Consequently, X *is a normal space.*

Proof. First note that $^\circ x \in X$ is defined for every $x \in \,^*X$, as X is compact.

Let $c = \,^\circ a$. If $c \in C$, then, for every open $U \supset C$, $a \in \mu(c) \subset \,^*U$.

Conversely, suppose $c \notin C$. Then by X being Hausdorff, for each $x \in C$ there are $U_x, V_x \in \mathcal{T}$ such that $c \in U_x, x \in V_x$ and $U_x \cap V_x = \emptyset$. In particular, $C \subset \bigcup_{x \in C} V_x$. Since C is a closed subset of the compact X, C is compact, so there is finite $C_0 \subset C$ such that $C \subset \cup_{x \in C_0} V_x$. Let $U := \cap_{x \in C_0} U_x$, and $V := \cup_{x \in C_0} V_x$, so $U, V \in \mathcal{T}$ and $U \cap V = \emptyset$, hence $^*U \cap {}^*V = \emptyset$. Moreover, $^*V \supset C$, but $a \notin {}^*V$, as $\mu(c) \subset {}^*U$. Therefore the first statement of the theorem is proved.

Next let $C, D \subset X$ be closed and disjoint. As a consequence of the above, for any $a \in {}^*X$ we have

$$°a \in C \Leftrightarrow a \in \bigcap_{C \subset U \in \mathcal{T}} {}^*U \quad \text{and} \quad °a \in D \Leftrightarrow a \in \bigcap_{D \subset U \in \mathcal{T}} {}^*U.$$

Hence $\displaystyle\bigcap_{C \subset U \in \mathcal{T}} {}^*U \cap \bigcap_{D \subset U \in \mathcal{T}} {}^*U = \emptyset$.

So we conclude from Prop. 1.16(ii) that X is normal. $\square$

In a metric space (X, d), given a subset $Y \subset X$ and $\epsilon \in \mathbb{R}^+$, we write

$$Y^\epsilon := \bigcup_{y \in Y} B_d(y, \epsilon).$$

We say that Y is **totally bounded** if for every $\epsilon \in \mathbb{R}^+$ there is finite $Y_0 \subset Y$ such that $Y \subset Y_0^\epsilon$.

If there is some $r \in \mathbb{R}^+$ such that $\forall y_1, y_2 \in Y \left(d(y_1, y_2) \leq r \right)$ we say that Y is **bounded**. Then it is clear that totally boundedness implies boundedness.

The following is also clear from saturation:

Proposition 1.25. *Let Y be a subset of a metric space. Then Y is totally bounded iff $^*Y \subset H^\epsilon$ for some hyperfinite $H \subset {}^*Y$ and $\epsilon \approx 0$.* $\square$

Let $K \subset X$ be compact in a metric space (X, d). By saturation, let H be hyperfinite such that $K \subset H \subset {}^*K$. Then by Robinson's characterization of compactness, $\forall x \in {}^*K \, \exists y \in H \, (x \approx y)$. By H hyperfinite, $\min_{y \in H} {}^*d(x, y)$ exists and is infinitesimal for each $x \in {}^*K$. Let

$$\epsilon := \sup_{x \in {}^*K} \min_{y \in H} {}^*d(x, y).$$

Then $\epsilon < n^{-1}$ for every $n \in \mathbb{N}$, hence $\epsilon \approx 0$.

From this point of view, compactness is close to being finite, at least in a metric space. In fact in many aspects, compactness is a kind of generalization of finiteness.

The classical **Heine-Borel Theorem** says that in an Euclidean space $\mathbb{R}^n$, $n \in \mathbb{N}$, a subset $Y \subset \mathbb{R}^n$ is compact iff it is closed and bounded. (Equivalently $^*Y \subset \mathrm{ns}({}^*\mathbb{R}^n) \wedge \mathrm{st}[{}^*Y] = Y$.). While this does not hold in

NONSTANDARD ANALYSIS   69

a general metric space (see Cor. 2.22, however), we do have the following result. Part of it was already proved, the rest is left as an exercise.

Theorem 1.22. (Generalized Heine-Borel Theorem) *In a metric space, a subset is compact iff it is complete and totally bounded.* □

An important feature of compactness is that the property is preserved by continuous functions.

Proposition 1.26. *Let* $f : X_1 \to X_2$ *be continuous then* $f[K]$ *is compact for any compact* $K \subset X_1$.

Proof. Let $K \subset X_1$ be compact. Then by Thm. 1.21, $^*K \subset \text{ns}(^*X_1)$ and $\text{st}[^*K] = K$. Together with Cor. 1.6, we have

$$^*f[^*K] \subset \bigcup_{x \in \text{st}[^*K]} \mu(f(x)) = \bigcup_{x \in K} \mu(f(x)) \subset \text{ns}(^*X_2).$$

Note that $^*(f[K]) = {}^*f[^*K]$, hence $\text{st}[^*(f[K])] \subset f[K]$ follows from the above and continuity, giving $\text{st}[^*(f[K])] = f[K]$. Therefore $f[K]$ is compact, by Thm. 1.21 again. □

Given topological spaces $(X_i, \mathcal{T}_i)$, indexed by $i \in I$, we form a product space $(\prod_{i \in I} X_i, \mathcal{T})$, where on the Cartesian product $\prod_{i \in I} X_i$ the topology $\mathcal{T}$ is generated by sets of the form

$$\prod_{i \in I} U_i, \quad \text{where } U_i \in \mathcal{T}_i \text{ and } U_i = X_i \text{ for all but finitely many } i \in I.$$

$(\prod_{i \in I} X_i, \mathcal{T})$, is called the **Tychonoff product** of the X_i. Unless specified, product space always refers to Tychonoff product only.

Let $X = \prod_{i \in I} X_i$, then $^*X = \prod_{i \in {}^*I} {}^*X_i$. Notice that under the above $\mathcal{T}$, we have

$$\text{ns}(^*X) = \{x \in {}^*X \mid \forall i \in I \, (x_i \in \text{ns}(^*X_i))\} \quad \text{and}$$

$$\text{st} : x = (x_i)_{i \in {}^*I} \mapsto \left(^\circ x_i\right)_{i \in I} \quad \text{for every } x \in \text{ns}(^*X).$$

Now we have a very simple proof of the following classical result, due to Robinson.

Theorem 1.23. (Tychonoff's Theorem) *Let* X_i, $i \in I$, *be compact, then* $\prod_{i \in I} X_i$ *is compact.*

Proof. Since the X_i's are compact, $\text{ns}(^*X) = {}^*X$. Moreover,

$$\text{st}[^*X] = \{(^\circ x_i)_{i \in I} \mid x \in {}^*X\} = \prod_{i \in I} X_i = X.$$

Therefore X is compact by Thm. 1.21. □

70   *Nonstandard Methods in Functional Analysis*

1.6.5   The Baire Category Theorem

Given $(X, \mathcal{T})$, the **interior** of a set $A \subset X$ is defined as:

$$\text{int}(A) := \bigcup \{U \mid U \in \mathcal{T} \wedge U \subset A\}.$$

A space X is called **locally compact** if

$$\forall x \in X \left(\exists \text{ compact } K \subset X \left(x \in \text{int}(K) \right) \right).$$

So locally compact spaces generalize compact spaces. For example, discrete spaces are locally compact and $\mathbb{R}$ is locally compact but not compact. We can see immediately the following consequence of the proof of Tychonoff's Theorem:

Corollary 1.9. *Let* X_i, $i \in I$, *be locally compact, then* $\prod_{i \in I} X_i$ *is locally compact.*   □

We leave as an exercise to check that open sets are determined by compact sets in a locally compact space.

Proposition 1.27. *In a locally compact space* X, $U \subset X$ *is open iff* $\forall$ *compact* $K \subset X$ $\left(K \setminus U \text{ is compact} \right)$.   □

A subset $A \subset X$ is called **dense** if $\bar{A} = X$. If X has a countable dense subset, it is called **separable**. So, since $\bar{\mathbb{Q}} = \mathbb{R}$, $\mathbb{R}$ is separable.

The following is straightforward.

Proposition 1.28. $A \subset X$ *is dense iff* $\forall x \in X \left(\mu(x) \cap {}^*A \neq \emptyset \right)$.   □

Corollary 1.10. *If* $A, B \subset X$ *are open dense, then* $A \cap B$ *is open dense.*

Proof. $A \cap B$ is clearly open. Suppose it is not dense, let $c \in X$ such that $\mu(c) \cap {}^*(A \cap B) = \emptyset$, by Prop. 1.28. Then by saturation, for some $U \in \mathcal{T}$, $c \in U$ and ${}^*U \cap {}^*A \cap {}^*B = {}^*U \cap {}^*(A \cap B) = \emptyset$.

By A being dense open, $U \cap A \neq \emptyset$ and open; then by B being dense open, $U \cap A \cap B \neq \emptyset$, a contradiction.   □

The following will be an important topological tool for the coming chapters.

Theorem 1.24.   (Baire Category Theorem) *Let* $(X, \mathcal{T})$ *be either a complete metric space or a locally compact space. Let* $V_n \subset X$ *be open dense,* $n \in \mathbb{N}$, *then* $\bigcap_{n \in \mathbb{N}} V_n$ *is dense.*

Proof. Let $f : \mathbb{N} \to \mathcal{T}$ be given by $f(n) = V_n$. Then $^*f : {}^*\mathbb{N} \to {}^*\mathcal{T}$ and we denote $^*f(n)$ by W_n.

In particular, ${}^*\left(\bigcap_{n \in \mathbb{N}} V_n \right) = \bigcap_{n \in {}^*\mathbb{N}} W_n$.

Fix an arbitrary $a \in X$, we will show that $\mu(a) \cap {}^*\bigcap_{n \in \mathbb{N}} V_n \neq \emptyset$, i.e. $\mu(a) \cap \bigcap_{n \in {}^*\mathbb{N}} W_n \neq \emptyset$. Then the result follows from Prop. 1.28.

Fix any $N \in {}^*\mathbb{N} \setminus \mathbb{N}$.

The complete metric space case:

Since the hyperfinite intersection $\bigcap_{n < N} W_n$ is *open dense, by Cor. 1.10, there is a nonempty *open ball $B_N \subset \left(B(a, N^{-1}) \cap \bigcap_{n < N} W_n \right)$. In particular, $B_N \subset \left(\mu(a) \cap \bigcap_{n < N} W_n \right)$.

Suppose $M > N$ and B_{M-1} was chosen. Since $\bigcap_{n < M} W_n$ is *open dense, we can find a nonempty *open ball $B_M \subset \left(\bigcap_{n < M} W_n \cap B_{M-1} \right)$ such that diameter$(B_M) < M^{-1}$.

By $^*X$ being *complete, there is a unique $b \in \bigcap_{N < M \in {}^*\mathbb{N}} B_M$. Then $b \in B_N \subset \mu(a)$ and $b \in \bigcap_{n \in {}^*\mathbb{N}} W_n$. Therefore $\mu(a) \cap \bigcap_{n \in {}^*\mathbb{N}} W_n \neq \emptyset$.

The locally compact space case:

Let $\mathcal{F} = \{ {}^*U \mid a \in U \in \mathcal{T} \}$. (So $\mu(a) = \bigcap \mathcal{F}$.) For any finite $\mathcal{F}_0 \subset \mathcal{F}$, by $^*X$ being *locally compact and $\bigcap_{n < N} W_n$ *open *dense, there is a *compact $K \subset \left(\bigcap \mathcal{F}_0 \cap \bigcap_{n < N} W_n \right)$ with int$(K) \neq \emptyset$. Then by saturation, we fix some *compact $K_N \subset \left(\bigcap \mathcal{F} \cap \bigcap_{n < N} W_n \right)$ with int$(K_N) \neq \emptyset$.

We are going to define for $M > N$ some $K_M \subset {}^*X$ satisfying the following property ϕ_M :

$$\left(K_M \text{ is } {}^*\text{compact} \right) \wedge \left({}^*\text{int}(K_M) \neq \emptyset \right) \wedge \left(K_M \subset \left(\bigcap_{n < M} W_n \cap {}^*\text{int}(K_{M-1}) \right) \right).$$

By $^*X$ being *local compact, $\bigcap_{n < N+1} W_n$ being *open *dense and *int(K_N) being nonempty *open, clearly such K_{N+1} can be found.

Now apply induction internally to $M > N$ in $^*\mathbb{N}$. For $M > N$, if K_M is defined so that ϕ_M is satisfied, then from ϕ_M, *int(K_M) is nonempty *open, together with the *local compactness of $^*X$ and the *openness and *denseness of $\bigcap_{n < M+1} W_n$ there is K_{M+1} satisfying ϕ_{M+1}.

The such constructed K_M's, $M > N$, are *compact, decreasing and nonempty, so $\bigcap_{M > N} K_M \neq \emptyset$. Therefore

$$\emptyset \neq \left(\bigcap_{M > N} K_M \right) \subset \left(K_N \cap \bigcap_{n \in {}^*\mathbb{N}} W_n \right) \subset \left(\mu(a) \cap \bigcap_{n \in {}^*\mathbb{N}} W_n \right).$$

$\square$

A subset $A \subset X$ is said to be *nowhere dense* if $\text{int}(\overline{A}) = \emptyset$. A set is said to be of the *first category*, or *meager*, if it is a countable union of nowhere dense sets. Otherwise it is said to be of the *second category*. A topological space X is called a *Baire space* if every nonempty open subset of X is of the second category.

It is easy to check that the condition in the statement of Thm. 1.24 is equivalent to being a Baire space. Hence complete metric spaces and locally compact spaces are Baire spaces.

1.6.6   *Stone-Čech compactification*

Given a topological space X, the *Stone-Čech compactification* of X is a compact space, denoted by βX, extending X, such that

- X is dense in βX;
- for every continuous $f : X \to [0, 1]$, there is a unique extension of f to a continuous function $\beta f : \beta X \to [0, 1]$;
- βX is unique up to homeomorphism *w.r.t* above properties.

It is with the above properties that Stone-Čech compactification is considered to be the largest compactification of a space.

As an illustration, consider $X = \mathbb{N}$ with the discrete topology.

Let $\beta \mathbb{N} := \{\mu(n) \,|\, n \in {}^{*}\mathbb{N}\}$, where we define

$$\mu(n) = \bigcap \big\{ \, {}^{*}U \,|\, (n \in {}^{*}U) \wedge (U \subset \mathbb{N}) \big\}.$$

Then the $\mu(n)$'s form a partition of ${}^{*}\mathbb{N}$ and behave like monads. Moreover, there is a correspondence between $\beta \mathbb{N}$ and the set of ultrafilters over $\mathbb{N}$ by mapping each $\mu(n)$ to $\{U \subset \mathbb{N} \,|\, n \in {}^{*}U\}$. It is easy to check that this correspondence is a bijection. For this reason, $\beta \mathbb{N}$ is also regarded as the set of ultrafilters over $\mathbb{N}$. Observe that the $\mu(n)$'s for $n \in \mathbb{N}$ correspond precisely to the principal ultrafilters over N.

The topology on $\beta \mathbb{N}$ is generated by open sets of the form

$$\{\mu(n) \,|\, n \in {}^{*}U\}, \, U \subset \mathbb{N}.$$

The mapping $\mathbb{N} \ni n \mapsto \mu(n)$ is clearly a homeomorphism onto its image, so we identify $\mathbb{N}$ with $\{\mu(n) \,|\, n \in \mathbb{N}\}$. Then for each $U \subset \mathbb{N}$, we simply let ${}^{*}U$ stand for $\{\mu(n) \,|\, n \in {}^{*}U\}$.

$\mathbb{N}$ is dense in $\beta \mathbb{N}$, since it holds for each $U \subset \mathbb{N}$ that ${}^{*}U \neq \emptyset$ iff $U \neq \emptyset$ iff ${}^{*}U \cap \mathbb{N} \neq \emptyset$.

For any family $U_i \subset \mathbb{N}$, $i \in I$, such that $^*\mathbb{N} = \bigcup_{i \in I} {}^*U_i$, there is a finite $I_0 \subset I$ such that $^*\mathbb{N} = \bigcup_{i \in I_0} {}^*U_i$. For otherwise it follows from saturation that there is some $n \in {}^*\mathbb{N} \setminus \bigcup_{i \in I} {}^*U_i$. Therefore $\beta\mathbb{N}$ is a compact space.

Now let $f : \mathbb{N} \to [0,1]$. (So f is automatically continuous.) Let $n \in {}^*\mathbb{N}$. So there is $y \in [0,1]$ such that $^*f(n) \approx y$. Then for each open $V \subset [0,1]$, if $y \in V$, then $^*f(n) \in {}^*V$ and hence $n \in {}^*(f^{-1}[V])$. Suppose $n' \in {}^*\mathbb{N}$ and $\mu(n') = \mu(n)$. Then $n \in {}^*(f^{-1}[V])$ iff $n' \in {}^*(f^{-1}[V])$ for any open $V \subset [0,1]$, by the continuity of f. Hence $^\circ\big(\,^*f(n)\big) = y = {}^\circ\big(\,^*f(n')\big)$. Therefore the function $\beta f : \beta\mathbb{N} \to [0,1]$ given by

$$(\beta f)(\mu(n)) := {}^\circ\big(\,^*f(n)\big), \quad \text{for each } \mu(n) \in \beta\mathbb{N},$$

is well-defined. It is clear that βf extends f.

Moreover, for each open $V \subset [0,1]$, we have

$$(\beta f)^{-1}[V] = \big\{\mu(n) \mid (n \in {}^*\mathbb{N}) \wedge (\beta f)\big(\mu(n)\big) \in V\big\}$$
$$= \big\{\mu(n) \mid (n \in {}^*\mathbb{N}) \wedge {}^*f(n) \in {}^*V\big\} = {}^*(f^{-1}[V]),$$

which is open in $\beta\mathbb{N}$. *i.e.* βf is continuous.

The fact that this $\beta\mathbb{N}$ is unique up to homeomorphism is a consequence of the following whose proof is left as an exercise.

Proposition 1.29. *Let X be a Tychonoff space. Let $\beta_1 X$ and $\beta_2 X$ be compact spaces having X as a dense subspace such that every continuous function $f : X \to [0,1]$ extends uniquely to a continuous function from $\beta_i X \to [0,1]$, for both $i = 1, 2$. Then $\beta_1 X$ and $\beta_2 X$ are homeomorphic.* $\square$

We will give a short proof in §3.2.2 that every Tychonoff space has a Stone-Čech compactification.

Note that in the above, if we replace $\mathbb{N}$ by any discrete space X, then the same proof shows that then βX, Stone-Čech compactification of X, is simply the space of ultrafilters over X.

We remark that the Stone-Čech compactification can be equivalently defined by using any compact spaces instead of the unit interval for the range of the continuous functions. The proof is left as an exercise.

Proposition 1.30. *Let X be a Tychonoff space and βX its Stone-Čech compactification. Then for any compact topological space Y, any continuous $f : X \to Y$ extends to a continuous function $\beta f : \beta X \to Y$.* $\square$

1.6.7 *Notes and exercises*

When the analytic and geometric features of analysis are removed, the foundation left is just topology, hence its particular relevance to functional analysis.

The nonstandard approach to the theory of topology was initiated and developed by Robinson.

The theory of monads was conceived by Leibniz and has philosophical meanings beyond topology. Both Robinson and Luxemburg are credited for the modern nonstandard formulation and study of monads.

It seems that so far Brouwer's Fixed Point Theorem (Thm. 1.20) has eluded a proof based on nonstandard techniques. However, in the reverse mathematics systems of Friedman and Simpson ([Simpson (2009)]), Brouwer's Fixed Point Theorem has the same strength as infinite combinatorial principals such as the Weak König's Lemma, so a proof based on properties of hyperfinite sets should be plausible.

The characterization of compactness by Robinson is a very simple but yet powerful tool having a large variety of applications not only in topology but in functional analysis and stochastic analysis as well.

The notion of Stone-Čech compactification as well as methods of constructing it are due to M.H. Stone and E. Čech for their work in the 1930's.

For more application of nonstandard methods in the construction of compactification, see [Salbany and Todorov (2000)].

Consult [Munkres (2000)] for a more thorough coverage of topics in topology.

EXERCISES

(1) For every $x \in$ ns($^*X$), show that there is $U \in {}^*\mathcal{T}$ such that $x \in U \subset \mu(x)$.
(2) Prove Prop. 1.16.
(3) Let $f : X_1 \to X_2$, where (X_1, d_1) and (X_2, d_2) are metric spaces. f is said to be **uniformly continuous** if

$$\forall \epsilon \in R^+ \ \exists \delta \in \mathbb{R}^+ \ \forall x, y \in X_1 \ \big(d_1(x, y) < \delta \Rightarrow d_2(f(x), f(y)) < \epsilon\big).$$

Show that f is uniformly continuous iff

$$\forall x, y \in {}^*X \ \big(x \approx y \Rightarrow {}^*f(x) \approx {}^*f(y)\big).$$

(4) Verify Prop. 1.25 and prove Thm. 1.22. Give an example of a complete and bounded but not compact subset of a metric space.

(5) Show that a function $f : X_1 \to X_2$ between two topological space is continuous iff graph$(f) := \{(x, f(x)) \mid x \in X_1\}$ is closed in the product space $X_1 \times X_2$.

(6) Let $f : X_2 \to X_3$ and $g : X_1 \to X_2$ be continuous functions between topological spaces X_1, X_2 and X_3. Show that the composition $f \circ g$ is continuous.

(7) Prove Prop. 1.22.

(8) Show that a space $(X, \mathcal{T})$ is regular iff $\forall x \in \mathrm{ns}(\,^*X)$

$$\mu(x) = \bigcap \{\,^*C \mid x \in C \wedge (X \setminus C) \in \mathcal{T}\,\}.$$

(9) Prove that locally compact spaces are completely regular and metric spaces are normal.

(10) Show that the uncountable product of $\mathbb{R}$, i.e. $\mathbb{R}^\lambda$ for some $\lambda \geq \omega_1$, is not a normal space

(11) Show that locally compact spaces are regular.

(12) Prove Prop. 1.27 and Prop. 1.28.

(13) Show that compact metric spaces are separable.

(14) Show that a complete pseudometric space is Baire.

(15) Let X_1, X_2 be topological spaces with X_1 a Baire space. Suppose $f_n : X_1 \to X_2$, $n \in \mathbb{N}$, are continuous and $f : X_1 \to X_2$ is given by $x \mapsto \lim_{n \to \infty} f_n(x)$.
Show that $\{x \in X \mid f$ is continuous at $x\}$ is dense in X_1.

(16) A *net* is some $\{x_i\}_{i \in I} \subset X$, where $(I, \leq)$ is a directed set.
The *net convergence*, $x_i \to x$, is defined as

$$\forall U \in \mathcal{T}\left((x \in U) \Rightarrow \left(\exists i \in I \,\forall j > i\,(x_j \in U)\right)\right).$$

Show that $f : X_1 \to X_2$ is continuous at $a \in X_1$ iff for every net $\{a_i\}_{i \in I} \subset X_1$, if $a_i \to a$ then $f(a_i) \to f(a)$.

(17) Prove Prop. 1.29.

(18) Prove Prop. 1.30.

Chapter 2

Banach Spaces

2.1 Norms and Nonstandard Hulls

The nonstandard hull method creates Banach spaces from internal ones.
Finite dimensional Banach spaces admit simple characterizations.

We always deal with a vector space X over a scalar field $\mathbb{F}$, where $\mathbb{F}$ is either $\mathbb{R}$ or $\mathbb{C}$. Unless it is necessary, the underlying field is normally not explicitly specified. A vector space is also referred to as a *linear space*. A subset of X forms a basis (or a *Hamel basis*, to distinguish it from another notion that will be used later on) if it is a maximal subset of linearly independent elements and the cardinality of the basis is $\dim(X)$, the *dimension* of X. The existence of a basis is guaranteed by AC.

In this section, we introduce basic results about seminorms, norms, seminormed/normed linear spaces and Banach spaces. Then we give the nonstandard hull construction, which is the most important construction used throughout this book. After some familiarization with the finite dimensional case, well-known examples of Banach spaces are provided.

2.1.1 *Seminormed linear spaces and quotients*

A *pseudo-seminorm* on a vector space X is a *subadditive positively homogeneous* function $p : X \to \mathbb{R}$, *i.e.*

$$\forall x, y \in X \ \forall \lambda \in \mathbb{R}^+ \left(\big(p(x+y) \leq p(x) + p(y)\big) \wedge \big(p(\lambda x) = \lambda p(x)\big)\right).$$

Note it follows that $p(0) = 0$.

A *seminorm* (usually written as $\|\cdot\|$,) on X is a *subadditive homogeneous* function with nonnegative range, *i.e.* $\|\cdot\| : X \to [0, \infty)$ satisfying

$$\forall x, y \in X \ \forall \alpha \in \mathbb{F} \left(\big(\|x+y\| \leq \|x\| + \|y\| \big) \wedge \big(\|\alpha x\| = |\alpha| \, \|x\| \big)\right).$$

All seminorms are pseudo-seminorms but the converse could fail as there are pseudo-seminorms that take negative values.

In particular, for a seminorm, the function on X^2 given by $(x, y) \mapsto \|x - y\|$ is a pseudometric. With the topology (not necessarily Hausdorff) given by this pseudometric, X forms a **seminormed linear space**. Trivially it is a topological vector space in the sense that the functions $x \mapsto \alpha x$, $\alpha \in \mathbb{F}$, are continuous and $+ : X^2 \to X$ is continuous *w.r.t.* the product topology on X^2. Note also that $\|\cdot\|$ is a continuous function *w.r.t.* the topology generated.

Occasionally we consider **infinite-valued seminorms**, *i.e.* some functions $\|\cdot\| : X \to [0, \infty]$ satisfying the seminorm axiom, where we make the requirements that

$$\forall r \in [0, \infty] \, (r + \infty = \infty = \infty + r), \quad \forall r \in (0, \infty] \, (r \cdot \infty = \infty),$$

but $0 \cdot \infty = 0$.

If $\|\cdot\|$ satisfies a further condition

- $\forall x \in X \, \big((\|x\| = 0) \Rightarrow (x = 0) \big)$,

$\|\cdot\|$ is called a **norm** and X is called a **normed linear space** and it is necessarily Hausdorff since the function $(x, y) \mapsto \|x - y\|$ becomes a metric.

For simplicity, the same symbol $\|\cdot\|$ may be used to denote different norms on the same or different spaces. If emphasis is needed, we would write the symbols with subscripts such as $\|\cdot\|_1$, $\|\cdot\|_2$, or $\|\cdot\|_X$ *etc.*

A normed/pseudonormed linear space is called a **real** normed/pseudonormed linear space or a **complex** normed/pseudonormed linear space according to whether $\mathbb{F}$ is $\mathbb{R}$ or $\mathbb{C}$.

The completeness of a seminormed linear space refers to the completeness as a pseudometric space.

Two seminorms on the same linear space are said to be **equivalent** iff the corresponding pseudometrics are equivalent, *i.e.* generating the same topology.

A complete normed linear space is called a **Banach space**. It is called a **real Banach space** if $\mathbb{F} = \mathbb{R}$ and a **complex Banach space** if $\mathbb{F} = \mathbb{C}$. So every Cauchy sequence in a Banach space converges to a limit in the space. It is worthwhile to notice that a normed linear space is Banach iff every absolutely convergent series is convergent in it.

Recall from the comment on p.17 the usage of *vector space, *normed linear space, *Banach space, etc. (*i.e. internal vector space, internal normed linear space, internal Banach space, etc.*)

For example, an internal vector space is some internal set X with an internal binary function $+$ on it, $0 \in X$ and scalar multiplication by each $\alpha \in {}^*\mathbb{F}$ so that all vector space axioms are satisfied. It is an internal normed linear space if there is an internal $\|\cdot\| : X \to {}^*[0, \infty)$ satisfying the internal counterpart of the norm axioms. It is an internal Banach space, if it is *complete.

Let A be a subset of a vector space X, then for $x \in X$ and $\alpha \in \mathbb{F}$, $x + A$ denotes the set $\{x + y \mid y \in A\}$ and αA the set $\{\alpha x \mid x \in A\}$.

Given a vector space space X and subspace $Y \subset X$, we write

$$X/Y := \{x + Y \mid x \in X\}.$$

Elements $x + Y$ are called **cosets** of Y.

It is easy to check that for $x_1, x_2 \in X$ and $\alpha \in \mathbb{F}$ the operation

$$(x_1 + Y) + \alpha(x_2 + Y) := (x_1 + \alpha x_2) + Y$$

is well-defined and X/Y forms a vector space, the **quotient space**, under these operations.

The following is left as an exercise.

Proposition 2.1. *Let X be a seminormed linear space and $Y \subset X$ a subspace. Define on X/Y a mapping*

$$(x + Y) \mapsto \|x + Y\| := \inf\{\|x + y\| \mid y \in Y\}.$$

Then

(i) *The mapping defines a seminorm on X/Y, the quotient seminorm.*
(ii) *If $Y \supset \{x \in X \mid \|x\| = 0\}$ and is closed, then X/Y becomes a normed linear space under the quotient norm.*
(iii) *If X is a Banach space and Y is closed, then X/Y is also a Banach space under the quotient norm.* $\square$

In particular, Prop. 2.1(ii) says that if X is a normed linear space with closed subspace $Y \subset X$, then the quotient space X/Y always forms a normed linear space.

When $\dim(X/Y) = 1$, we say that Y is a **hyperplane** of X.

The **distance** between an element $x \in X$ and some $A \subset X$ is defined as

$$\mathrm{dist}(x, A) := \inf\{\|x - y\| \mid y \in A\}.$$

More generally, for $A, B \subset X$,

$$\mathrm{dist}(A, B) := \inf\{\|x - y\| \mid x \in A, y \in B\}.$$

Note that in Prop. 2.1, since Y is a subspace, $\|x + Y\| = \mathrm{dist}(x, Y)$.

Proposition 2.2. *Let X be a normed linear space and $Y \subset X$ a closed subspace. Then*

(i) *For any $x \in {}^*X$, if $x \approx 0$ in ${}^*X$ then $(x + Y) \approx 0$ in ${}^*(X/Y)$.*
(ii) *Suppose Z/Y, is closed in X/Y, where $Y \subset Z \subset X$ is some subspace. Then Z is a closed subspace of X.*

Proof.   For (i), if $x \approx 0$, then $\|(x + Y)\| = \mathrm{dist}(x, Y) \approx 0$, as $0 \in Y$. Hence $(x + Y) \approx 0$ in ${}^*(X/Y)$.

For (ii), let $a \in X$ such that $a \approx c$ for some $c \in {}^*Z$. Then by (i), $(a + Y) \approx (c + Y) \in {}^*(Z/Y)$. Since Z/Y is closed, by Prop.1.14, $(a + Y) \in Z/Y$. Hence $a \in Z$, since Z is a subspace such that $Y \subset Z \subset X$.

Therefore Z is closed, by Prop.1.14. $\square$

2.1.2   *Internal spaces and nonstandard hulls*

Consider an internal seminormed linear space X (with an internal seminorm $\|\cdot\|$). The finite part is defined as

$$\mathrm{Fin}(X) := \{ x \in X \mid \|x\| \in \mathrm{Fin}({}^*\mathbb{R}) \}.$$

In particular,

$$\mathrm{Fin}(X) = \bigcup_{n \in \mathbb{N}} {}^*B(0, n).$$

(Recall that $B(a, r)$ denotes the open ball $\{ x \in X \mid \|x - a\| < r \}$.)

For any $A \subset X$ we sometime write $\mathrm{Fin}(A)$ for $A \cap \mathrm{Fin}(X)$.

Note also that for a seminormed linear space X, $\mathrm{ns}({}^*X) \subset \mathrm{Fin}({}^*X)$ always holds. (See Cor. 2.1 below for criteria for the other inclusion.) Of course if the internal seminormed linear space is not of the form ${}^*X$, the nearstandard part is undefined.

For an internal seminormed linear space X, the monad of any $x \in X$, is given by

$$\mu(x) = \{ y \in X \mid \|y - x\| \approx 0 \} = \{ y \in X \mid y \approx x \}.$$

Hence $\forall x \in X \left(\mu(x) = \mu(0) + x \right)$, i.e. $\{ y + x \mid y \in \mu(0) \}$. In particular, the topology is determined by $\mu(0)$:

Proposition 2.3. *Let $\|\cdot\|_1$ and $\|\cdot\|_2$ be seminorms on a linear space X. Then $\|\cdot\|_1$ and $\|\cdot\|_2$ are equivalent iff $\mu_1(0) = \mu_2(0)$, i.e. their corresponding monads in ${}^*X$ are equal.* $\square$

Proof.   Let $\mathcal{T}_1$ and $\mathcal{T}_2$ denote the corresponding topologies and B_1 and B_2 the corresponding unit open balls centered at 0.

$(\Rightarrow)$: $\mathcal{T}_1 \subset \mathcal{T}_2$, implies $\forall r \in \mathbb{R}^+ \, \exists s \in \mathbb{R}^+ \left(B_2(0,s) \subset B_1(0,r)\right)$, hence for any such r, s,

$$\mu_2(0) = \bigcap_{n \in \mathbb{N}} {}^*B_2(0, n^{-1}s) \subset \bigcap_{n \in \mathbb{N}} {}^*B_1(0, n^{-1}r) = \mu_1(0).$$

Likewise $\mathcal{T}_2 \subset \mathcal{T}_1$, implies $\mu_1(0) \subset \mu_2(0)$, therefore $\mu_1(0) = \mu_2(0)$.

$(\Leftarrow)$: If $\mu_1(0) \subset \mu_2(0)$, then

$$\mu_1(0) = \bigcap_{n \in \mathbb{N}} {}^*B_1(0, n^{-1}) \subset \mu_2(0) \subset {}^*B_2(0,1),$$

so saturation implies that ${}^*B_1(0, n^{-1}) \subset {}^*B_2(0,1)$, for some $n \in \mathbb{N}^+$. Then $\left(x + {}^*B_1(0, n^{-1}r)\right) \subset \left(x + {}^*B_2(0,r)\right)$ for any $x \in {}^*X$ and $r \in \mathbb{R}^+$, therefore $\mathcal{T}_2 \subset \mathcal{T}_1$.

Similarly $\mu_2(0) \subset \mu_1(0)$ implies $\mathcal{T}_1 \subset \mathcal{T}_2$. Hence $\mathcal{T}_1 = \mathcal{T}_2$. $\square$

A useful fact to note is the following.

Proposition 2.4. *Let X be a linear space quipped with equivalent seminorms $\|\cdot\|_1$ and $\|\cdot\|_2$. Then there is $k \in \mathbb{N}^+$ such that*

$$\forall x \in X \left(k^{-1} \|x\|_1 \leq \|x\|_2 \leq k \|x\|_1\right).$$

Proof.   Let μ_i and B_i denote the monads and unit open balls centered at 0 *w.r.t.* the seminorm $\|\cdot\|_i$.

Then $\mu_1(0) = \mu_2(0)$. As in the proof of Prop. 2.3, we have for some $n, m \in \mathbb{N}^+$ that ${}^*B_1(0, n^{-1}) \subset {}^*B_2(0,1)$ and ${}^*B_2(0, m^{-1}) \subset {}^*B_1(0,1)$.

Then $\forall x \in X \left((\|x\|_2 \leq n \|x\|_1) \wedge (\|x\|_1 \leq m \|x\|_2)\right)$. Hence, by taking $k = \max\{n, m\}$,

$$k^{-1} \|x\|_1 \leq mk^{-1} \|x\|_2 \leq \|x\|_2 \leq n \|x\|_1 \leq k \|x\|_1, \quad x \in X.$$ $\square$

When emphasis is needed, we write μ_X for the monads in X. If X is a seminormed linear space, we sometimes write μ_X instead of μ_{*X}, for monads in ${}^*X$.

For an internal seminormed linear space X and $x \in \mathrm{Fin}(X)$, monads $\mu(x)$ are also denoted by $\widehat{x}$.

The **nonstandard hull** of an internal seminormed linear space X is defined as

$$\widehat{X} := \{\widehat{x} \mid x \in \mathrm{Fin}(X)\}.$$

For convenience, if $A \subset \mathrm{Fin}(X)$,

$$\widehat{A} \quad \text{denotes} \quad \{\widehat{x} \mid x \in A\}.$$

In particular, $\widehat{A}$ and $\widehat{X}$ are the same thing when $A = \mathrm{Fin}(X)$.

We also sometimes write

$$(\cdots \quad \cdots)^{\wedge}$$

when dealing with a long expression.

We define for $x, y \in \mathrm{Fin}(X)$ and $\alpha \in \mathrm{Fin}(\,^*\mathbb{F})$ (*i.e.* $\mathrm{ns}(\,^*\mathbb{F})$) the following operations:

- $(\widehat{x}, \widehat{y}) \mapsto \widehat{x + y}$;
- $\widehat{x} \mapsto \widehat{\alpha x}$;
- $\widehat{x} \mapsto \|\widehat{x}\| := {}^{\circ}\|x\|$.

Then it is simple to check that they are well-defined. For example, if $\widehat{x_1} = \widehat{x_2}$ and $\widehat{y_1} = \widehat{y_2}$, then $(x_1 - x_2) \approx (y_1 - y_2)$, hence $(x_1 + y_1) - (x_2 + y_2) \approx 0$ and so $\widehat{x_1 + y_1} = \widehat{x_2 + y_2}$.

For convenience, the same symbols are used for the above operations and the corresponding internal ones in X. So $0 = \widehat{0}$, $\widehat{x} + \widehat{y} = \widehat{x + y}$, ${}^{\circ}\alpha\widehat{x} = \widehat{\alpha x}$ and $\|\widehat{x}\| = {}^{\circ}\|x\|$.

Theorem 2.1. *Let X be an internal seminormed linear space. Then the nonstandard hull $\widehat{X}$ forms a Banach space.*

Proof. The needed properties of a normed linear space are easy to check. For example, $\widehat{x} + \widehat{y} = \widehat{x + y} = \widehat{y + x} = \widehat{y} + \widehat{x}$ and

$$\|\widehat{x} + \widehat{y}\| = \left\|\widehat{x + y}\right\| \approx \|x + y\| \le \|x\| + \|y\| \approx \|\widehat{x}\| + \|\widehat{y}\|,$$

hence $\|\widehat{x} + \widehat{y}\| \le \|\widehat{x}\| + \|\widehat{y}\|$, *etc.*

The completeness of $\widehat{X}$ follows from saturation. Let $\{\widehat{x}_n\}_{n \in \mathbb{N}}$ be a Cauchy sequence in $\widehat{X}$, where $x_n \in \mathrm{Fin}(X)$. Extend $\{x_n\}_{n \in \mathbb{N}}$ to an internal sequence $\{x_n\}_{n < \,^*\mathbb{N}}$. (Recall Prop. 1.1.)

For any $k \in \mathbb{N}^+$, there is $n \in \mathbb{N}$ so that for all $n < m \in \mathbb{N}$, $\|x_n - x_m\| \approx \|\widehat{x}_n - \widehat{x}_m\| < k^{-1}$. So it follows from overspill that for some $M_k \in \,^*\mathbb{N}$ that $\forall m \left((n < m < M_k) \Rightarrow (\|\widehat{x}_n - \widehat{x}_m\| \le k^{-1})\right)$. In particular, $x_m \in \mathrm{Fin}(X)$, for all $m < M_k$. Now let $M \in \,^*\mathbb{N} \setminus \mathbb{N}$ and $M < M_k$, $k \in \mathbb{N}$. Then $x_M \in \mathrm{Fin}(X)$ and $\lim_{n \to \infty} \|\widehat{x}_n - \widehat{x}_M\| = 0$, hence $\widehat{x}_n \to \widehat{x}_M \in \widehat{X}$. $\square$

Similarly, by saturation, we also have the following.

Proposition 2.5. *Let X be an internal seminormed linear space and let $A \subset \mathrm{Fin}(X)$ be an internal subset. Then $\widehat{A}$ is closed.* $\square$

Note that the above conclusion remains valid for $A \subset \mathrm{Fin}(X)$ which is the intersection of fewer than κ internal sets, where κ is the cardinality of saturation.

Most of time we only deal with the nonstandard hull of an internal normed linear space, instead of an internal seminormed linear space, as the resulting space is the same if we replace the internal seminormed linear space by its quotient by the kernel of the seminorm, which is an internal normed linear space.

Here is an alternative way to view the nonstandard hull $\widehat{X}$.

Proposition 2.6. *Regard* $\mathrm{Fin}(X)$ *as a seminormed linear space under the seminorm* $^{\circ}\|\cdot\|$. *Then* $\mu(0)$ *is a closed subspace of* $\mathrm{Fin}(X)$ *and* $\widehat{X}$ *is identical to the seminormed quotient space* $\mathrm{Fin}(X)/\mu(0)$, *which in fact forms a normed linear space.*

Proof. Trivially, $\mu(0)$ is closed under the seminorm $^{\circ}\|\cdot\|$.

For each $x \in \mathrm{Fin}(X)$, $\widehat{x} = \mu(x) = x + \mu(0)$, hence $\widehat{X} = \mathrm{Fin}(X)/\mu(0)$ as linear spaces.

Moreover, from the definition, the norm on the nonstandard hull $\widehat{X}$ is the same as the quotient seminorm on $\mathrm{Fin}(X)/\mu(0)$. Furthermore, by Prop. 2.1(ii), the quotient seminorm is actually a norm. $\square$

A bijection between linear spaces that preserves linear operations is called a *(linear) isomorphism*. When it exists, the linear spaces are called *(linearly) isomorphic*. A bijection between seminormed linear spaces (over the same field $\mathbb{F}$) is called an *isometric isomorphism* if it is a linear isomorphism that preserves the seminorms. When it exists, the spaces are called *isometrically isomorphic*. A linear subspace of a seminormed linear space is automatically a seminormed linear space. Likewise a linear subspace of a normed linear space is a normed linear space. A closed subspace of a Banach space is automatically a Banach space. An isometric isomorphism of a seminormed linear space onto a linear subspace of a seminormed linear space is called an *isometric embedding*.

In Thm. 2.1, if the X is of the form $^{*}X$ for some normed linear space X, we have a natural isometric embedding of X into a nonstandard hull.

Proposition 2.7. *Every normed linear space* X *embeds isometrically into* $\widehat{^{*}X}$ *via the mapping* $x \mapsto \widehat{x}$.

(*Henceforth, we regard* $X \subset \widehat{X}$ *under this identification.*) $\square$

Note that for a normed linear space X, we have $X = \mathrm{ns}\big[\widehat{^{*}X}\big]$.

For a subset $Y \subset X$ we let

$$\mathrm{Lin}(Y) := \Big\{ \sum_{e \in Y_0} \alpha_e \, e \mid Y_0 \subset Y \text{ is finite}, \; \alpha : Y_0 \to \mathbb{F} \Big\},$$

called the **linear span** of Y. Normally the scalar field $\mathbb{F}$ used is the same as the one for X. If they are allowed to be different, we would specify **real linear span** or **complex linear span** and write $\mathrm{Lin}_{\mathbb{R}}$ or $\mathrm{Lin}_{\mathbb{C}}$ accordingly.

So $\mathrm{Lin}(Y)$ is a linear subspace of X. The topological closure of $\mathrm{Lin}(Y)$ in the Banach space $^*\widehat{X}$ is called the **closed linear span** of X, denoted by $\overline{\mathrm{Lin}}(Y)$, and forms a Banach space.

In particular, given any normed linear space X, $\overline{\mathrm{Lin}}(X)$, is the minimal Banach space with X isometrically embedded into. $\overline{\mathrm{Lin}}(X)$, is called the **completion** of X and is simply denoted as $\overline{X}$, same as the topological closure.

Proposition 2.8. *Let X be an internal seminormed linear space, $n \in \mathbb{N}$ and $x_1, \ldots, x_n \in \mathrm{Fin}(X)$. Suppose $\widehat{x}_1, \ldots, \widehat{x}_n$ are linearly independent then $x_1, \ldots, x_n$ are *linearly independent.*

Proof. Suppose $x_1, \ldots, x_n$ are not *linearly independent, then let $\alpha_1, \ldots, \alpha_n \in \,^*\mathbb{F}$ with some of them nonzero and $\sum_{i=1}^{n} \alpha_i x_i = 0$. Without loss of generality, let $|\alpha_1| = \max_{i=1,\ldots,n} |\alpha_i| \neq 0$. Then $\sum_{i=1}^{n} \alpha_i \alpha_1^{-1} x_i = 0$. But then all $\alpha_i \alpha_1^{-1} \in \mathrm{Fin}(\,^*\mathbb{F})$ with one of them $= 1$ and $\sum_{i=1}^{n} {}^{\circ}(\alpha_i \alpha_1^{-1}) \widehat{x}_i = 0$, therefore $\widehat{x}_1, \ldots, \widehat{x}_n$ are not linearly independent. $\square$

It should be noted that Prop. 2.8 gives an internal reflection of an algebraic condition such as linear independence in the nonstandard hull, even though whose construction involves topology.

2.1.3   *Finite dimensional Banach spaces*

Proposition 2.9. *Let X be a finite dimensional normed linear space. Then X is isometrically isomorphic to $^*\widehat{X}$, hence $X = {}^*\widehat{X}$ through the natural identification $x \mapsto \widehat{x}$.*

Proof. Let $\theta : X \to \,^*\widehat{X}$ be defined by $\theta(x) = \widehat{x}$.

Then clearly θ preserves linear operations. Moreover, for $x \in X$,

$$\|\theta(x)\| = \|\widehat{x}\| = {}^\circ(\,^*\|x\|) = \|x\|.$$

(Note that $\,^*\|\cdot\|$ is the internal norm on $\,^*X$.)

In particular, θ is an isometric isomorphism onto a subspace of $\widehat{\,^*X}$.

Let the $\dim(X) = n \in \mathbb{N}$. Let $\{e_i\}_{i=1,\ldots,n}$ be a basis of X. Then since θ is a linear isomorphism onto a subspace of $\,^*\widehat{X}$, the set $\{\theta(e_i)\}_{i=1,\ldots,n} = \{\widehat{e}_i\}_{i=1,\ldots,n}$ is linearly independent.

Suppose θ is not onto $\,^*\widehat{X}$, let $\widehat{e} \in \,^*\widehat{X} \setminus \theta[X]$. Then $\{\widehat{e}_1,\ldots,\widehat{e}_n,\widehat{e}\}$ is linearly independent. So it follows from Prop. 2.8 that $\{e_1,\ldots,e_n,e\}$ is linearly independent. Hence $\dim(\,^*X) > n$. Then, by transfer, $\dim(X) > n$, a contradiction.

Therefore θ is a bijective. $\square$

The converse of Prop. 2.9 can be proved and is left as an exercise.

Proposition 2.10. *Let X be a finite dimensional normed linear space over $\mathbb{F}$ with a basis $\{e_i\}_{i=1,\ldots,n}$. Then*

$$\mathrm{Fin}(\,^*X) = \Big\{ \sum_{i=1}^n \alpha_i e_i \mid \alpha_1,\ldots,\alpha_n \in \mathrm{Fin}(\,^*\mathbb{F}) \Big\}.$$

Proof. The inclusion $(\supset)$ is clear.

For the other direction, let $a \in \mathrm{Fin}(\,^*X)$. Write $a = \sum_{i=1}^n \beta_i e_i$ for some $\beta_i \in \,^*\mathbb{F}$, as $\{e_i\}_{i=1,\ldots,n}$ is a basis of $\,^*X$.

By Prop. 2.9, $\widehat{a} = \sum_{i=1}^n \alpha_i e_i$ for some $\alpha_i \in \mathbb{F}$. Therefore

$$b := \sum_{i=1}^n (\beta_i - \alpha_i) e_i \approx 0.$$

Now, since each $\widehat{e}_i = e_i$,

$$\sum_{i=1}^n {}^\circ(\beta_i - \alpha_i) e_i = \sum_{i=1}^n {}^\circ(\beta_i - \alpha_i)\widehat{e}_i = \widehat{b} = 0.$$

Then, by the linear independence of $\{e_i\}_{i=1,\ldots,n}$, all ${}^\circ(\beta_i - \alpha_i) = 0$, hence all $\beta_i \in \mathrm{Fin}(\,^*\mathbb{F})$. $\square$

Proposition 2.11. *There is no isometric isomorphism between an infinite dimensional normed linear space X and $\widehat{\,^*X}$.* $\square$

Corollary 2.1. *Let X be a normed linear space, then $\mathrm{ns}(\,^*X) = \mathrm{Fin}(\,^*X)$ iff $\,^*B(0,1) \subset \mathrm{ns}(\,^*X)$ iff $\dim(X) < \infty$.* $\square$

Corollary 2.2. *A normed linear space X is finite dimensional iff the closed unit ball $\overline{B(0,1)}$ is compact.*

Proof.   Note that $\overline{B(0,1)} = \mathrm{st}\big({}^*B(0,1)\big)$ if ${}^*B(0,1) \subset \mathrm{ns}({}^*X)$. Then use Prop. 1.23 and Cor. 2.1. $\square$

Notice from Thm. 2.1 that finite dimensional normed linear spaces are automatically Banach spaces.

Note also from Cor. 2.2 that for a *Euclidean space ${}^*\mathbb{R}^N$, where $N \in$ $(\,{}^*\mathbb{N} \setminus \mathbb{N})$, the closed unit ball in ${}^*\mathbb{R}^N$ is not compact although it is clearly bounded and closed. (See Ex. 4 on p.74.)

Proposition 2.12. *Let X be a finite dimensional normed linear space. Then any two norms on X are equivalent.*

Proof.   For $i = 1, 2$, let $\|\cdot\|_i$ be a norm on X generating a topology $\mathcal{T}_i$ with monads denoted by $\mu_i(x)$ and the finite part by $\mathrm{Fin}_i({}^*X)$.

By an earlier remark, if $\mathcal{T}_1 \neq \mathcal{T}_2$ we would have $\mu_1(0) \neq \mu_2(0)$. Clearly, this is impossible if $\dim(X) = 1$, so we assume that $\dim(X) > 1$.

Suppose without loss of generality that $\mu_1(0) \setminus \mu_2(0) \neq \emptyset$. So there is $a \in {}^*X$ so that $\|a\|_1 \approx 0$ but $\|a\|_2 \not\approx 0$.

Let $\{e_i\}_{i=1,\ldots,n}$ be a basis of X and write $a = \sum_{i=1}^{n} \alpha_i e_i$. Since $a \in \mathrm{Fin}_1({}^*X)$, we have $\alpha_i \in \mathrm{Fin}({}^*\mathbb{F})$, by Prop. 2.10

Then $\|a\|_2 \leq \max\limits_{i=1,\ldots,n} |\alpha_i| \sum_{i=1}^{n} \|e_i\|_2$. But $\sum_{i=1}^{n} \|e_i\|_2 \in \mathbb{F}$, therefore it follows from $\|a\|_2 \not\approx 0$ that $\max_{i=1,\ldots,n} |\alpha_i| \not\approx 0$.

But then, in the nonstandard hull *w.r.t.* $\mathcal{T}_1$, $0 = \widehat{a} = \sum_{i=1}^{n} {}^\circ\alpha_i \widehat{e}_i$ with one of the coefficients ${}^\circ\alpha_i \neq 0$, since $\max_{i=1,\ldots,n} |\alpha_i| \not\approx 0$. So $\{\widehat{e}_i\}_{i=1,\ldots,n}$ is linearly dependent, leading to a contradiction by Prop. 2.9. $\square$

As a result, note that any two normed linear spaces over the same $\mathbb{F}$ of the same finite dimension are homeomorphic as topological spaces.

In particular, we have:

Corollary 2.3. *A finite dimensional normed linear space over $\mathbb{F}$ is homeomorphic to $\mathbb{F}^n$ for some $n \in \mathbb{N}$. (Hence also homeomorphic to the Euclidean space $\mathbb{R}^n$ for some $n \in \mathbb{N}$.)* $\square$

2.1.4   Examples of Banach spaces

A set C in a linear space is **convex** if it is closed under convex combinations, *i.e.* $\big(tx + (1-t)y\big) \in C$ whenever $x, y \in C$ and $t \in [0,1]$.

In the case $X = \mathbb{R}^n$ for some $n \in {}^*\mathbb{N}$, any bounded convex set C with 0 as an interior point w.r.t. the usual topology produces a norm given by $x \mapsto \sup\{r \mid rx \in C\}$. For example, define $\|(a_1, \ldots, a_n)\|_\infty := \max\{|a_i| \mid i = 1, \ldots, n\}$. Then $\|\cdot\|_\infty$ is a norm on $\mathbb{R}^n$.

Take $X = \mathbb{F}^n$ for some $n \in \mathbb{N}^+$, and let $\{e_i\}_{i=1,\ldots,n}$ be a basis. Then the following are norms on X :

$$a \mapsto \max_{i=1,\ldots,n} |\alpha_i|, \quad a \mapsto \Big(\sum_{i=1}^n |\alpha_i|^p \Big)^{1/p}, \ p \in [1, \infty), \quad \text{where } a = \sum_{i=1}^n \alpha_i e_i,$$

and hence they are all equivalent according to Prop. 2.12. Of course, the geometry corresponding to each one of them is different.

When the above n is replaced by an infinite set, we get examples of infinite dimensional Banach spaces.

In the following, the addition and scalar multiplications for functions from a set Ω to $\mathbb{F}$ are inherited from the pointwise definition in the linear space $\mathbb{F}^\Omega$, i.e.

$$\forall x \in \Omega \ \Big[(f + \alpha g) : x \mapsto f(x) + \alpha g(x) \Big], \quad \text{where } f, g \in \mathbb{F}^\Omega, \ \alpha \in \mathbb{F}.$$

For $\Omega = \mathbb{N}$, the following Banach spaces are referred to as the **classical sequence spaces**.

- ℓ_p, where $p \in [1, \infty)$: the elements are sequences $a : \mathbb{N} \to \mathbb{F}$ with $\|a\|_p < \infty$, where the norm, called the *p-norm*, is defined as

$$\|a\|_p := \Big(\sum_{n \in \mathbb{N}} |a_n|^p \Big)^{1/p}.$$

Note in particular that ℓ_1 consists of *absolutely convergent* sequences.

- ℓ_∞ : the elements are sequences $a : \mathbb{N} \to \mathbb{F}$ with $\|a\|_\infty < \infty$, where the norm, called the **supremum norm**, is defined as

$$\|a\|_\infty := \sup_{n \in \mathbb{N}} |a_n|.$$

Convergence of functions in ℓ_∞ is the same as uniform convergence, hence the supremum norm is also called the **uniform norm**.

- c, or $c(\mathbb{N})$: the elements are convergent sequences $a : \mathbb{N} \to \mathbb{F}$, with the supremum norm $\|\cdot\|_\infty$. Hence c forms a closed subspace of ℓ_∞.

- c_0 : the elements are sequences $a : \mathbb{N} \to \mathbb{F}$, such that $a_n \to 0$. The norm is still the supremum norm $\|\cdot\|_\infty$. So c_0 forms a closed subspace of c.

The Banach spaces ℓ_p for $p \in (1, \infty)$, c and c_0 are separable, while ℓ_∞ is nonseparable.

Observe also the set $X := \{ f \in \ell_\infty \mid |f[\mathbb{N}]| \in \mathbb{N} \}$ is a dense but incomplete subspace of ℓ_∞. However, it follows from Thm. 2.1 that $^*\widehat{X} = \widehat{\ell_\infty}$ through a natural identification of the monads and both form the same Banach space.

Now let Ω be a topological space and

$$C(\Omega) := \{ f : \Omega \to \mathbb{F} \mid f \text{ is continuous} \}.$$

Then under the pointwise addition and scalar multiplications, $C(\Omega)$ forms a linear space.

- $C_{\mathrm{b}}(\Omega)$: this is the subspace of $C(\Omega)$ consisting of elements $f : \Omega \to \mathbb{F}$ such that $\|f\|_\infty < \infty$, where the supremum norm is given by
$$\|f\|_\infty := \sup_{x \in \Omega} |f(x)|.$$
$C_{\mathrm{b}}(\Omega)$ is complete under the supremum norm, by the property of uniform limit of continuous functions.

For a locally compact Ω, we define

- $C_0(\Omega)$: this is a closed subspace of $C_{\mathrm{b}}(\Omega)$ consisting of elements $f : \Omega \to \mathbb{F}$ such that $\{ x \mid |f(x)| \geq \epsilon \}$ is compact for all $\epsilon \in \mathbb{R}^+$, with the supremum norm inherited from $C_{\mathrm{b}}(\Omega)$. Elements in $C_0(\Omega)$ are called **continuous functions vanishing at infinity**.

Of course, if Ω is compact, then $C_0(\Omega) = C_{\mathrm{b}}(\Omega) = C(\Omega)$.

For $\Omega = [0, 1]$, the Banach space $C([0, 1])$ is separable, by the Weierstrass Approximation Theorem using polynomials over $\mathbb{Q}$.

Note that when $\Omega = \mathbb{N}$ with the discrete topology, we have $C_{\mathrm{b}}(\mathbb{N}) = \ell_\infty$ and $C_0(\mathbb{N}) = c_0$.

Let Ω be a locally compact space.

- $\mathbb{M}(\Omega)$: this is the collection of regular σ-additive complex Borel measures on Ω under the obvious addition and scalar multiplication, with the norm of each measure μ given by $|\mu|\,(\Omega)$, where $|\mu|$ is the total variation of μ. If Ω is a countable union of compact subsets, then any σ-additive complex Borel measures on Ω is automatically regular, as mentioned on p.37.

Let $N \in {}^*\mathbb{N}$, then $^*\mathbb{F}^N = \{ a \mid a : \{0, \ldots, N\} \to {}^*\mathbb{F} \}$ forms an internal linear space under the pointwise addition and scalar multiplication. We have the following example of nonstandard hulls.

- $\widehat{\ell_p(N)}$, where $p \in {}^*[1, \infty)$: here $\ell_p(N) := \{a \in {}^*\mathbb{F}^N \mid \|a\|_p \in {}^*\mathbb{R}^+\}$, with the internal p-norm given by

$$\|a\|_p := \Big(\sum_{0 \le n \le N} |a_n|^p \Big)^{1/p}, \qquad a \in \ell_p(N).$$

- $\widehat{\ell_\infty(N)}$: here $\ell_\infty(N) := \{a \in {}^*\mathbb{F}^N \mid \|a\|_\infty \in {}^*\mathbb{R}^+\}$, with the internal supremum norm given by

$$\|a\|_\infty := \max_{0 \le n \le N} |a_n|, \qquad a \in \ell_\infty(N).$$

Note that the above internal space $\ell_p(N)$ and $\ell_\infty(N)$ are hyperfinite dimensional and in the definition of the norms, the sum is a hyperfinite sum (in comparison with the convergence requirement for ℓ_p and ℓ_∞) and the maximum is attained.

Given $\mathbb{F}$-valued finitely additive measures μ_1, μ_2 on $\mathcal{P}(\mathbb{N})$ and $\alpha \in \mathbb{F}$, we let $\mu_1 + \alpha\mu_2$ be the $\mathbb{F}$-valued finitely additive measure $\mathbb{N} \supset A \mapsto \mu_1(A) + \alpha\mu_2(A)$. Then the set of $\mathbb{F}$-valued finitely additive measures forms a linear space under these operations.

The following are Banach spaces.

- $ba(\mathbb{N})$: here the elements consist of $\mathbb{F}$-valued finitely additive measures $\mu : \mathcal{P}(\mathbb{N}) \to \mathbb{F}$, with $\|\mu\| < \infty$, where the norm is the total variation of μ, with the definition on p.35 restricted to finite partitions. Elements of $ba(\mathbb{N})$ are also called **bounded additive measures**.
- $ca(\mathbb{N})$: this is a closed subspace of $ba(\mathbb{N})$ consisting of $\mathbb{F}$-valued σ-additive measures—**bounded countably additive measures**.

Both $ba(\mathbb{N})$ and $ca(\mathbb{N})$ are nonseparable.

Instead of $\mathbb{N}$, and hence $\mathcal{P}(\mathbb{N})$, one can use a Boolean algebra of sets or a σ-algebra of sets and produce more examples of Banach spaces similar to $ba(\mathbb{N})$ and $ca(\mathbb{N})$.

Consider a set Ω, a σ-algebra $\mathcal{B} \subset \mathcal{P}(\Omega)$ and a positive σ-additive measure μ on $\mathcal{B}$. (Recall the arithmetic of $[0, \infty]$ given on p.78.) For such $(\Omega, \mathcal{B}, \mu)$, we have the following Banach spaces, called the **Lebesgue spaces**.

- $L_p(\mu)$ (or as $L_p(\Omega)$, or more precisely $L_p(\Omega, \mathcal{B}, \mu)$), where $p \in [1, \infty)$: first we define a seminormed linear space (under the pointwise addition and scalar multiplications) and a closed subspace:

$$X := \{ f \mid f : \Omega \to \mathbb{F} \ \text{ is } \mu\text{-measurable and } \|f\|_p < \infty \}$$

and $\mathcal{N} := \{ f \in X \mid \|f\|_p = 0 \}$,

where $\|\cdot\|_p$ is the seminorm given by the Lebesgue integral

$$\|f\|_p := \left(\int_\Omega |f|^p \, d\mu \right)^{1/p}.$$

Then the quotient space $X/\mathcal{N}$ forms a Banach space denoted by $L_p(\mu)$. In our convention, elements $L_p(\mu)$ are just written as f instead of as equivalent classes, *i.e.* cosets $f + \mathcal{N}$, and $\|\cdot\|_p$ is identified with the corresponding quotient norm $\|\cdot\|_p$, also called the *p-norm*.

- $L_\infty(\mu)$ (or $L_\infty(\Omega)$, or $L_\infty(\Omega, \mathcal{B}, \mu)$) : this is defined similarly using the *essential supremum (semi)norm* given by

$$\|f\|_\infty := \inf \{ r \in \mathbb{R}^+ \mid \mu(|f|^{-1} [[r, \infty)]) = 0 \},$$

where $f : \Omega \to \mathbb{F}$ is μ-measurable. (By convention, $\inf \emptyset := \infty$.)

Note that if we take $\Omega = \mathbb{N}$, $\mathcal{B} = \mathcal{P}(\mathbb{N})$ and μ the counting measure, then the above $L_p(\mu)$ and $L_\infty(\mu)$ are just ℓ_p and ℓ_∞.

If we take $\Omega = [0,1]$, $\mathcal{B}$ the Lebesgue measurable subsets and μ the Lebesgue measure, then the above $L_p(\mu)$ are separable while $L_\infty(\mu)$ is non-separable.

Suppose μ is the Loeb measure of an internal positive measure ν and $\mathcal{B}$ the corresponding Loeb algebra, then for $p \in [1, \infty]$, $L_p(\mu)$ embeds isometrically as a closed subspace of the nonstandard hull $\widehat{L_p(\nu)}$ via liftings of the Loeb measurable functions. In general, this embedding is proper. Take for example the internal counting measure ν on $\{0, \ldots, N\}$, where $N \in {}^*\mathbb{N} \setminus \mathbb{N}$, then $\widehat{L_p(\nu)}$ contains elements having an infinite coordinate while this is not the case for elements of $L_p(\mu)$.

Another important class of Banach spaces are the *Sobolev Spaces* $W^{k,p}(\Omega)$, where for some $n \in \mathbb{N}$, $\Omega \subset \mathbb{R}^n$ is open, $k \in \mathbb{N}$, and $p \in [1, \infty]$. These are linear subspaces of $L_p(\Omega)$ (with μ being the Lebesgue measure) equipped with the following norm for each $f \in W^{k,p}(\Omega)$:

$$\|f\|_{W^{k,p}(\Omega)} := \left(\sum_{m=0}^{k} \int_\Omega \left| f^{(m)} \right|^p \, d\mu \right)^{1/p},$$

where the $f^{(m)}$'s are derivatives in the distributional sense.

Finally we mention that more examples of Banach spaces can be constructed by the **direct sum** method.

Let X_i, $i \in I$ be a family of normed linear spaces over the same $\mathbb{F}$. Then for $p \in [1, \infty)$, the L_p-direct sum of the family is given by

$$\bigoplus_p X_i := \{ x \in \prod_{i \in I} X_i \mid \|x\|_p < \infty \},$$

where

$$\|x\|_p := \left(\sum_{i \in I} \|x_i\|_{X_i}^p \right)^{1/p}.$$

(We define $\sum_{i \in I} r_i$ as $\sup\{\sum_{i \in J} r_i \mid J \subset I \text{ is finite}\}$.)

Likewise, we have $\bigoplus_\infty X_i$ under the norm $\|x\|_\infty := \sup_{i \in I} \|x_i\|_{X_i}$.

Usually, we simply write $\bigoplus_{i \in I} X_i$ for $\bigoplus_1 X_i$.

It can be checked that such direct sums of normed linear spaces are normed linear spaces and it forms a Banach space when all X_i's are Banach spaces.

Note that in any of the above direct sums, the factors X_i are closed subspaces under the obvious identification.

Note also that for finite I, we have

$$\widehat{\bigoplus_p X_i} = \bigoplus_p \widehat{X_i} \quad \text{for} \quad p \in \{0\} \cup [1, \infty].$$

2.1.5 Notes and exercises

The nonstandard hull construction was invented by Luxemburg in [Luxemburg (1969)].

The ultrapower construction of Banach spaces has been used in [McDuff (1970)] and [Dacunha-Castelle and Krivine (1970)], it can be viewed as a special case of the nonstandard hull construction when the underlying nonstandard universe is produced from an ultrapower. However the nonstandard hull approach gives better access to other nonstandard objects (such as the hyperfinite numbers) and is therefore much more versatile than the Banach space ultrapower approach.

In a sense, nonstandard hulls are generalizations of finite dimensional Banach spaces. Other than finite dimensional cases, most well-studied examples of normed linear spaces are not nonstandard hulls. However, all of them are embeddable into such.

Regard $\mathbb{C}$ as a linear space over $\mathbb{R}$. Let $e_1 = 1$ and $e_2 = -e^{i\theta}$, where $0 \neq \theta \approx 0$. Then $\|e_1\| = 1 = \|e_2\|$ and e_1 and e_2 are *linearly independent in *$\mathbb{C}$. Moreover, $0 \approx e_1 + e_2 \in \text{Fin}(\,^*\mathbb{C})$, so $\widehat{e}_1 + \widehat{e}_2 = 0$. In particular, one doesn't have a reasonable version of a converse of Prop. 2.8. Furthermore, let $\epsilon = \|e_1 + e_2\|$, then $\|\epsilon^{-1}(e_1 + e_2)\| = 1$, so the statement in Prop. 2.10 fails for a 2−dimensional internal normed linear space.

Here is another example. Let $X = \,^*\mathbb{R}^2$, fix some $\epsilon \in \,^*\mathbb{R}$ with $0 \neq \epsilon \approx 0$ and define a norm $\|(x,y)\|$ as $|x| + \epsilon\,|y|$. Then X forms a finite dimensional internal normed linear space with basis $(1,0), (1,1)$ such that $\|(1,0)\| = 1 \approx \|(1,1)\|$. But $\widehat{(1,0)} = \widehat{(1,1)}$, so the converse of Prop. 2.8 fails. Moreover, $(0, \epsilon^{-1}) \in \text{Fin}(X)$, so Prop. 2.10 cannot be generalized for arbitrary finite dimensional internal normed linear spaces.

Most of the results here generalize to topological vector space setting. See, for example, [Rudin (1991)].

For more examples of classical Banach spaces, see [Dunford and Schwartz (1988a)].

EXERCISES

(1) Prove Prop. 2.1.
(2) Let X be a normed linear space with a Hamel basis $\{e_i\}_{i \in I}$. Let $a \in \,^*X$ and write $a = \sum_{i \in J} \alpha_i e_i$ for some hyperfinite $J \subset \,^*I$.
    Suppose $a \in \text{Fin}(\,^*X)$, is it necessarily true that $\forall i \in J\left(\alpha_i \in \text{Fin}(\mathbb{F})\right)$?
    Suppose $a \approx 0$, is it necessarily true that $\forall i \in J\left(\alpha_i \approx 0\right)$?
    (Compare these with Prop.2.10.)
(3) Let X be a normed linear space. Show that $\text{ns}(\,^*X) = \text{Fin}(\,^*X)$ iff X is finite dimensional.
(4) Let X be an infinite dimensional normed linear space. Show that for any countable list of finite dimensional subspaces of X, the union of their bases is never a Hamel basis of X. In particular any Hamel basis of X is uncountable.
(5) A normed linear space X is said to have the **Heine-Borel property** if every bounded closed subset is compact. Show that X is finite dimensional iff it has the Heine-borel property.
(6) Prove Prop. 2.11 and Cor. 2.1.
(7) Show that a nonstandard hull of an internal normed linear space is separable iff it is finite-dimensional.
(8) Prove the Minkowski Inequality:

Let $n \in \mathbb{N}$, for any $\alpha_0, \ldots, \alpha_n, \beta_0, \ldots, \beta_n \in \mathbb{F}$ and $p \in (0, \infty)$,

$$\Big(\sum_{i=0}^{n} |\alpha_i + \beta_i|^p \Big)^{1/p} \leq \Big(\sum_{i=0}^{n} |\alpha_i|^p \Big)^{1/p} + \Big(\sum_{i=0}^{n} |\beta_i|^p \Big)^{1/p}.$$

(9) Prove the corresponding Minkowski Inequality for functions measurable *w.r.t.* a positive measure.

(10) Check that the spaces in §2.1.4 are Banach spaces. In particular, check that the defined functions are norms and the spaces are complete under those norms.

(11) Verify the separability and nonseparability of the above spaces.

(12) Let $p \in [1, \infty]$. Find a closed subspace $X \subsetneq \ell_p$ such that the quotient space ℓ_p / X is isometrically isomorphic to ℓ_p.

(13) Can ℓ_p and ℓ_q be isometrically isomorphic for some $1 \leq p < q \leq \infty$?

(14) Show that the set of elements in $C([0,1])$ which are nowhere differentiable on $(0,1)$ is of second category.

(15) Is the nonstandard hull $\widehat{ba(\mathbb{N})}$ isometrically embeddable into some space of measures?

2.2   Linear Operators and Open Mappings

> *Bounded linear operators form a Banach space. Open mappings are those matching up the finite parts.*

Given linear spaces X, Y over $\mathbb{F}$, a function $f : X \to Y$ is a **linear operator** if it preserves linearity, *i.e.*

$$\forall x_1, x_2 \in X \; \forall \alpha \in \mathbb{F} \; \Big(f(x_1 + \alpha x_2) = f(x_1) + \alpha f(x_2) \Big).$$

The **kernel** of f refers to the subspace $f^{-1}[0]$ and is also denoted by $\mathrm{Ker}(f)$.

If X, Y are seminormed linear spaces, $f : X \to Y$ is called **bounded**, if

$$\forall s \in \mathbb{R}^+ \; \exists r \in \mathbb{R}^+ \; \forall x \in X \; \Big(\|x\| \le s \Rightarrow \|f(x)\| \le r \Big).$$

So, for bounded linear f, $\mathrm{Ker}(f)$ is a closed subspace of X.

We write B_X for the **open unit ball** $B(0,1)$ in X and hence the closure $\bar{B}_X$ corresponds to the **closed unit ball** $\{ x \in X \mid \|x\| \le 1 \}$. The symbol S_X is used for the **unit sphere** $\{ x \in X \mid \|x\| = 1 \}$. Sometimes we write $B_X(x,r)$, $S_X(x,r)$ *etc* instead of $B(x,r)$, $S(x,r)$ to emphasize the underlying space.

For $A \subset X$ and $\alpha \in \mathbb{F}$, the set $\{ \alpha x \mid x \in A \}$ is denoted by αA. So a linear operator $f : X \to Y$ is bounded if

$$\exists r \in \mathbb{R}^+ \; \big(f[\bar{B}_X] \subset r\bar{B}_Y \big).$$

Or equivalently, $\exists r \in \mathbb{R}^+ \; \big(f[B_X] \subset rB_Y \big)$.

This section covers properties of the space of bounded linear operators, its nonstandard hull, the Open Mapping Theorem and its well-known consequences and then there is a discussion on projections in Banach spaces.

2.2.1   Bounded linear operators and dual spaces

The following are useful characterizations of continuity for linear operators.

Proposition 2.13. *Let X, Y be normed linear spaces over the same $\mathbb{F}$. Suppose $f : X \to Y$ is linear. Then the following are equivalent:*

(i) *f is continuous.*
(ii) *f is continuous at 0.*
(iii) *f is uniformly continuous.*
(iv) *$^*f[\mu_X(0)] \subset \mu_Y(0)$.*
(v) *$\exists r \in \mathbb{R}^+ \; \big(f[B_X] \subset rB_Y \big)$. (i.e. f is bounded.)*
(vi) *$\exists n \in \mathbb{N} \; \big(f[\bar{B}_X] \subset n\bar{B}_Y \big)$.*

(vii) $\exists n \in \mathbb{N} \left({}^*f[{}^*B_X] \subset n\,{}^*B_Y \right)$.

(viii) ${}^*f[{}^*B_X] \subset \mathrm{Fin}({}^*Y)$.

(ix) ${}^*f[{}^*\bar{B}_X] \subset \mathrm{Fin}({}^*Y)$.

(x) ${}^*f[\mathrm{Fin}({}^*X)] \subset \mathrm{Fin}({}^*Y)$.

Proof. We only prove the equivalence of (ii) and (v). The rest is left as an exercise.

$((\mathrm{ii}) \Rightarrow (\mathrm{v}))$:

Suppose $\forall r \in \mathbb{R}^+ \left(f[B_X] \setminus rB_Y \neq \emptyset \right)$. Then by saturation, there is $N \in {}^*\mathbb{N} \setminus \mathbb{N}$ and $x \in {}^*B_X$ such that ${}^*f(x) \notin N\,{}^*B_Y$, i.e. $\| {}^*f(x) \| \geq N$. Then $N^{-1}x \approx 0$ but ${}^*f(N^{-1}x) \not\approx 0$, so f is not continuous at 0 by Prop. 1.18.

$((\mathrm{v}) \Rightarrow (\mathrm{ii}))$:

We use Prop. 1.18 again. (ii) follows if any $0 \approx x \in {}^*X$ satisfies ${}^*f(x) \approx 0$. Since $f(0) = 0$, we assume that $x \neq 0$. Then by transferring (v), there is $r \in \mathbb{R}^+$ such that ${}^*f\left(\|x\|^{-1} x \right) \in r\,{}^*B_Y$, i.e. ${}^*f(x) \in \|x\|\, r\,{}^*B_Y$ and hence ${}^*f(x) \approx 0$ as required. $\square$

The set of bounded linear operators from X to Y is denoted by $\mathcal{B}(X,Y)$. When $X = Y$, we use the symbol $\mathcal{B}(X)$ instead.

$\mathcal{B}(X,Y)$ forms a linear space over $\mathbb{F}$ via pointwise addition and scalar multiplications, i.e. for $f, g \in \mathcal{B}(X,Y)$ and $\alpha \in \mathbb{F}$,

$$f + \alpha g : x \mapsto \left(f(x) + \alpha g(x) \right).$$

If we define on $\mathcal{B}(X,Y)$ a function

$$f \mapsto \|f\| := \sup\{ \, \|f(x)\|_Y \mid x \in \bar{B}_X \, \},$$

then it is easy to check that $\|\cdot\|$ is a norm, called the **operator norm**, on the linear space $\mathcal{B}(X,Y)$. Moreover, if Y is a Banach space, $\mathcal{B}(X,Y)$ is also a Banach space. Unless stated otherwise, we always work with the operator norm on $\mathcal{B}(X,Y)$.

It follows from linearity that for $f \in \mathcal{B}(X,Y)$,

$$\|f\| = \sup\left\{ \, \|x\|_X^{-1}\, \|f(x)\|_Y \mid 0 \neq x \in X \, \right\}.$$

For the special case $Y = \mathbb{F}$, where $\mathbb{F}$ is regarded naturally as a normed linear space over itself, $\mathcal{B}(X, \mathbb{F})$ is called the **dual** of X and is denoted by X'. Elements in X' are called **bounded linear functionals**. They are called **real bounded linear functionals** if $\mathbb{F} = \mathbb{R}$ and **complex bounded linear functionals** if $\mathbb{F} = \mathbb{C}$.

Note that even if X is an incomplete normed linear space, X' is always complete.

Note also that $^*(X')$ has the same meaning as $(^*X)'$ and is simply written as $^*X'$.

Example 2.1.

- Let μ be a positive measure on a σ-algebra of subsets of Ω. For $p \in (1, \infty)$, let q be such that $p^{-1} + q^{-1} = 1$. Then we have the Hölder's Inequality (see [Rudin (1991)]):

$$\forall f \in L_p(\Omega) \; \forall g \in L_q(\Omega) \left[\int_\Omega |fg| \, d\mu \leq \left(\int_\Omega |f|^p \, d\mu \right)^{1/p} \left(\int_\Omega |g|^q \, d\mu \right)^{1/q} \right].$$

From this, it can be shown that the mapping $\pi : L_q(\mu) \to L_p(\mu)'$ given by $\pi(g) : f \mapsto \int_\Omega fg d\mu$ is an isometric isomorphism. We simply regard $L_q(\mu) = L_p(\mu)'$. Likewise, $L_p(\mu) = L_q(\mu)'$.

- It can be shown that $L_\infty(\mu) = L_1(\mu)'$. However $L_1(\mu) \neq L_\infty(\mu)'$ except for trivial cases.

- As special cases, $\ell_q = \ell_p'$ and $\ell_p = \ell_q'$ for $p, q \in (1, \infty)$ such that $p^{-1} + q^{-1} = 1$ and $\ell_\infty = \ell_1'$.

- One also has $\ell_1 = c_0'$ and $ba(\mathbb{N}) = \ell_\infty'$.

$\square$

Proposition 2.14. *Let X be a normed linear space and $Y \subset X$ a closed subspace. Let $f \in (X/Y)'$ and define $f_0 : X \to \mathbb{F}$ by $x \mapsto f(x + Y)$.*
Then $f_0 \in X'$ and $\|f_0\| \leq \|f\|$.
Let $g \in X'$, then $g = f_0$ as above for some $f \in (X/Y)'$ iff $g[Y] = \{0\}$.

Proof. We will apply the equivalence of (ii) and (v) in Prop. 2.13.
Clearly f_0 is linear.
For any $x \in X$ since $0 \in Y$, we have

$$\|x + Y\| = \inf\{\|x + y\| \mid y \in Y\} \leq \|x\|.$$

Let $x \in {}^*X$ with $x \approx 0$. Then $\|x\| \approx 0$, so $\|x + {}^*Y\| \approx 0$. Now by Prop. 2.13, $^*f(x + {}^*Y) \approx 0$, *i.e.* $^*f_0(x) \approx 0$, implying $f_0 \in X'$ by Prop. 2.13 again. $\|f_0\| \leq \|f\|$ follows from the above inequality.
Now let $g \in X'$.
If $g[Y] = \{0\}$, then the function $f : (x + Y) \mapsto g(x)$ is well-defined, linear and $g = f_0$ as above. Furthermore, if $x \in {}^*X$ and $(x + {}^*Y) \approx {}^*Y$, then $x \approx y$ for some $y \in {}^*Y$, so $^*g(x) \approx {}^*g(y) = 0$ by Prop. 2.13, *i.e.* $^*f(x + {}^*Y) \approx 0$, therefore $f \in (X/Y)'$ by Prop. 2.13 again.
The converse is clear.

$\square$

If X is a normed linear space, $Y \subset X$ a subspace and $f \in X'$, then it is easy to see that the restriction $(f \restriction_Y) \in Y'$ with $\|f \restriction_Y\|_{Y'} \leq \|f\|_{X'}$.

Proposition 2.15. *Let X be a normed linear space and $Y \subset {}^*X$ an internal subspace such that $X \subset Y$. Suppose $f_1, \ldots, f_n \in X'$, $n \in \mathbb{N}$, are linearly independent. Let $Z := {}^*\mathrm{Lin}({}^*f_1 \restriction_Y, \ldots, {}^*f_n \restriction_Y) \subset Y'$. Then*

(i) *${}^*f_1 \restriction_Y, \ldots, {}^*f_n \restriction_Y$ are *linearly independent in Y';*

(ii) *$\mathrm{Fin}(Z) = \left\{ \sum_{i=1}^n \alpha_i {}^*f_i \restriction_Y \mid \alpha_i \in \mathrm{Fin}({}^*\mathbb{F}) \right\}$ and $\widehat{Z} = \mathrm{Lin}(f_1, \ldots, f_n)$.*

Proof. (i) This is similar to the proof of Prop. 2.8. Suppose $\sum_{i=1}^n \alpha_i {}^*f_i \restriction_Y = 0$, for some $\alpha_i \in {}^*\mathbb{F}$, not all zero. Assume without loss of generality that $\alpha_1 \neq 0$. Then $\sum_{i=1}^n \alpha_i \alpha_1^{-1} {}^*f_i \restriction_Y = 0$ and therefore $\sum_{i=1}^n {}^\circ(\alpha_i \alpha_1^{-1}) f_i(x) = 0$ for each $x \in X$, since $X \subset Y$.

Hence $f_1 + \sum_{i=2}^n {}^\circ(\alpha_i \alpha_1^{-1}) f_i = 0$, a contradiction.

(ii) By the same proof in (i), we see that $\widehat{{}^*f_1 \restriction_Y}, \ldots, \widehat{{}^*f_n \restriction_Y}$, are linearly independent. Now define

$$\theta : \mathrm{Lin}(f_1, \ldots, f_n) \to \widehat{Z} \quad \text{by} \quad \theta\left(\sum_{i=1}^n \alpha_i f_i \right) = \sum_{i=1}^n \alpha_i \widehat{{}^*f_i \restriction_Y},$$

where $\alpha_1, \ldots \alpha_n \in \mathbb{F}$.

Similar to the proofs of Prop. 2.9 and Prop. 2.10, the conclusions follow from $\dim\big(\theta[\mathrm{Lin}(f_1, \ldots, f_n)]\big) = n = \dim(Z)$. $\qquad\qquad\square$

For normed linear spaces X, Y, Z, it is clear that given $g \in \mathcal{B}(X, Y)$ and $f \in \mathcal{B}(Y, Z)$, then $(f \circ g) \in \mathcal{B}(X, Z)$.

If X, Y are internal normed linear spaces, we see that

$$\mathrm{Fin}({}^*\mathcal{B}(X, Y)) = \left\{ f \in {}^*\mathcal{B}(X, Y) \mid \|f\| \in \mathrm{Fin}({}^*\mathbb{F}) \right\}.$$

Note also that

$$\mathrm{ns}({}^*\mathcal{B}(X, Y)) = \left\{ f \in {}^*\mathcal{B}(X, Y) \mid \exists g \in \mathcal{B}(X, Y) \, (f \approx {}^*g) \right\}.$$

For $f \in \mathcal{B}(X, Y)$, $f \approx 0$ means $\|f\| \approx 0$, *i.e.* ${}^\circ f$ is the zero operator.

The following gives a useful identification of the nonstandard hull $\widehat{{}^*\mathcal{B}(X, Y)}$. Its proof is left as an exercise.

Proposition 2.16. *Let X, Y be internal normed linear spaces over ${}^*\mathbb{F}$.*

(i) *Let $f \in \mathrm{Fin}({}^*\mathcal{B}(X, Y))$. Then the function $\tilde{f} : \widehat{x} \mapsto \widehat{f(x)}$, where $x \in X$, is well-defined and $\tilde{f} \in B(\widehat{X}, \widehat{Y})$.*

Moreover, $\tilde{f} = \tilde{g}$ whenever $f \approx g$.

(ii) *Define* $\theta : {}^*\widehat{\mathcal{B}(X,Y)} \to \mathcal{B}(\widehat{X},\widehat{Y})$ *by* $\theta(\widehat{f}) = \tilde{f}$.
   Then θ *is an isometric embedding into* $\mathcal{B}(\widehat{X},\widehat{Y})$.   □

From now on, we identify $\widehat{f}$ with $\theta(\widehat{f})$ and ${}^*\widehat{\mathcal{B}(X,Y)}$ as a subspace of $\mathcal{B}(\widehat{X},\widehat{Y})$. In general, ${}^*\widehat{\mathcal{B}(X,Y)}$ is a proper subspace of $\mathcal{B}(\widehat{X},\widehat{Y})$. We will discuss this in § 2.5.4.

Corollary 2.4. *Let* X, Y *be normed linear spaces over* $\mathbb{F}$.
   Then $\mathcal{B}(X,Y)$ *isometrically embeds into* $\mathcal{B}({}^*\widehat{X}, {}^*\widehat{Y})$ *via the mapping* $f \mapsto \widehat{{}^*f}$.
   Moreover, $f = \widehat{{}^*f}\!\restriction_X$, *the restriction of* $\widehat{{}^*f}$ *on* X.   □

From now on, we identify f with $\widehat{{}^*f}\!\restriction_X$.

2.2.2   Open mappings

We now turn our attention to open mappings. The following can be viewed as the dual version of Prop. 2.13, as all the inclusions are reversed.

Proposition 2.17. *Let* X, Y *be normed linear spaces over* $\mathbb{F}$. *The following are equivalent for any* $f \in \mathcal{B}(X,Y)$:

 (i) f *is an open mapping.*
 (ii) ${}^*f[\mu_X(0)] \supset \mu_Y(0)$.
 (iii) ${}^*f[\mu_X(0)] = \mu_Y(0)$.
 (iv) $\exists r \in \mathbb{R}^+ \left(f[B_X] \supset r B_Y \right)$.
 (v) $\exists n \in \mathbb{N} \left(n\,{}^*f[\,{}^*B_X] \supset {}^*B_Y \right)$.
 (vi) ${}^*f[\mathrm{Fin}({}^*X)] \supset \mathrm{Fin}({}^*Y)$.

Proof.   We only prove the equivalence of (i), (iv), (vi). The rest is left as an exercise using Prop. 1.20 and Prop. 1.18.

$((i) \Rightarrow (iv))$:
Since f is open, $\mu_Y(0) \subset {}^*f[\mu_X(0)]$. Then $\forall N \in {}^*\mathbb{N} \setminus \mathbb{N}$

$$\frac{1}{N}\,{}^*B_Y \subset \mu_Y(0) \subset {}^*f[\mu_X(0)] \subset {}^*f[\,{}^*B_X].$$

Hence, the underspill implies that for some $n \in \mathbb{N}$, $n^{-1}\,{}^*B_Y \subset {}^*f[\,{}^*B_X]$, and so $n^{-1}B_Y \subset f[B_X]$ by transfer.

$((iv) \Rightarrow (vi))$:

Let $r \in \mathbb{R}^+$ so that $f[B_X] \supset rB_Y$. Then

$$*f[\mathrm{Fin}(*X)] = *f\Big[\bigcup_{n\in\mathbb{N}} n\,*B_X\Big] = \bigcup_{n\in\mathbb{N}} n\,*f[\,*B_X] \supset \bigcup_{n\in\mathbb{N}} nr\,*B_Y = \mathrm{Fin}(*Y).$$

$((\mathrm{vi}) \Rightarrow (\mathrm{i}))$:

From $\mathrm{Fin}(*Y) \subset *f[\mathrm{Fin}(*X)]$, we have

$$*B_Y \subset *f[\mathrm{Fin}(*X)] = \bigcup_{n\in\mathbb{N}} n\,*f[\,*B_X],$$

i.e. $\bigcap_{n\in\mathbb{N}} \big(*B_Y \setminus n\,*f[\,*B_X]\big) = \emptyset$, hence, by saturation, $*B_Y \subset n\,*f[\,*B_X]$ for some $n \in \mathbb{N}^+$. Then

$$\mu_Y(0) = \bigcap_{m\in\mathbb{N}} m^{-1}\,*B_Y \subset \bigcap_{m\in\mathbb{N}} m^{-1}n\,*f[\,*B_X]$$

$$= *f\Big[\bigcap_{m\in\mathbb{N}} m^{-1}n\,*B_X\Big] = *f[\mu_X(0)]. \qquad \square$$

We remark that if $f \in \mathcal{B}(X,Y)$ is an open mapping, then so is αf for any $0 \neq \alpha \in \mathbb{F}$.

For a bounded linear operator, having a nonstandard version that takes the finite part onto the finite part is precisely the property for being an open mapping. Note that in the following (as compared with the classical version Thm. 2.3) the completion of the spaces is not required.

Theorem 2.2. (Nonstandard Open Mapping Theorem) *Let* $f \in \mathcal{B}(X,Y)$, *where* X, Y *are normed linear spaces over* $\mathbb{F}$.

Then $*f[\mathrm{Fin}(*X)] = \mathrm{Fin}[*Y]$ *iff* f *is an open mapping.*

Proof. We give a proof using Prop. 2.17 (i) $\Leftrightarrow$ (ii) only.

$(\Rightarrow)$: Assume $*f[\mathrm{Fin}(*X)] = \mathrm{Fin}[*Y]$ and let $0 \neq b \in \mu_Y(0)$. Then $\dfrac{b}{\|b\|} = *f(a)$ for some $a \in \mathrm{Fin}(*X)$, hence $b = *f(\|b\|\,a)$ with $\|b\|\,a \approx 0$. i.e. $b \in *f[\mu_X(0)]$, therefore $\mu_Y(0) \subset *f[\mu_X(0)]$ and so f is open by Prop. 2.17.

$(\Leftarrow)$: $*f[\mathrm{Fin}(*X)] \subset \mathrm{Fin}[*Y]$ always holds as f is bounded. Suppose f is an open mapping, i.e. $*f[\mu_X(0)] \supset \mu_Y(0)$, but there is some

$$b \in \Big(\mathrm{Fin}[*Y] \setminus *f[\mathrm{Fin}(*X)]\Big) = \Big(\mathrm{Fin}[*Y] \setminus \Big(\bigcup_{n\in\mathbb{N}} f[n\,*B_X]\Big)\Big).$$

Then $b \notin *f[N\,*B_X]$, i.e. $N^{-1}b \notin *f[\,*B_X]$, for all small $N \in *\mathbb{N} \setminus \mathbb{N}$, by overspill. But for such N, $N^{-1}b \in \mu_Y(0) \subset *f[\mu_X(0)] \subset *f[\,*B_X]$, a contradiction. $\qquad \square$

In particular, taking $Y = \mathbb{F}$, it follows that every bounded linear functional is an open mapping.

The following technical lemma is needed for the classical Open Mapping Theorem.

Lemma 2.1. *Let $f \in \mathcal{B}(X, Y)$, where X is a Banach space and Y a normed linear space, over the same $\mathbb{F}$.*
Suppose $\forall y \in B_Y \, \exists x \in B_X \, \big(\|y - f(x)\| < 2^{-1} \big)$.
Then $B_Y \subset 2f[\bar{B}_X]$.

Proof.   Let $b \in B_Y$.
Then there is some $a_0 \in B_X$ such that $\|b - f(a_0)\| < 2^{-1}$.
Let $n \in \mathbb{N}$ and assume inductively that $a_m \in B_X$, $m \le n$, are defined such that $\left\| b - f\left(\sum_{m=0}^{n} 2^{-m} a_m \right) \right\| < 2^{-n-1}$.

Then $\left(2^{n+1} b - 2^{n+1} f\left(\sum_{m=0}^{n} 2^{-m} a_m \right) \right) \in B_Y$, so $a_{n+1} \in B_x$ can be found satisfying $\left\| 2^{n+1} b - 2^{n+1} f\left(\sum_{m=0}^{n} 2^{-m} a_m \right) - a_{n+1} \right\| < 2^{-1}$, therefore, by linearity, we can write $\left\| b - f\left(\sum_{m=0}^{n+1} 2^{-m} a_m \right) \right\| < 2^{-n-2}$, *i.e.* the inductive hypothesis is satisfied.

By completeness, since $\sum_{m=0}^{\infty} 2^{-m} a_m$ is absolute convergent, it converges in X to some a. Note $a \in 2\bar{B}_X$ and, by continuity, $b = f(a)$. $\square$

The completeness of a Banach space plays a crucial role in the proof of the following fundamental result—for its reliance on the Baire Category Theorem.

Theorem 2.3.   (Open Mapping Theorem) *Let X, Y be Banach spaces over $\mathbb{F}$ and $f \in \mathcal{B}(X, Y)$ be surjective. Then*

(i) $\,^*\!f\big[\mathrm{Fin}(\,^*\!X)\big] = \mathrm{Fin}[\,^*\!Y]$.
(ii) f *is an open mapping.*
(iii) $\widehat{\,^*\!f}$ *is an open mapping in $\mathcal{B}(\widehat{\,^*\!X}, \widehat{\,^*\!Y})$.*

Proof.   (i): By assumption, $Y = \bigcup_{n \in \mathbb{N}^+} \overline{f[nB_X]}$. Then the Baire Category Theorem (Thm. 1.24) implies that $\mathrm{int}\big(\overline{f[nB_X]}\big) \ne \emptyset$ for some $n \in \mathbb{N}^+$.
Take $m \in \mathbb{N}^+$ and $b \in Y$ so that $B(b, m^{-1}) \subset \overline{f[nB_X]}$.
That is, $b + m^{-1} B_Y \subset \overline{f[nB_X]}$.

By Prop. 1.14(iii), let $a \in n\,{}^*B_X$ so that $b \approx {}^*f(a)$, *i.e.* $b = {}^*f(a) - \epsilon$ for some $0 \approx \epsilon \in {}^*Y$.

Now by transfer, we have $m^{-1}\,{}^*B_Y \subset {}^*\!\left(\overline{f[nB_X]}\right) - {}^*f(a) + \epsilon$.

Since $a \in n\,{}^*B_X$, it follows that

$$ {}^*B_Y \subset {}^*\!\left(\overline{f[2mnB_X]}\right) + m\epsilon = {}^*\!\left(\overline{2mnf[B_X]}\right) + m\epsilon. $$

Apply transfer to Lem. 2.1 with $2mnf$ in place of f, we have

$$ {}^*B_Y \subset 4mn\,{}^*f[\bar{B}_X] \subset (4mn+1)\,{}^*f[B_X]. $$

Finally,

$$ {}^*f\left[\mathrm{Fin}({}^*X)\right] = {}^*f\Big[\bigcup_{k \in \mathbb{N}} k\,{}^*B_X\Big] = \bigcup_{k \in \mathbb{N}} k\,{}^*f[{}^*B_X] \subset \bigcup_{k \in \mathbb{N}} k\,\|f\|\,{}^*B_Y $$

$$ \subset \bigcup_{k \in \mathbb{N}} k(4mn+1)\,\|f\|\,f[{}^*B_X] = {}^*f\left[\mathrm{Fin}({}^*X)\right] $$

and, since $\mathrm{Fin}({}^*Y) = \bigcup_{k \in \mathbb{N}} k\,\|f\|\,{}^*B_Y$, we conclude that

$$ {}^*f\left[\mathrm{Fin}({}^*X)\right] = \mathrm{Fin}({}^*Y). $$

(ii): From (i) and Thm. 2.2, it follows that f is an open mapping.

(iii): By Cor. 2.4, $\widehat{{}^*f} \in \mathcal{B}(\widehat{{}^*X}, \widehat{{}^*Y})$. By (i), the condition ${}^*f\left[\mathrm{Fin}({}^*X)\right] = \mathrm{Fin}[{}^*Y]$ holds, implying that $\widehat{{}^*f}$ is surjective. Hence it follows from (ii) that $\widehat{{}^*f}$ is an open mapping.   □

Immediately, we have the following easy result which is often useful in checking equivalent norms.

Corollary 2.5. *Let X form Banach spaces under norms $\|\cdot\|_1$ and $\|\cdot\|_2$. Suppose the identity mapping $(X, \|\cdot\|_1) \to (X, \|\cdot\|_2)$ is continuous. Then $\|\cdot\|_1$ and $\|\cdot\|_2$ are equivalent.*

Proof. Let $X_1 := (X, \|\cdot\|_1)$, $X_2 := (X, \|\cdot\|_2)$ and μ_1, μ_2 be the corresponding monads. Let $f : X_1 \to X_2$ be the identity mapping. Then $f \in \mathcal{B}(X_1, X_2)$ and is surjective, so f is an open mapping, by Thm. 2.3(ii). Hence, by Prop. 2.17(iii), we have ${}^*f[\mu_1(0)] = \mu_2(0)$, meaning $\mu_1(0) = \mu_2(0)$, which implies that $\|\cdot\|_1$ and $\|\cdot\|_2$ are equivalent, by Prop. 2.3.   □

A typical corollary of the Open Mapping Theorem is the following.

Corollary 2.6. (Inverse Mapping Theorem) *Let X, Y be Banach spaces over $\mathbb{F}$ and $f \in \mathcal{B}(X, Y)$ be bijective. Then $f^{-1} \in \mathcal{B}(Y, X)$ and is an open mapping.*

Proof. Note that $^*(f^{-1}) = (\,^*f)^{-1}$, so we simply write it as $^*f^{-1}$.

Clearly f^{-1} is linear. By Thm. 2.3(ii), f is an open mapping, so Thm. 2.2 gives that $^*f^{-1}[\text{Fin}(\,^*Y)] = \text{Fin}(\,^*X)$. Hence, by Prop. 2.13, f^{-1} is continuous, *i.e.* $f^{-1} \in \mathcal{B}(Y, X)$.

From $^*f^{-1}[\text{Fin}(\,^*Y)] = \text{Fin}(\,^*X)$, and Thm. 2.2, we also conclude that f^{-1} is an open mapping. □

Note that by Prop. 2.13, the above bijection f has the property that $f[\mu_X(0)] = \mu_Y(0)$, hence it is a linear isomorphism that preserves the topology, *i.e.* a **linear homeomorphism**.

Recall the direct sum construction on p.91.

The following is a useful classical consequence of the Inverse Mapping Theorem, hence of the Open Mapping Theorem.

Corollary 2.7. (Closed Graph Theorem) *Let X, Y be Banach spaces over $\mathbb{F}$. Suppose $f : X \to Y$ is a linear operator with* $\text{graph}(f)$ *closed in $X \oplus_1 Y$. Then f is continuous.*

Proof. Clearly, by the linearity of f, $\text{graph}(f)$ is a vector subspace of $X \oplus_1 Y$ with the inherited L_1-norm. Since it is closed, it forms a Banach space.

Let $\pi_X : \text{graph}(f) \to X$ be given by $(x, y) \mapsto x$. Likewise, we define $\pi_Y : \text{graph}(f) \to Y$ by $(x, y) \mapsto y$.

Then $\pi_X \in \mathcal{B}\big(\text{graph}(f), X\big)$ is bijective. Hence, by applying Cor. 2.6, $(\pi_X)^{-1} \in \mathcal{B}\big(X, \text{graph}(f)\big)$. Moreover, $\pi_Y \in \mathcal{B}\big(\text{graph}(f), Y\big)$.

Therefore $f = \big(\pi_Y \circ (\pi_X)^{-1}\big) \in \mathcal{B}(X, Y)$. □

Recall that a homeomorphism between two topological spaces is a bijective continuous open mapping. In such case, the inverse is also a bijective continuous open mapping. We now have the following application of the Inverse Mapping Theorem

Proposition 2.18. *Let X be a Banach space and $Y, Z \subset X$ be closed subspaces such that $X = Y + Z$ and $Y \cap Z = \{0\}$.*

Then X and $Y \oplus_1 Z$ are homeomorphic.

Proof. Let $f : Y \oplus_1 Z \to X$ be given by $(y, z) \mapsto y + z$. Then f is linear and bijective. Moreover, $\|y + z\|_X \leq \|y\|_X + \|z\|_X = \|(y, z)\|_{Y \oplus_1 Z}$, so $f \in \mathcal{B}(Y \oplus_1 Z, X)$. Then by the Inverse Mapping Theorem (Cor. 2.6), f^{-1} is a bijective continuous open mapping. □

When the above conditions are satisfied in a normed linear space X, namely $Y, Z \subset X$ are *closed* subspaces such that

$$X = Y + Z \quad \text{and} \quad Y \cap Z = \{0\},$$

we say that the closed subspace Y is **complemented** in X and Z is **complementary** to Y. Hence we also call Y, Z **complementary to each other**.

It is not true in general that given a closed subspace Y of a Banach space X, there is a closed subspace $Z \subset X$ complementary to Y. A deep result is that Banach spaces for which each closed subspace is complemented are precisely Hilbert spaces (to be discussed later).

An application of the Inverse Mapping Theorem similar to the proof of Prop.2.18 shows the following.

Proposition 2.19. *Let X be a Banach space and $Y \subset X$ be a complemented closed subspace. Suppose $Z \subset X$ is a closed subspace complementary to Y. Then Z and X/Y are isometrically isomorphic.* □

In Prop. 2.18, if we define $p : X \to X$ by $(y + z) \mapsto y$, where we decompose each $x \in X$ as $y + z$ for some unique $(y, z) \in Y \times Z$, then $p[X] = Y$ and $p \in \mathcal{B}(X)$ satisfies $p^2 = p$, *i.e.* $p \circ p = p$, an **idempotent**.

In general, for a normed linear space X, if an operator $p \in \mathcal{B}(X)$ is such that $p^2 = p$, we say that p is a **projection**—in the linear space sense, as we will later redefine projection in other context by imposing some sort of orthogonality.

Let 1 denote the identity mapping in $\mathcal{B}(X)$. Then whenever $p \in \mathcal{B}(X)$ is a projection, $(1 - p) \in \mathcal{B}(X)$ is a projection, as

$$(1 - p)^2 = 1 - p - p + p^2 = 1 - p.$$

Moreover, it can be seen that $\text{Ker}(p) = (1-p)[X]$. So, as p is continuous, $\text{Ker}(p)$ is closed, hence $(1 - p)[X]$ is closed. By a dual argument, *i.e.* replacing p by $(1 - p)$, it follows that $p[X]$ is closed.

Clearly $p[X] \cap (1 - p)[X] = \{0\}$. This is one half of the following result. The other is not hard to check.

Proposition 2.20. *Let X be a Banach space and $p \in \mathcal{B}(X)$ be a projection. Then $p[X]$ and $(1-p)[X]$ are closed subspaces complementary to each other in X.*

Conversely, if $Y \subset X$ is a complemented closed subspace, there is a projection $p \in \mathcal{B}(X)$ such that $Y = p[X]$. □

Now an observation about internal projections. Recall the nonstandard hull of linear operators in Prop.2.16.

Proposition 2.21. *Let X be an internal normed linear space. Suppose $p \in \mathrm{Fin}\big(\,^*\mathcal{B}(X)\big)$ is a *projection, then $\widehat{p} \in \mathcal{B}(\widehat{X})$ and is a projection.*

Proof. Let $p \in \mathrm{Fin}\big(\,^*\mathcal{B}(X)\big)$ be such that $p^2 = p$. Let $a \in \mathrm{Fin}(X)$, then

$$(\widehat{p})^2(\widehat{a}) = \widehat{p}(\widehat{p(a)}) = \widehat{p^2(a)} = \widehat{p(a)} = \widehat{p}(\widehat{a}).$$

Hence $\widehat{p}$ is a projection.   □

Note in particular that if X is a normed linear space and $p \in \mathcal{B}(X)$ is a projection, then $\widehat{^*p} \in \mathcal{B}(\widehat{X})$ and is also a projection.

In a normed linear space X, for nonzero projection $p \in \mathcal{B}(X)$, if we let $a \in p[X] \cap S_X$, then $p(a) = a$ and so $\|p(a)\| = \|a\| = 1$. Therefore, $\|p\| \geq 1$. We leave it as an exercise to show that, in general, for any $r \in \mathbb{R}^+$, there are examples such that $\|p\| \geq r$.

Corollary 2.8. *Let X be a normed linear space and $Y \subset X$ be a complemented closed subspace.*
   Then $\widehat{^*Y}$ is a complemented closed subspace of $^*\widehat{X}$.
   Moreover, if $Z \subset X$ is a closed subspace complementary to Y, then $\widehat{^*Z} \subset \widehat{^*X}$ is a closed subspace complementary to $\widehat{^*Y}$.

Proof. Let $p \in \mathcal{B}(X)$ be a projection given by Prop.2.20 so that $p[X] = Y$. Then by Prop.2.21, $\widehat{^*p} \in \mathcal{B}(\widehat{^*X})$ is a projection. So, by Prop.2.20 again, $\widehat{^*p}[\widehat{^*X}]$ is a complemented closed subspace of $\widehat{^*X}$.

On the other hand, $\widehat{^*p}[\widehat{^*X}] = \widehat{^*p[^*X]} = \widehat{^*Y}$ i.e. $\widehat{^*Y}$ is a complemented closed subspace of $\widehat{^*X}$.

Replacing p by $(1-p)$, we have $^*\widehat{(1-p)}[\widehat{^*X}] = \widehat{^*(1-p)[^*X]}$.

Let $Z \subset X$ be a closed subspace complementary to Y. Then we have $(1-p)[X] = Z$, hence $\widehat{^*(1-p)[^*X]} = \widehat{^*Z}$.

Therefore $\widehat{^*Z}$ is a closed subspace complementary to $\widehat{^*Y}$.   □

2.2.3   *Uniform boundedness*

Another typical application of the Baire category Theorem is the following important result relating pointwise boundedness to uniform boundedness.

Theorem 2.4. (The Uniform Boundedness Principle) *Let X be a Banach space and Y be a normed linear space over the same field.*

Suppose $A \subset \mathcal{B}(X, Y)$ is such that $\forall x \in X \left(\sup_{f \in A} \|f(x))\| < \infty \right)$.
Then $\sup_{f \in A} \|f\| < \infty$.

Proof. For $n \in \mathbb{N}$, let

$$C_n := \{x \in X \mid \sup_{f \in A} \|f(x)\| \le n\} = \bigcap_{f \in A} f^{-1} \left[\overline{B_Y}(0, n) \right],$$

so the C_n's are closed and $X = \cup_{n \in \mathbb{N}} C_n$.

By applying the Baire Category Theorem (Thm. 1.24) to the Banach space X, we see that for some $n \in \mathbb{N}$, $\mathrm{int}(C_n) \ne \emptyset$. Then for some $a \in X$ and $r \in \mathbb{R}^+$, we have $B_X(a, r) \subset C_n$, hence

$$\forall x \in X \left(\|x - a\| < r \Rightarrow \sup_{f \in A} \|f(x)\| \le n \right),$$

$$i.e. \quad \forall x \in X \left((\|x\| < r) \Rightarrow (\sup_{f \in A} \|f(x)\| \le 2n) \right),$$

$$i.e. \quad \forall x \in B_X \left(\sup_{f \in A} \|f(x)\| \le 2nr^{-1} \right),$$

so we conclude that $\sup_{f \in A} \|f\| \le 2nr^{-1}$. $\qquad\qquad\qquad \Box$

Corollary 2.9. (Banach-Steinhaus Theorem) *Let X be a Banach space and Y a normed linear space over the same field. Let $f_n \in \mathcal{B}(X, Y)$, $n \in \mathbb{N}$. Suppose $\lim_{n \to \infty} f_n(x)$ exists for each $x \in X$. Define $f : X \to Y$ by $f(x) := \lim_{n \to \infty} f_n(x)$. Then $f \in \mathcal{B}(X, Y)$.*

Proof. It is clear that such f is linear.

By the Uniform Boundedness Principle, for some $r \in \mathbb{R}^+$, we have $\sup_{n \in \mathbb{N}} \|f_n\| = r$. Then for each $x \in B_X$ we have

$$\|f(x)\| \le \|f(x) - f_n(x)\| + \|f_n(x)\| \le \|f(x) - f_n(x)\| + r,$$

i.e. $\|f\| = \sup_{x \in B_X} \|f(x)\| \le r$, hence $f \in \mathcal{B}(X, Y)$. $\qquad\qquad \Box$

2.2.4   Notes and exercises

The Open Mapping Theorem was proved by Schauder in the 1930's. It remains still valid in the more general setting of a kind of metrizable topological vector space called F-spaces. (See [Rudin (1991)].)

For a linear operator $f : X \to Y$, Prop. 2.13 shows that continuity means $*f[\mathrm{Fin}(\,^*X)] \subset \mathrm{Fin}(\,^*Y)$. From Prop. 2.17, intuitively, being an open mapping means the dual notion that $\mathrm{Fin}(\,^*Y) \subset f[\mathrm{Fin}(\,^*X)]$. Hence the

Nonstandard Open Mapping Theorem (Thm. 2.2) is obtained by putting both together.

In the Open Mapping Theorem (Thm. 2.3), the assumption can be weakened by requiring the target space Y to be any normed linear space, as the proof did not use the completeness of Y. However this weakening is simply superficial: whenever there is a Banach space X with a surjective $f \in \mathcal{B}(X,Y)$, Y is automatically a Banach space. This can be proved from Thm. 2.3 (with Y weakened to a normed linear space) as follows:

- $\widehat{{}^*f} \in \mathcal{B}(\widehat{{}^*X}, \widehat{{}^*Y})$ is a surjective open mapping, by Thm. 2.3 (i) and (iii).
- X is a closed subspace of $\widehat{{}^*X}$, since X is Banach.
- Therefore $Y = f[X] = \widehat{{}^*f}[X]$ is closed in the Banach space $\widehat{{}^*Y}$.
- Hence Y is a Banach space.

It is worth-noticing from Prop. 2.1 and the Open Mapping Theorem (Thm. 2.3) that if X is a Banach space with a closed subspace Y, the mapping $X \to X/Y$ given by $x \mapsto (x + Y)$ is an open mapping.

The linear homeomorphism in the Inverse mapping Theorem (Cor. 2.6) is not necessarily an isometry: just think of the identity functions on a space with two distinct but equivalent norms.

The converse of the Closed Graph Theorem holds in general, *i.e.* continuity implies closed graph. (See Ex. 5 on p.75.)

EXERCISES

(1) Complete the proof of Prop. 2.13.
(2) Verify the representation of the dual spaces in Example 2.1.
(3) Prove Prop. 2.16 and Cor. 2.4.
(4) Let X, Y be normed linear spaces and $\bar{X}$, $\bar{Y}$ their completions. Show that for any $f \in \mathcal{B}(X,Y)$, there is an extension $\bar{f} \in \mathcal{B}(\bar{X}, \bar{Y})$ with $\|f\| = \|\bar{f}\|$.
(5) Show that under the operator norm, $\mathcal{B}(X,Y)$ is a Banach space whenever Y is a Banach space.
(6) Let X, Y be Banach spaces and $f \in \mathcal{B}(X,Y)$. Show that if there is $r \in \mathbb{R}^+$, $\forall x \in \bar{B}_X \left(\|f(x)\|_Y \geq r \right)$, then $\mathrm{Ker}(f) = \{0\}$ and $f[X]$ is closed. Show that the converse also holds.
(7) Prove the following version of Thm. 2.2:
Let X, Y be internal normed linear spaces. Let $f \in \mathrm{Fin}({}^*\mathcal{B}(X,Y))$ be such that $f\left[\mathrm{Fin}(X)\right] = \mathrm{Fin}[Y]$. Then f is an *open mapping and $\widehat{f}$ is an open mapping.

(8) Given an example of a normed linear space X such that for some $a \in \bar{B}_X$ and $b \in \bar{B}_{*X}$, $a \not\approx b$ but $\forall \phi \in X' \left(\phi(a) \approx {}^*\phi(b) \right)$.

(9) Verify in Example 2.1 that $L_q(\mu) = L_p(\mu)'$ and $L_\infty(\mu) = L_1(\mu)'$.

(10) Show that the statement in the Closed Graph Theorem (Cor. 2.7) could fail if X is replaced by an incomplete normed linear space.

(11) Let $r \in \mathbb{R}^+$ be given. Find a normed linear space X and a projection $p \in \mathcal{B}(X)$ such that $\|p\| \geq r$.

(12) Give a direct proof of the Uniform Boundedness Principle (Thm. 2.4) without quoting the Baire Category Theorem.

2.3   Helly's Theorem and the Hahn-Banach Theorem

> *To norm or not to be normed, that is the question leading to the*
> *extension of functionals without increase in norm.*

We begin with a remark that any functional is almost normed by elements in the space and precisely normed by an element in the nonstandard hull:

Proposition 2.22. *Let X be a normed linear space and $f \in X'$. Then there is $a \in {}^*B_X$ such that $\|f\| \approx {}^*f(a) \in {}^*[0, \infty)$, hence $\widehat{{}^*f}(\hat{a}) = \left\| \widehat{{}^*f} \right\| = \|f\|$.*

Proof.   Since $\forall \epsilon \in \mathbb{R}^+ \, \exists x \in B_X \big(\|f\| \leq |f(x)| + \epsilon \big)$, it follows from saturation that $| {}^*f(a_0)| \approx \|f\|$, for some $a_0 \in {}^*B_X$. Now let $\alpha \in {}^*\mathbb{F}$ be such that $\alpha \, {}^*f(a_0) = | {}^*f(a_0)|$ and $a := \alpha a_0$, then $a \in {}^*B_X$ and $\|f\| \approx {}^*f(a) \in {}^*[0, \infty)$. Because $\| {}^*f\| = \|f\|$, we have $\widehat{{}^*f}(\hat{a}) = \left\| \widehat{{}^*f} \right\| = \|f\|$. $\square$

This section provides the major tools in Banach space theory, namely Helly's Theorem, Hahn-Banach Theorem and the Hahn-Banach Separation Theorem. The introduction of these tools comes with some related norming properties. The Hahn-Banach Theorem follows from the proof of Helly's Theorem, so one can regard the latter as the most fundamental result.

2.3.1   *Norming and Helly's Theorem*

In general, the space X itself may not have enough norming elements. We will characterize this in a later section.

Before exploring further these norming properties in dual form, we need the following geometric property from finite dimensional Euclidean spaces showing that every nonzero element attains the maximal possible distance from a hyperplane.

Lemma 2.2. *Let X be a normed linear space with $1 < \dim(X) < \infty$. Then for any $c \in X$ with $\|c\| = 1$, there is a subspace $Z \subset X$ such that $X = \mathrm{Lin}(Z \cup \{c\})$ and $\mathrm{dist}(c, Z) = 1$.*

Proof.   Since $\|c\| = 1$, we only need to find such Z with $\mathrm{dist}(c, Z) \geq 1$.

CASE 1: $\mathbb{F} = \mathbb{R}$.

Let $\dim(X) = n + 1$ and fix a list of subspaces

$$\mathrm{Lin}(\{c\}) = X_0 \subsetneqq X_1 \subsetneqq \cdots \subsetneqq X_n = X.$$

We are going to construct subspaces $\{0\} = Y_0 \subsetneq Y_1 \subsetneq \cdots \subsetneq Y_n$ such that

$$\forall k = 1, \ldots, n \left(\left(X_k = \mathrm{Lin}(Y_k \cup \{c\}) \right) \wedge \left(\mathrm{dist}(c, Y_k) \geq 1 \right) \right).$$

Assume inductively that Y_k is constructed with above properties. To continue with the construction, first let

$$V := \bigcup_{\lambda \in \mathbb{R}^+, \, y \in Y_k} \lambda B_{X_{k+1}}(c + y, 1),$$

where $B_{X_k}(x, r)$ denotes the open ball $B(x, r) \cap X_k$ in X_k.

CLAIM: V is an open convex set and $V \cap (-V) = \emptyset$.

As a union of open balls, V is clearly open. Let $a_1, a_2 \in V$ and $t \in [0, 1]$. Then for some $\lambda_1, \lambda_2 \in \mathbb{R}^+$ and some $y_1, y_2, y_3, y \in Y_k$,

$$\begin{aligned}
\left(ta_1 + (1-t)a_2 \right) &\in \left(t\lambda_1 B_{X_{k+1}}(c + y_1, 1) \right) + \left((1-t)\lambda_2 B_{X_{k+1}}(c + y_2, 1) \right) \\
&= t B_{X_{k+1}}(\lambda_1(c + y_1), \lambda_1) + (1-t) B_{X_{k+1}}(\lambda_2(c + y_2), \lambda_2) \\
&\subset B_{X_{k+1}}\left((t\lambda_1 + (1-t)\lambda_2)c + y_3, \, (t\lambda_1 + (1-t)\lambda_2) \right) \\
&= (t\lambda_1 + (1-t)\lambda_2) B_{X_{k+1}}(c + y, \, 1) \subset V,
\end{aligned}$$

by $(t\lambda_1 + (1-t)\lambda_2) \in \mathbb{R}^+$. Hence the first part of the Claim is proved.

For the second part, if there is $a \in V \cap (-V)$, then $a, -a \in V$ and, by convexity, the line segment $[-a, a] \subset V$. But $0 \in [-a, a]$, hence $0 \in V$ and consequently $0 \in B_{X_{k+1}}(c + y, 1)$ for some $y \in Y_k$, contradicting to the assumption $\mathrm{dist}(c, Y_k) \geq 1$.

Since X_{k+1} is homeomorphic to $\mathbb{R}^m$ for some $m > 1$, $X_{k+1} \setminus \{0\}$ is connected and it follows from the Claim that $X_{k+1} \neq \left(V \cup (-V) \cup \{0\} \right)$.

Now let $a \in X_{k+1} \setminus \left(V \cup (-V) \cup \{0\} \right)$. By assumption,

$$X_k = \mathrm{Lin}(Y_k \cup \{c\}) \subset \bigcup_{\lambda \in \mathbb{R}, \, y \in Y_k} \lambda B_{X_{k+1}}(c + y, 1) = \left(V \cup (-V) \cup \{0\} \right),$$

so $a \in X_{k+1} \setminus X_k$ and therefore $X_{k+1} = \mathrm{Lin}(Y_k \cup \{a, c\})$.

By $a \notin \bigcup_{\lambda \in \mathbb{R}, \, y \in Y_k} \lambda B_{X_{k+1}}(c + y, 1)$ and $\mathrm{dist}(c, Y_k) \geq 1$, we have $\|c + y - \lambda a\| \geq 1$ for any $\lambda \in \mathbb{R}$ and $y \in Y_k$.

That is, $\mathrm{dist}\left(c, \mathrm{Lin}(Y_k \cup \{a\}) \right) \geq 1$.

Hence the required properties are satisfied by $Y_{k+1} := \mathrm{Lin}(Y_k \cup \{a\})$.

With all Y_k's constructed, let $Z = Y_n$, then the proof for Case 1 is completed.

CASE 2: $\mathbb{F} = \mathbb{C}$.

Regarding X as a normed linear space over $\mathbb{R}$ under the same norm, the above shows that there is a real subspace $Y \subset X$ such that

$$X = \mathrm{Lin}_{\mathbb{R}}(Y \cup \{c\}) \quad \wedge \quad \mathrm{dist}(c, Y) \geq 1.$$

In the above construction, we may assume without loss of generality that $ic \in X_1$ and put $ic \in Y_1 \subset Y$, because $\|c + \lambda ic\| = |1 + \lambda i| \, \|c\| \geq 1$, for any $\lambda \in \mathbb{R}^+$.

Now let $Z \subset Y$ be a real subspace such that $Y = \mathrm{Lin}_{\mathbb{R}}(Z \cup \{ic\})$. Then clearly $\mathrm{dist}(c, Z) \geq \mathrm{dist}(c, Y) \geq 1$.

Let $X_0 := \mathrm{Lin}_{\mathbb{R}}(\{c, ic\}) = \mathrm{Lin}_{\mathbb{C}}(\{c\})$.

So $X = \mathrm{Lin}_{\mathbb{R}}(X_0 \cup Z)$ with $iX_0 = X_0$ and $Z \cap X_0 = \{0\}$. Finally,

$$Z \subset X = iX = \mathrm{Lin}_{\mathbb{R}}(iX_0 \cup iZ) = \mathrm{Lin}_{\mathbb{R}}(X_0 \cup iZ),$$

implying $Z \subset iZ$. But then $iZ \subset (-Z) = Z$, therefore $Z = iZ$, *i.e.* Z is a complex subspace of X, completing the proof for case 2. $\qquad \square$

With the hyperplane treated as a level curve of a bounded linear functional, the corresponding element is normed. As a result, the dual to Prop. 2.22 has a stronger form.

Theorem 2.5. *Let X be a normed linear space.*

Then $\forall x \in X \left(\exists f \in X' \left((\|f\| = 1) \wedge (f(x) = \|x\|) \right) \right).$

Proof. To avoid triviality, we assume that $\dim(X) > 1$ and consider some $0 \neq c \in X$. Furthermore, it suffices to norm $\|c\|^{-1} c$. So we assume that $\|c\| = 1$.

Apply Lem. 1.1, let $Y \subset {}^*X$ be a hyperfinite dimensional subspace such that $X \subset Y \subset {}^*X$. In particular, $c \in Y$.

By transferring Lem. 2.2 and applying it to c and Y, there is an internal subspace $Z \subset Y$ such that $Y = {}^*\mathrm{Lin}(Z \cup \{c\})$ and ${}^*\mathrm{dist}(c, Z) = 1$.

Let $g : Y \to {}^*\mathbb{F}$ be defined internally by $g(x) = \alpha$, where each $x \in Y$ is represented as $\alpha c + z$ for some unique $\alpha \in {}^*\mathbb{F}$ and $z \in Z$.

Clearly, g is linear. By assumption, for any $0 \neq \alpha \in {}^*\mathbb{F}$ and $z \in Z$, we have $\|\alpha c + z\|_Y = |\alpha| \, \|c + \alpha^{-1} z\|_Y \geq |\alpha|$. Moreover, $g[Z] = \{0\}$, hence $g \in Y'$ and $\|g\|_{Y'} \leq 1$.

Note that $g(c) = 1$, which shows also that $\|g\|_{Y'} = 1$.

Now define $f := {}^\circ g \!\restriction_X$, *i.e.* $f : X \ni x \mapsto {}^\circ g(x)$.

From the definition, we have $f(c) = {}^\circ g(c) = 1$.

Obviously, f is linear. Moreover,

$$1 = f(c) \leq \|f\|_{X'} \leq {}^\circ \|g\|_{Y'} = 1,$$

since X is a subspace of Y. So $f \in X'$, $\|f\|_{X'} = 1$ and $f(c) = 1$. $\qquad \square$

In particular, the above Theorem implies that X' separates elements in X : for if $a \neq b$ in X, then there is $f \in X'$ with $f(a - b) = \|a - b\| \neq 0$, hence $f(a) \neq f(b)$.

As an application of Thm. 2.5, we digress a little bit with a continuation of the discussion of the projections on p.103.

Theorem 2.6. *Let X be a normed linear space and $Y \subset X$ be a finite dimensional subspace. Then Y is complemented.*

In particular, there is a projection $p \in \mathcal{B}(X)$ such that $Y = p[X]$.

Proof. First we need the following.

CLAIM: Let $a \in X$. Then $\mathrm{Lin}(\{a\}) = p[X]$ for some projection $p \in \mathcal{B}(X)$.

If $a = 0$, we simply take $p = 0$. So we assume that $a \neq 0$. Since we get the same conclusion if a is replaced by $\|a\|^{-1} a$, we further assume that $\|a\| = 1$.

Let $f \in S_{X'}$ be such that $f(a) = \|a\| = 1$, as given by Thm. 2.5.

Define $p : X \to X$ by $X \ni x \mapsto f(x)a$. Then immediately p is linear and $\|p\| \leq \|a\| = 1$, hence $p \in \mathcal{B}(X)$. For any $x \in X$, $p^2(x) = p(f(x)a) = f(x)p(a) = f(x)a = p(x)$. i.e. $p^2 = p$.

Clearly we also have $p[X] = \mathrm{Lin}(\{a\})$. Therefore the Claim is proved.

By Prop. 2.20, the Claim implies that Y is complemented if $\dim(Y) = 0, 1$. Hence the theorem holds when the subspace is of dimension $0, 1$.

Now assume inductively that the theorem holds for subspaces of dimension $\leq n$, where $n \in \mathbb{N}$.

Let $Y \subset X$ be a subspace with $\dim(Y) = n + 1$. So there is some $a \in Y$ and a subspace $Y_0 \subset Y$ with $\dim(Y_0) = n$ such that $Y = \mathrm{Lin}(Y_0 \cup \{a\})$. Let $L := \mathrm{Lin}(\{a\})$. Then since $\dim(Y/L) = n$, it follows from the inductive hypothesis that some closed subspace $V \subset X/L$ is complementary to Y/L.

Let $W = \{x \in X \mid (x + L) \in V\}$. Then $L \subset W$ and $V = W/L$. Clearly W forms a subspace of X. In fact it is a closed subspace by Prop. 2.2(ii).

Now apply the inductive hypothesis (or the Claim), let $Z \subset W$ be a closed subspace complementary to L in W. So $W = L + Z$. Then from $X/L = Y/L + W/L$, we have $X = Y + L + W = Y + Z$.

Let $c \in Y \cap Z$. Then $(c + L) \in \left((Y/L) \cap (W/L) \right) = \{L\}$, showing that $c \in L$. But $L \cap Z = \{0\}$, so $c = 0$. i.e. $Y \cap Z = \{0\}$.

Moreover, Z is closed in W and W is closed in X so Z is a closed subspace of X. Therefore Z is a closed subspace complementary to Y.

Finally, by Prop. 2.20, $Y = p[X]$ for some projection $p \in \mathcal{B}(X)$. $\square$

Let X be a normed linear space. The dual of X' is denoted by X'', called the **bidual** of X. The following shows that X can be viewed as closed subspace of its bidual.

Proposition 2.23. *Let X be a normed linear space. For each $x \in X$, let $\theta(x)$ be the evaluation map $\theta(x) : X' \ni f \mapsto f(x)$.*

Then $\theta : X \to X''$ and is an isometric embedding.

Proof.   Clearly, $\theta(x) : X' \to \mathbb{F}$ is linear. Note that
$$\|\theta(x)\|_{X''} = \sup_{f \in \bar{B}_{X'}} |\theta(x)(f)| = \sup_{f \in \bar{B}_{X'}} |f(x)| \leq \|x\|_X .$$
By Thm. 2.5, there is $f \in \bar{B}_{X'}$ such that $f(x) = \|x\|_X$.

Therefore $\|\theta(x)\|_{X''} = \|x\|_X$ and consequently $\forall x \in X \left(\theta(x) \in X'' \right)$ and θ is an isometric isomorphism into a subspace of X''.   □

In general, the above identification of X is not onto X''. We will discuss this in more details in later sections, but here is a quick application that should be compared with the norming in Thm. 2.5.

Proposition 2.24. *Let X be a normed linear space and $S \subset X$. Then S is bounded in norm iff $\forall f \in X' \left(\sup_{x \in S} |f(x)| < \infty \right)$.*

Proof.   ($\Rightarrow$) : is trivial.

($\Leftarrow$) : Use Prop. 2.23 to view $S \subset X''$. Then apply the Uniform Boundedness Principle (Thm. 2.4) to the Banach space X' (see p.95) and $\mathbb{F}$, so the given condition implies that
$$\sup_{a \in S} \|a\|_X = \sup_{a \in S} \|a\|_{X''} < \infty.$$
   □

Given $0 \neq f \in X'$ and $\gamma \in \mathbb{F}$, trivially there exists $x \in X$ so that $f(x) = \gamma$. The following generalization will be an important tool for us.

(For convenience, we use $\forall^{\mathrm{fin}} \cdots$ as an abbreviation for $\forall$ *finite* $\cdots$.)

Theorem 2.7.   (Helly's Theorem—nonstandard version) *Let X be a normed linear space, $f_i \in X'$ and $\gamma_i \in \mathbb{F}$, $i \in I$, where I is an index set. Then for any $r \in \mathbb{R}^+$,*
$$\exists x \in {}^*X \left[\left(\|x\| \lesssim r \right) \wedge \left(\forall i \in I \left({}^*f_i(x) \approx \gamma_i \right) \right) \right] \tag{2.1}$$
iff
$$\exists x \in \widehat{{}^*X} \left[\left(\|x\| \leq r \right) \wedge \left(\forall i \in I \left(\widehat{{}^*f_i}(x) = \gamma_i \right) \right) \right] \tag{2.2}$$
iff
$$\forall^{\mathrm{fin}} J \subset I \left[\forall \{\alpha_i\}_{i \in J} \subset \mathbb{F} \left(\left| \sum_{i \in J} \alpha_i \gamma_i \right| \leq r \left\| \sum_{i \in J} \alpha_i f_i \right\| \right) \right]. \tag{2.3}$$

Proof. By the nonstandard hull construction, (2.1) and (2.2) are equivalent.

$((2.2) \Rightarrow (2.3))$: Let $a \in {}^*\widehat{X}$ with $\|a\| \leq r$ and $\widehat{{}^*f_i}(a) = \gamma_i$, $i \in I$. Then for any finite $J \subset I$ and $\alpha_i \in \mathbb{F}$, $i \in J$, we have

$$\left| \sum_{i \in J} \alpha_i \gamma_i \right| = \left| \sum_{i \in J} \alpha_i \widehat{{}^*f_i}(a) \right| \leq \|a\| \left\| \sum_{i \in J} \alpha_i \widehat{{}^*f_i} \right\| = \|a\| \left\| \sum_{i \in J} \alpha_i f_i \right\|$$

by Prop. 2.7.

$((2.3) \Rightarrow (2.1))$: Assume (2.3), fix a finite $J \subset I$.

List $\{f_i \mid i \in J\}$ as $f_1, \ldots, f_m, \ldots, f_n$, where $f_1, \ldots, f_m$ is a maximal linearly independent subsequence, and list $\{\gamma_i \mid i \in J\}$ as $\gamma_1, \ldots, \gamma_n$ accordingly.

Let $e_1, \ldots, e_N \in {}^*X$ with $\|e_1\| = \cdots = \|e_N\| = 1$, where $N \in {}^*\mathbb{N}$, be a Hamel basis extending that of X. Note that $N \geq m$ by the linear independence of $f_1, \ldots, f_m$.

Let $Y = {}^*\mathrm{Lin}(e_1, \ldots, e_N)$, then $X \subset Y \subset {}^*X$.

For $i = 1, \ldots, m$, let $g_i = {}^*f_i \upharpoonright_Y$ and extend $g_1, \ldots, g_m$ to a linearly independent sequence $g_1, \ldots, g_N$ in Y'. So $Y' = {}^*\mathrm{Lin}(g_1, \ldots, g_N)$.

Define an $N \times N$ matrix $M := \begin{bmatrix} g_1(e_1) & \cdots & g_N(e_1) \\ \vdots & \cdots & \vdots \\ g_1(e_N) & \cdots & g_N(e_N) \end{bmatrix}$.

Note that M has full rank and hence is invertible.

Now let $[\alpha_1, \ldots, \alpha_N] := [\gamma_1, \ldots, \gamma_m, 0, \ldots, 0] \cdot M^{-1}$ and define

$$a_0 := \sum_{i=1}^{N} \alpha_i e_i \in Y.$$

Then

$$\forall i = 1, \ldots m \; \left({}^*f_i(a_0) = g_i(a_0) = \gamma_i \right). \tag{2.4}$$

Let $Z := {}^*f_1^{-1}(0) \cap \cdots \cap {}^*f_m^{-1}(0) \cap Y$. Note that Z forms a closed subspace of Y.

By transfer and Thm. 2.5, there is some $h \in (Y/Z)'$ such that

$$\left(\|h\|_{(Y/Z)'} = 1 \right) \wedge \left(h(a_0 + Z) = \|a_0 + Z\|_{Y/Z} \right).$$

Define $h_0 \in Y'$ as in Prop. 2.14. *i.e.*

$$h_0 : x \mapsto h(x + Z) \quad \text{with} \; \|h_0\|_{Y'} \leq \|h\|_{(Y/Z)'} = 1.$$

In particular,

$$h_0[Z] = \{0\}. \tag{2.5}$$

Write $h_0 = \sum_{i=1}^{N} \beta_i g_i$ for some $\beta_i \in {}^*\mathbb{F}$. For each k with $m < k \leq N$, let $\zeta_1, \ldots, \zeta_N \in {}^*\mathbb{F}$ satisfy $[\zeta_1, \ldots, \zeta_N] = [0, \ldots, 0, \underset{k^{\text{th}}}{1}, 0, \ldots, 0] \cdot M^{-1}$ and set $c = \sum_{i=1}^{N} \zeta_i e_i \in Y$, then $\forall i \neq k \left({}^*f_i(c) = g_i(c) = 0 \right) \wedge g_k(c) = 1$. In particular $c \in Z$ hence $0 = h_0(c) = \sum_{i=1}^{N} \beta_i g_i(c) = \beta_k$.

Therefore $h_0 = \sum_{i=1}^{m} \beta_i g_i$.

By $\|h_0\| \leq 1$ and Prop. 2.15, $\beta_1, \ldots, \beta_m \in \text{Fin}({}^*\mathbb{F})$. Moreover, apply the isometric isomorphism in Prop. 2.15,

$$\left\| \sum_{i=1}^{m} {}^{\circ}\beta_i f_i \right\|_{X'} = \left\| \sum_{i=1}^{m} {}^{\circ}\beta_i \widehat{g_i} \right\|_{\widehat{Y'}} \approx \left\| \sum_{i=1}^{m} \beta_i g_i \right\|_{Y'} = \|h_0\|_{Y'} \leq 1. \qquad (2.6)$$

On the other hand,

$$h_0(a_0) = h(a_0 + Z) = \inf\{ \|a_0 + z\|_Y \mid z \in Z \} \approx \|a_0 + z_0\| \qquad (2.7)$$

for some $z_0 \in Z$, by saturation.

Now define $a = a_0 + z_0$. Then for $k = 1, \ldots, m$, we have ${}^*f_k(a) = \gamma_k$, by (2.4) and (2.5). For $k = (m+1), \ldots, n$, by linear dependence, write $f_k = \sum_{i=1}^{m} \lambda_i f_i$ for some $\lambda_i \in \mathbb{F}$, then (2.3) gives

$$|\gamma_k - {}^*f_k(a)| = \left| \gamma_k - \sum_{i=1}^{m} \lambda_i \gamma_i \right| \leq r \left\| f_k - \sum_{i=1}^{m} \lambda_i f_i \right\| = 0.$$

Therefore we obtain

$$\forall i = 1, \ldots, n \left({}^*f_i(a) = \gamma_i \right). \qquad (2.8)$$

By transferring (2.3) and combining it with (2.6) and (2.7), we have

$$\|a\| = \|a_0 + z_0\| \approx h_0(a_0) = \sum_{i=1}^{m} \beta_i g_i(a_0) = \sum_{i=1}^{m} \beta_i \gamma_i$$
$$\leq r \left\| \sum_{i=1}^{m} \beta_i {}^*f_i \right\|_{({}^*X)'} \approx r \left\| \sum_{i=1}^{m} {}^{\circ}\beta_i f_i \right\|_{X'} \lessapprox r. \qquad (2.9)$$

Consequently, by (2.8) and (2.9), $\forall^{\text{fin}} J \subset I \quad \forall n \in \mathbb{N} \quad \exists x \in {}^*X$

$$\left(\|x\| \leq r + n^{-1} \right) \wedge \left(\forall i \in J \left({}^*f_i(x) = \gamma_i \right) \right).$$

Finally, by saturation, let $b \in {}^*X$ with $\|b\| \lessapprox r$ and $\forall i \in I \left({}^*f_i(b) = \gamma_i \right)$, therefore (2.1) is proved. $\qquad \square$

As a corollary, we have the usual version of Helly's Theorem.

Corollary 2.10. (Helly's Theorem) *Let X be a normed linear space, $f_i \in X'$ and $\gamma_i \in \mathbb{F}$, $i \in I$, where I is a finite set. Then for any $r \in \mathbb{R}^+$,*

$$\forall \epsilon \in \mathbb{R}^+ \exists x \in X \left[\left(\|x\| \leq r + \epsilon \right) \wedge \left(\forall i \in I \left(f_i(x) = \gamma_i \right) \right) \right] \qquad (2.10)$$

iff

$$\forall \{\alpha_i\}_{i \in I} \subset \mathbb{F} \left(\left| \sum_{i \in I} \alpha_i \gamma_i \right| \leq r \left\| \sum_{i \in J} \alpha_j f_i \right\| \right). \qquad (2.11)$$

Proof. The easy direction $(\Rightarrow)$ is similar to the proof before.

For the harder direction$(\Leftarrow)$, by Thm. 2.7, we have $x \in {}^*X$ satisfying $\big(\|x\| \leq r + \epsilon \big) \wedge \big(\forall i \in I \, ({}^*\!f_i(x) = \gamma_i) \big)$, and therefore (2.10) follows by transfer. $\qquad\square$

However, the dual version bears a closer resemblance to Thm. 2.7.

Corollary 2.11. *Let X be a normed linear space, $a_i \in X$ and $\gamma_i \in \mathbb{F}$, $i \in I$, where I is any index set. Then for any $r \in \mathbb{R}^+$,*

$$\exists f \in X' \Big[\big(\|f\| \leq r \big) \wedge \big(\forall i \in I \, (f(a_i) = \gamma_i) \big) \Big] \tag{2.12}$$

iff

$$\forall^{\mathrm{fin}} J \subset I \Big[\forall \{\alpha_i\}_{i \in J} \subset \mathbb{F} \, \Big(\Big| \sum_{i \in J} \alpha_i \gamma_i \Big| \leq r \Big\| \sum_{i \in J} \alpha_i a_i \Big\| \Big) \Big]. \tag{2.13}$$

Proof. Let $\theta : X \to X''$ be as in Prop. 2.23 and consider Thm. 2.7 for $\theta(a_i) \in X''$ and γ_i, $i \in I$.

Then (2.12) is the same as

$$\exists f \in X' \Big[\big(\|f\| \leq r \big) \wedge \big(\forall i \in I \, (\theta(a_i)(f) = \gamma_i) \big) \Big]$$

implying $\forall^{\mathrm{fin}} J \subset I \quad \forall \{\alpha_i\}_{i \in J} \subset \mathbb{F}$

$$\Big| \sum_{i \in J} \alpha_i \gamma_i \Big| \leq r \Big\| \sum_{i \in J} \alpha_j \theta(a_i) \Big\|_{X''} = r \Big\| \sum_{i \in J} \alpha_j a_i \Big\|_X,$$

as in the proof of Thm. 2.7 and by Prop. 2.23.

Conversely, from the proof of Thm. 2.7, the above condition implies for some $g \in {}^*X'$ that

$$\big(\|g\|_{*X'} \lessapprox r \big) \wedge \big(\forall i \in I \, ({}^*\theta(a_i)(g) = \gamma_i) \big).$$

Now let $f = {}^\circ g \!\upharpoonright_X$, then $\|f\|_{X'} \leq {}^\circ \|g\|_{*X} \leq r$, in particular $f \in X'$. Moreover, for each $i \in I$, we have $f(a_i) = {}^\circ g(a_i) = {}^\circ \big({}^*\theta(a_i)(g) \big) = \gamma_i$. Therefore (2.12) is satisfied. $\qquad\square$

Corollary 2.12. *Let Y be a subspace of a normed linear space X. Suppose $a \in X$ is such that $\mathrm{dist}(a, Y) = r \in \mathbb{R}^+$. Then there is $f \in \bar{B}_{X'}$ such that $f[Y] = \{0\}$ and $f(a) = r$.*

Proof. By assumption, for any finite subsets $\{a_i\}_{i \in I} \subset Y$, $\{\alpha_i\}_{i \in I} \subset \mathbb{F}$ and $\alpha \in \mathbb{F}$, we have $|\alpha| \, r = |\alpha| \, \mathrm{dist}(a, Y) \leq \big\| \sum_{i \in I} \alpha_i a_i + \alpha a \big\|$. Therefore

$$\Big| \sum_{i \in I} \alpha_i \cdot 0 + \alpha r \Big| = |\alpha| \, r \leq \Big\| \sum_{i \in I} \alpha_i a_i + \alpha a \Big\|$$

and the conclusion follows from Cor. 2.11. $\qquad\square$

Further elaboration of Thm. 2.7 will yield the following result, whose proof is left as an exercise.

Theorem 2.8. (Extended Helly's Theorem) *Let X be a normed linear space and I, J index sets with some $\{f_i\}_{i \in I} \subset X'$, $\{\gamma_{ij}\}_{i \in I, j \in J} \subset \mathbb{F}$ and $r : \bigcup \{\mathbb{F}^{J_0} \mid J_0 \subset J \text{ is finite}\} \to [0, \infty)$. Then*

$$\exists \{x_j\}_{j \in J} \subset {}^*X \left[\left[\forall^{\text{fin}} J_0 \subset J \; \forall \{\beta_j\}_{j \in J_0} \subset \mathbb{F} \left(\left\| \sum_{j \in J_0} \beta_j x_j \right\| \lessapprox r(\{\beta_j\}_{j \in J_0}) \right) \right] \right.$$

$$\left. \wedge \; \left[\forall (i,j) \in (I \times J) \left(\gamma_{ij} \approx {}^*f_i(x_j) \right) \right] \right]$$

iff

$$\forall^{\text{fin}} I_0 \subset I \quad \forall^{\text{fin}} J_0 \subset J \quad \forall \{\alpha_i\}_{i \in I_0} \subset \mathbb{F} \quad \forall \{\beta_j\}_{j \in J_0} \subset \mathbb{F}$$

$$\left(\left| \sum_{i \in I_0, j \in J_0} \alpha_i \beta_j \gamma_{ij} \right| \leq r(\{\beta_j\}_{j \in J_0}) \left\| \sum_{i \in I_0} \alpha_i f_i \right\| \right).$$

$\square$

2.3.2   The Hahn-Banach Theorem

From the exercises in the last section, we see that if X is a normed linear space over $\mathbb{F}$ and $f \in X'$ (*i.e.* $\mathcal{B}(X, \mathbb{F})$), then $\bar{f} := \left({}^*\widehat{f} \upharpoonright_{\bar{X}} \right) \in \bar{X}'$ is the unique extension of f in $\bar{X}'$ with $\|\bar{f}\| = \|f\|$.

Actually, such f extends to some functional in Y' with the norm preserved whenever X is a subspace of Y, although generally there is no uniqueness. This is the essence of the Hahn-Banach Theorem which is now shown to be a corollary of Helly's Theorem.

Theorem 2.9. (Hahn-Banach Theorem) *Let X be a subspace of a normed linear space Y over $\mathbb{F}$. Then for any $f \in X'$ there is $\bar{f} \in Y'$ such that $f = \bar{f} \upharpoonright_X$ and $\|f\| = \|\bar{f}\|$.*

Proof.   Apply Cor. 2.11 to the space Y with $I = X$, and for each $x \in I$ let $a_x = x$ and $\gamma_x = f(x)$. Let $r = \|f\|$. Then (2.13) is satisfied and consequently (2.12) gives some $\bar{f} \in Y'$ such that $\|\bar{f}\| \leq r = \|f\|$ and $\forall x \in X \left(\bar{f}(x) = \bar{f}(a_x) = \gamma_x = f(x) \right)$. $\square$

However, it may be useful to see a second proof of the Hahn-Banach Theorem.

Proof.   (A direct proof of the Hahn-Banach Theorem.)

First we need the following:

CLAIM: *Suppose there is some $a \in Y \setminus X$, then f extends to some $\bar{f} \in \left(\mathrm{Lin}(X \cup \{a\})\right)'$ with $\|f\| = \|\bar{f}\|$.*

To prove the Claim, by replacing f by $\|f\|^{-1} f$, we can assume that $\|f\| = 1$. Write Y_0 for the real linear span of $X \cup \{a\}$ and Y_1 for the real linear span of $Y_0 \cup \{ia\}$. So, in the case $\mathbb{F} = \mathbb{R}$, we have $Y_0 = \mathrm{Lin}(X \cup \{a\})$; otherwise, we have $Y_1 = \mathrm{Lin}(X \cup \{a\})$.

Let $g := \mathrm{Re}(f)$, the real part of f. So in the case $\mathbb{F} = \mathbb{R}$, $f = g$. When X is viewed as a normed linear space over $\mathbb{R}$, $g : X \to \mathbb{R}$ is a real bounded linear functional. Clearly $\|g\| \leq \|f\|$. By Prop. 2.22, let $c \in {}^*B_X$ ${}^*f(c) \approx \|f\|$, then ${}^*g(c) = \mathrm{Re}({}^*f)(c) = {}^*f(c)$, therefore $\|f\| = \|g\| = 1$.

For any $x, y \in X$, $g(x) - g(y) = g(x - y) \leq \|x - y\| \leq \|x - a\| + \|y - a\|$, *i.e.* $-\|x - a\| + g(x) \leq \|y - a\| + g(y)$, hence

$$r := \sup_{x \in X} \left(-\|x - a\| + g(x) \right) \leq \inf_{y \in X} \left(\|y - a\| + g(y) \right).$$

Define $\bar{g}_0 : Y_0 \to \mathbb{R}$ by $\bar{g}_0(x + ka) = g(x) - kr$ for each element in Y_0 represented uniquely as $(x + ka)$ for some $x \in X$ and $k \in \mathbb{R}$. Clearly $\bar{g}_0 : Y_0 \to \mathbb{R}$ is linear and extends g. Moreover, for $x \in X$ and $0 \neq k \in \mathbb{R}$,

$$-\|k^{-1}x - a\| + g(k^{-1}x) \leq r \leq \|k^{-1}x - a\| + g(k^{-1}x),$$

that is, $\left|g(k^{-1}x) - r\right| \leq \|k^{-1}x + a\|$, hence $\left|\bar{g}_0(x + ka)\right| \leq \|x + ka\|$ and therefore $\|\bar{g}_0\| = \|g\| = 1$.

In the case $\mathbb{F} = \mathbb{R}$, we let $\bar{f} = \bar{g}_0$ and the proof is complete.

In the case $\mathbb{F} = \mathbb{C}$, by repeating the same procedure, we obtain a real bounded linear functional $\bar{g} : Y_1 \to \mathbb{R}$ which extends $\bar{g}_0$ and satisfies $\|\bar{g}\| = \|\bar{g}_0\| = 1$.

Finally, define $\bar{f} : Y_1 \to \mathbb{C}$ by $x \mapsto \left(\bar{g}(x) - i\bar{g}(ix)\right)$. It is straightforward to check that $\bar{f}$ is linear. Furthermore, by an application of Prop. 2.22 as before, $\|\bar{f}\| = \|\mathrm{Re}(\bar{f})\| = \|\bar{g}\|$ and therefore $\bar{f} \in Y_1'$ extends f and $\|\bar{f}\| = \|f\|$.

Therefore the Claim is proved.

Apply Lem. 1.1, let Z be an internal extension of ${}^*X$ which is hyperfinite dimensional over ${}^*X$ and $Y \subset Z \subset {}^*Y$.

By iterating the Claim, the same conclusion holds for any subspace of Y which is finite dimensional over X, *i.e.* a linear span of the union of X with a finite subset of Y.

Then by transferring this, ${}^*f$ extends to some $\tilde{f} \in Z'$ with $\left\|\tilde{f}\right\| = \|{}^*f\|$.

Now the conclusion follows if we let $\bar{f} = \widehat{\tilde{f}} \restriction_Y$. $\square$

The Hahn-Banach Theorem has the following generalization. The proof is based on similar techniques and is left as an exercise.

Theorem 2.10. (Hahn-Banach Theorem—extended versions)

(i) *Let X be a subspace of a linear space Y over $\mathbb{R}$ and $p : Y \to \mathbb{R}$ be a pseudo-seminorm. Then for any linear $f : X \to \mathbb{R}$ such that $\forall x \in X \left(f(x) \leq p(x) \right)$, there is a linear $\bar{f} : Y \to \mathbb{R}$ extending f such that $\forall x \in Y \left(\bar{f}(x) \leq p(x) \right)$.*

(ii) *Let Y be a linear space over $\mathbb{F}$ having a seminorm $\|\cdot\|$. On a subspace $X \subset Y$, suppose $f : X \to \mathbb{F}$ is linear and $\forall x \in X \left(|f(x)| \leq \|x\| \right)$. Then there is a linear $\bar{f} : Y \to \mathbb{F}$ which is an extension of f satisfying $\forall x \in Y \left(|\bar{f}(x)| \leq \|x\| \right)$.* □

2.3.3   The Hahn-Banach Separation Theorem

As an application of the above Hahn-Banach Theorem, we have the following important result:

Theorem 2.11. (Hahn-Banach Separation Theorem) *Let X be a normed linear space with nonempty disjoint convex subsets A and B, where $\operatorname{int}(A) \neq \emptyset$. Then there is $f \in X'$ such that*

$$\operatorname{Re}(f)[A] \subset (-\infty, \lambda] \quad and \quad \operatorname{Re}(f)[B] \subset [\lambda, \infty) \quad for\ some \quad \lambda \in \mathbb{R}.$$

(If A is open, we can require that $\operatorname{Re}(f)[A] \subset (-\infty, \lambda)$.)

Proof.   Let $C := (A - B) = \{ x - y \mid x \in A \wedge y \in B \}$, then the assumptions imply that C is convex, $\operatorname{int}(C) \neq \emptyset$ and $0 \notin C$.

Fix $c \in \operatorname{int}(C)$, let $K := (C - c)$ and define $p : X \to [0, \infty)$ by

$$p(x) := \inf \left\{ t \in \mathbb{R}^+ \mid x \in tK \right\}.$$

As $0 \in \operatorname{int}(K)$, p is well-defined. Clearly, p is positively homogeneous. Moreover, given $t_1, t_2 \in \mathbb{R}^+$, for any $x_1, x_2 \in K$, convexity implies that $\left(t_1(t_1 + t_2)^{-1} x_1 + t_2(t_1 + t_2)^{-1} x_2 \right) \in K$. Therefore $\left(t_1 K + t_2 K \right) \subset (t_1 + t_2) K$. It follows that p is subadditive, *i.e.* p forms a pseudo-seminorm.

Furthermore, $p(-c) \geq 1$, since $-c \notin K$ (because $0 \notin C$).

Regard X as a normed linear space over $\mathbb{R}$. Let $g_0 : \operatorname{Lin}(\{c\}) \to \mathbb{R}$ be given by $g_0(rc) = -rp(-c)$, $r \in \mathbb{R}$. Note that $g_0(-c) = p(-c) \geq 1$.

Clearly, g_0 is linear. For $r < 0$, we have $g_0(rc) = (-r)p(-c) = p(rc)$, by positive homogeneity. Hence $\forall x \in \operatorname{Lin}(\{c\}) \left(g_0(x) \leq p(x) \right)$.

Therefore, by Thm. 2.10, g_0 extends to some linear $g : X \to \mathbb{R}$ such that $\forall x \in X \left(g(x) \leq p(x) \right)$.

Note that $\forall x \in K \left(p(x) \leq 1\right)$, so $\forall x \in K \left(g(x) \leq 1\right)$. As $0 \in \mathrm{int}(K)$, it follows that $^*g(x) \approx 0$ whenever $^*X \ni x \approx 0$.

By Prop. 2.13, this means that g is a real bounded linear functional. By $\forall x \in K \left(g(x) \leq 1\right)$, it follows that $\forall x \in C \left(g(x - c) \leq 1\right)$, *i.e.*

$$\forall x \in A \, \forall y \in B \left(g(x) - g(y) + g(-c) \leq 1\right).$$

As $g(-c) = g_0(-c) \geq 1$, we have $\forall x \in A \, \forall y \in B \left(g(x) \leq g(y)\right)$.

Let $\lambda := \inf \left\{g(y) \,|\, y \in B \right\}$, we have $g[A] \subset (-\infty, \lambda]$ and $g[B] \subset [\lambda, \infty)$. If A is open, actually $g[A] \subset (-\infty, \lambda)$, since g is an open mapping (by Thm. 2.2).

If $\mathbb{F} = \mathbb{R}$, we simply let $f = g$ and the proof is complete.

If $\mathbb{F} = \mathbb{C}$, we define $f : X \to \mathbb{C}$ by $f(x) := g(x) - ig(ix)$. Then $g = \mathrm{Re}(f)$ and, as in the direct proof of the Hahn-Banach Theorem, f is linear and has the same norm as g, so $f \in X'$ and has all the required properties. $\square$

Note that, as a consequence, linear functionals separate points from a convex set in the following sense:

Corollary 2.13. *Let X be a normed linear space, $C \subset X$ be convex and $a \in X$ with $\mathrm{dist}(a, C) > 0$.*

Then $\exists f \in X' \left(\mathrm{Re}(f)(a) < \inf \left\{\mathrm{Re}(f)(x) \,|\, x \in C \right\}\right).$

Proof. Apply Thm. 2.11 by taking $B = C$ and A an open ball centered at a disjoint from B. $\square$

In particular, in the above, a and C are separated by an affine plane (*i.e.* a translated hyperplane).

Corollary 2.14. *Let X be a normed linear space, $Y \subset X$ a closed subspace and $a \in X$ with $\mathrm{dist}(a, Y) > 0$.*

Then $\exists f \in X' \left(f(a) = 1 \land f[Y] = \{0\}\right).$

Proof. Again, we apply Thm. 2.11 by taking $B = Y$ and A an open ball centered at a disjoint from Y.

Then since Y is closed under multiplication by real numbers, the conclusion of Thm. 2.11 implies that $\forall y \in Y \left(f(y) = 0\right)$.

Replacing f by a scaling and rotation of f if necessary, we can require that $f(a) = 1$. $\square$

It may be worthwhile to compare the above with Lem. 2.2.

2.3.4 *Notes and exercises*

What we call norming here is also called *norm-attaining* elsewhere.

The Hahn-Banach Theorem should have been named *Helly-Hahn-Banach Theorem*, for it was Eduard Helly who first prove a special version of it in 1912. (See [Hochstadt (1979/80)]. See also AMS Review MR595079 of the article by J. Dieudonné for a different view.)

The stronger assumptions in the usual version of Helly's Theorem (Cor. 2.10) are needed for the existence of the realization in the original space instead of the nonstandard hull. (See [Megginson (1998)] for details about this and a different proof of the theorem.)

The Hahn-Banach Theorem is essentially a phenomena of finite dimensional extension of real bounded linear functionals. Because of this, in the second proof, the basic tool is the same as everywhere in the literature—with a slight difference here the use of AC gets hidden behind the hyperfinite dimensional extension.

The Hahn-Banach Theorem remains valid for functionals taking values in certain types of partially ordered real linear spaces. See [Day (1973)] for results due to L.V. Kantorovič, W. Bonnice, R. Silverman *et al.*

Typically, the Hahn-Banach Separation Theorem (Thm. 2.11) is an application of the Hahn-Banach Theorem. Compare also with Cor. 2.12. See [Rudin (1991)] for a general version for topological vector spaces and variants.

EXERCISES

(1) Prove Thm. 2.5 by generalizing Lem. 2.2 using a transfinite induction argument.
(2) Prove Thm. 2.8.
(3) Prove Thm. 2.5 from the Hahn-Banach Theorem.
(4) Prove the general version of the Hahn-Banach Theorem (Thm. 2.10).
(5) Show that there is $\phi \in \bar{B}_{\ell'_\infty}$ such that
  - $\forall a \in c \left(\lim_{n \to \infty} a_n = \phi(a) \right)$;
  - $\forall a \in \ell_\infty \left(\left(\forall n \in \mathbb{N} \; a_n \in [0, \infty) \right) \Rightarrow \left(\phi(a) \in [0, \infty) \right) \right)$;
  - $\forall a, b \in \ell_\infty \left(\forall n \in \mathbb{N} \; (b_n = a_{n+1}) \Rightarrow \left(\phi(b) = \phi(a) \right) \right)$.

  Functionals of the above kind are called *Banach limits*.
(6) Give a direct proof of the following: Let X be an internal normed linear space and $Y \subsetneq X$ be an internal closed subspace. Then there is $a \in S_X$ such that ${}^*\text{dist}(a, Y) \approx 1$. (Compare with Cor. 2.14.)

2.4   General Nonstandard Hulls and Biduals

Biduals are nonstandard hulls w.r.t. the weak topology, Robinson makes Alaoglu's Theorem weak*ly apparent.

Using a family of internal seminorms, a Banach space can be constructed by a generalization of the nonstandard hull construction given in Section 2.1.

This section is about this construction and the special case of the weak nonstandard hull.

2.4.1   *Nonstandard hulls by internal seminorms*

Let X be an internal linear space over $^*\mathbb{F}$. Consider a family W of internal seminorms on X. (Implicitly, $|W| < \kappa$, the cardinality of saturation.) Then the following defines an equivalence relation on X :

$$x \approx_{\mathrm{w}} y \;\Leftrightarrow\; \forall p \in W \left(p(x - y) \approx 0 \right), \quad \text{where } x, y \in X.$$

Also, for $x \in X$, the monad *w.r.t.* W is defined as

$$\mu_{\mathrm{w}}(x) := \left\{ y \in X \mid x \approx_{\mathrm{w}} y \right\}.$$

The finite part *w.r.t.* W is given by:

$$\mathrm{Fin}_{\mathrm{w}}(X) := \left\{ x \in X \mid \sup_{p \in W} {}^{\circ}p(x) < \infty \right\}.$$

Note that $\forall x \in X \left(\mu_{\mathrm{w}}(x) = x + \mu_{\mathrm{w}}(0) \right)$ and

$$\forall x \in \mathrm{Fin}_{\mathrm{w}}(X)\, \forall y \in X \left((x \approx_{\mathrm{w}} y) \Rightarrow (y \in \mathrm{Fin}_{\mathrm{w}}(X)) \right).$$

So, if $x \in \mathrm{Fin}_{\mathrm{w}}(X)$, then $\mu_{\mathrm{w}}(x) = \left\{ y \in \mathrm{Fin}_{\mathrm{w}}(X) \mid x \approx_{\mathrm{w}} y \right\}$.

In particular, $\mu_{\mathrm{w}}(0) \subset \mathrm{Fin}_{\mathrm{w}}(X)$.

Clearly, both $\mathrm{Fin}_{\mathrm{w}}(X)$ and $\mu_{\mathrm{w}}(0)$ are closed under linear combination with scalar multiplications from $\mathrm{Fin}(\,^*\mathbb{F})$.

In particular, $\mathrm{Fin}_{\mathrm{w}}(X)$ can be viewed as a linear space over $\mathbb{F}$ with $\mu_{\mathrm{w}}(0)$ as a subspace.

We denote the quotient space $\mathrm{Fin}_{\mathrm{w}}(X)/\mu_{\mathrm{w}}(0)$ over $\mathbb{F}$ by

$$\widehat{X}^{\mathrm{w}} := \left\{ x + \mu_{\mathrm{w}}(0) \mid x \in \mathrm{Fin}_{\mathrm{w}}(X) \right\} = \left\{ \mu_{\mathrm{w}}(x) \mid x \in \mathrm{Fin}_{\mathrm{w}}(X) \right\}.$$

For convenience, when W is fixed, we write $\mu_{\mathrm{w}}(x)$, *i.e.* $x + \mu_{\mathrm{w}}(0)$, as $\widehat{x}$.

Notice that, for $\widehat{x}, \widehat{y} \in \widehat{X}^{\mathrm{w}}$ and $\mathbb{F} \ni \alpha \approx \beta \in {}^*\mathbb{F}$, we have $\widehat{x} + \alpha\widehat{y} = \overline{x + \beta y}$.

Now define $\|\cdot\|_{\mathrm{w}} : \widehat{X}^{\mathrm{w}} \to \mathbb{R}$ by

$$\|\widehat{x}\|_{\mathrm{w}} := \sup \left\{ {}^{\circ}p(x) \mid p \in W \right\}.$$

It is easy to see that $\|\cdot\|_w$ forms a norm on $\widehat{X}^w$.

Observe that the nonstandard hull $\widehat{X}$ in Section 2.1 corresponds to $\widehat{X}^w$ with $W = \{\|\cdot\|_X\}$.

Lemma 2.3. *Let X be an internal linear space and W a family of internal seminorms on X. Then $\widehat{X}^w$ forms a Banach space under $\|\cdot\|_w$.*

Proof. By the above remarks, we only need to verify completeness.

Consider a Cauchy sequence $\{\widehat{a}_n \mid n \in \mathbb{N}\}$ in $\widehat{X}^w$.

For $m \in \mathbb{N}$, fix $k_m \in \mathbb{N}$ so that

$$\forall n \in \mathbb{N} \left((n > k_m) \Rightarrow \left(\|\widehat{a}_n - \widehat{a}_{k_m}\|_w < (2m)^{-1} \right) \right).$$

Extend $\{a_n\}_{n \in \mathbb{N}}$ to a hyperfinite sequence $\{a_n\}_{n < N}$ in X, for some $N \in {}^*\mathbb{N}$. For $p \in W$ and $m \in \mathbb{N}$, define

$$\mathcal{F}_{p,m} := \{a_n \mid n < N \wedge p(a_n - a_{k_m}) < (2m)^{-1}\}.$$

Then the $\mathcal{F}_{p,m}$'s satisfy the finite intersection property. Hence, by saturation, there is $a \in \bigcap_{p \in W, m \in \mathbb{N}} \mathcal{F}_{p,m}$. Then the following holds:

$$\forall n, m \in \mathbb{N} \left((n > k_m) \Rightarrow \left(\sup_{p \in W} {}^\circ p(a - a_n) \le m^{-1} \right) \right).$$

In particular, $a \in \mathrm{Fin}_w(X)$ and $\lim_{n \to \infty} \|\widehat{a} - \widehat{a}_n\|_w = 0$.

Hence $\widehat{X}^w$ is a Banach space under $\|\cdot\|_w$. $\square$

The Banach space $\widehat{X}^w$ is called the **nonstandard hull** of X w.r.t. W. Similar to Prop. 2.6, we have the following view of $\widehat{X}^w$:

Proposition 2.25. *Let X be an internal linear space and W a family of internal seminorms on X.*

Under the seminorm $\mathrm{Fin}_w(X) \ni x \mapsto \|\widehat{x}\|_w$, $\mu_w(0)$ is a closed subspace of $\mathrm{Fin}_w(X)$ and $\widehat{X}^w$ is identical to the seminormed quotient space $\mathrm{Fin}_w(X)/\mu_w(0)$, which in fact forms a Banach space. $\square$

To emphasize the difference, the notation $\widehat{X}$ is reserved for the nonstandard hull w.r.t. the norm, called the **norm-nonstandard hull** of X.

2.4.2   Weak nonstandard hulls and biduals

Now we turn our attention to internal spaces of the form $^*X$, where X is a normed linear space over $\mathbb{F}$.

For each $\phi \in X'$, it is easy to check that the mapping $p_\phi : X \to [0, \infty)$ given by

$$p_\phi(x) := |\phi(x)|, \quad \text{where } x \in X,$$

defines a seminorm on X.

The topology on X generated by seminorms $\{p_\phi \mid \phi \in X'\}$ is called the **weak topology** on X, *i.e.* the weakest (*i.e.* the coarsest) topology that makes all $\phi \in X'$ continuous.

So a net $\{a_i\}_{i \in I} \subset X$ converges to $a \in X$ under the weak topology, written $a_i \to_w a$, if $\forall \phi \in X' \left(\lim_{i \in I} p_\phi(a_i - a) = 0 \right)$.

Under the weak topology, the infinitely close relation on X is given by

$$x \approx_w y \Leftrightarrow \forall \phi \in X' \left({}^*\phi(x) \approx {}^*\phi(y) \right), \quad x, y \in {}^*X.$$

The following is easy to check.

Proposition 2.26. *Let $x, y \in {}^*X$ where X is a normed linear space. Then $x \approx y \Rightarrow x \approx_w y$.*

Consequently, weakly closed subsets of X are norm-closed. $\square$

Moreover, we have a partial converse:

Proposition 2.27. *Let C be a closed convex subset of a normed linear space X. Then C is weakly closed.*

Proof. Let $c \in {}^*C$. Suppose $X \ni a \approx_w c$, we want to show that $a \in C$, then by the nonstandard characterization (Prop. 1.14(ii)), C is weakly closed.

But if $a \notin C$, then by Cor. 2.13, there is $\phi \in X'$ such that ϕ separates a from C. In particular, by transfer, $\phi(a) \not\approx {}^*\phi(c)$, contradicting to $a \approx_w c$. $\square$

Next we apply Lem. 2.2 with

$$W = \{ \, {}^*p_\phi \mid \phi \in \bar{B}_{X'} \, \}.$$

The resulted $\widehat{{}^*X}^{\mathrm{w}}$ is called the **weak nonstandard hull** of X. Then we have

$$\mathrm{Fin}_w({}^*X) = \left\{ x \in {}^*X \mid \sup\{ {}^\circ \mid {}^*\phi(x)\mid \mid \phi \in \bar{B}_{X'} \} < \infty \right\}$$

$$\mu_w(0) = \{ x \in \mathrm{Fin}_w({}^*X) \mid \forall \phi \in X' \left({}^*\phi(x) \approx 0 \right) \} \quad \text{and}$$

$$\|\widehat{x}\|_w = \sup \left\{ {}^\circ \mid {}^*\phi(x)\mid \mid \phi \in \bar{B}_{X'} \right\}.$$

Notice that $\mathrm{Fin}(^*X) \subset \mathrm{Fin}_w(^*X)$ and $\mu(0) \subset \mu_w(0)$. In general, both inclusions are proper.

By Prop. 2.25, we can view the above $\widehat{^*X^w}$ as the seminormed quotient space $\mathrm{Fin}_w(^*X)/\mu_w(0)$. The following theorem gives another quotient space identification for $\widehat{^*X^w}$ using the internal norm on $^*X$.

Firstly, $^*X$ can be regarded as an infinite-valued seminormed linear space under $^\circ\|\cdot\|_{*X} : {}^*X \to [0,\infty]$, where we assign $^\circ\|x\|_{*X} = \infty$ whenever $\|x\|_{*X} \notin \mathrm{Fin}(^*\mathbb{R})$.

Then secondly, $\mathrm{Fin}_w(^*X)$ is a subspace of $^*X$, hence $\mu_w(0)$ is also a subspace of $^*X$. On the quotient space $^*X/\mu_w(0)$, we let $\|\cdot\|_q$ denote the quotient seminorm, i.e.

$$\|x + \mu_w(0)\|_q := \inf\left\{ {}^\circ\|x + y\|_{*X} \,\middle|\, y \in \mu_w(0) \right\} \in [0,\infty], \text{ where } x \in {}^*X.$$

The finite part of $^*X/\mu_w(0)$, is defined as

$$\mathrm{Fin}_q\big({}^*X/\mu_w(0)\big) := \left\{ x + \mu_w(0) \mid x \in {}^*X \wedge \|x + \mu_w(0)\|_q < \infty \right\}$$

and forms a subspace of $^*X/\mu_w(0)$.

As an *obiter dictum*, note that $^*X \ni x \mapsto \|x + \mu_w(0)\|_q$ forms a seminorm on $^*X$.

Theorem 2.12.

(i) $\forall x \in \mathrm{Fin}_w(^*X) \big(\|\widehat{x}\|_w = \|x + \mu_w(0)\|_q \big)$.

(ii) *As Banach spaces,* $\widehat{^*X^w} = \mathrm{Fin}_q\big({}^*X/\mu_w(0)\big)$.

Proof.   (i) : We will show for $a \in \mathrm{Fin}_w(^*X)$ that $\|\widehat{a}\|_w = \|a + \mu_w(0)\|_q$.
For any $y \in \mu_w(0)$, since

$$\|\widehat{a}\|_w = \sup_{\phi \in \bar{B}_{X'}} {}^\circ|{}^*\phi(a)| = \sup_{\phi \in \bar{B}_{X'}} {}^\circ|{}^*\phi(a+y)| \leq {}^\circ\|a + y\|_{*X},$$

we have $\|\widehat{a}\|_w \leq \|a + \mu_w(0)\|_q$.

Now we prove $\|a + \mu_w(0)\|_q \leq \|\widehat{a}\|_w$. Without loss of generality, assume that $\|\widehat{a}\|_w = 1$. Then it suffices to show that $\|a + \mu_w(0)\|_q \leq 1$.

For each $\phi \in X'$, write $r_\phi := {}^\circ({}^*\phi(a))$.

CLAIM: For any finite subset $A \subset S_{X'}$ and $n, m \in \mathbb{N}$, there is some $c \in (1 + m^{-1})B_X$ such that

$$\forall \phi \in A\left(|\phi(c) - r_\phi| \leq n^{-1} \right).$$

Suppose the claim fails for some finite $A \subset S_{X'}$ and $n, m \in \mathbb{N}$. Then

$$\forall x \in X\left[\bigwedge_{\phi \in A} \left(|\phi(x) - r_\phi| \leq n^{-1} \right) \Rightarrow \left(x \notin (1 + m^{-1})B_X \right) \right].$$

i.e. $\left\{ x \in X \mid \bigwedge_{\phi \in A} |\phi(x) - r_\phi| \leq n^{-1} \right\}$ is disjoint from $(1 + m^{-1})B_X$. Moreover, both sets are convex with the latter open. So by the Hahn-Banach Separation Theorem (Thm. 2.11), for some $\theta \in S_{X'}$ and $\lambda \in \mathbb{R}$,

$$(1 + m^{-1})B_X \subset \{x \in X \mid \mathrm{Re}(\theta(x)) < \lambda\}$$

$$\left\{ x \in X \mid \bigwedge_{\phi \in A} |\phi(x) - r_\phi| \leq n^{-1} \right\} \subset \{x \in X \mid \mathrm{Re}(\theta(x)) \geq \lambda\}.$$

That is, for any $x, y \in X$, whenever $\|x\|_X < 1$ and $\bigwedge_{\phi \in A} |\phi(y) - r_\phi| \leq n^{-1}$,

$$\mathrm{Re}(\theta(x)) < \frac{\lambda}{1 + m^{-1}} < \lambda \leq \mathrm{Re}(\theta(y)). \tag{2.14}$$

(Note that $\lambda > 0$, since (2.14) holds for any x with $\|x\|_X < 1$.)

By re-scaling, we may assume that $\|\theta\|_{X'} = 1$.

By Prop. 2.22, for some $x \in {}^*B_X$, we have $\|\theta\|_{X'} \approx {}^*\theta(x) \in {}^*\mathbb{R}$. Then by transferring (2.14),

$$1 = \|\theta\|_{X'} \approx {}^*\theta(x) < \frac{\lambda}{1 + m^{-1}} < \lambda \leq \mathrm{Re}({}^*\theta(a)) \leq |{}^*\theta(a)|.$$

As $m \in \mathbb{N}$, the above gives $\|\widehat{a}\|_w > 1$, a contradiction.

Hence the Claim is proved.

By transferring the Claim for all those finite subset A and $n, m \in \mathbb{N}$, an application of the saturation shows that

$$\exists c \in {}^*X \left(\left({}^\circ \|c\|_{*X} \leq 1 \right) \wedge \left(\forall \phi \in X' \left({}^*\phi(a) \approx {}^*\phi(c) \right) \right) \right),$$

i.e. we have some $(c - a) \in \mu_w(0)$ such that ${}^\circ \|c\|_{*X} \leq 1$.

Therefore $\|a + \mu_w(0)\|_q \leq 1$ as required and (i) is proved.

(ii) : From (i), it follows immediately that $\widehat{{}^*X^w} \subset \mathrm{Fin}_q({}^*X/\mu_w(0))$.

To show the other inclusion, let $x \in {}^*X$ such that $\|x + \mu_w(0)\|_q < \infty$. So there is $y \in \mu_w(0)$ such that ${}^\circ \|x + y\|_{*X} < \infty$. Then, for any $\phi \in \bar{B}_{X'}$,

$${}^\circ |{}^*\phi(x)| = {}^\circ |{}^*\phi(x + y) - {}^*\phi(y)| = {}^\circ |{}^*\phi(x + y)| \leq {}^\circ \|x + y\|_{*X} < \infty.$$

Hence $x \in \mathrm{Fin}_w({}^*X)$ and we have $\mathrm{Fin}_q({}^*X/\mu_w(0)) \subset \widehat{{}^*X^w}$.

Therefore, by (i), as Banach spaces, $\widehat{{}^*X^w} = \mathrm{Fin}_q({}^*X/\mu_w(0))$. $\square$

By the definition of $\|\cdot\|_q$ and saturation, we have from Thm. 2.12(i) that

Corollary 2.15. $\forall \widehat{x} \in \widehat{{}^*X^w} \left(\exists y \in \mu_w(0) \left(\|\widehat{x}\|_w \approx \|x + y\|_{*X} \right) \right).$ $\square$

In particular, for each $a \in \mathrm{Fin}_{\mathrm{w}}(*X)$, there is $b \in \mathrm{Fin}(*X)$ such that $a \approx_{\mathrm{w}} b$ and $\|\widehat{a}\|_{\mathrm{w}} \approx \|b\|_{*X}$, so Thm. 2.12(ii) further implies the following.

Corollary 2.16. $\widehat{*X^{\mathrm{w}}} = \mathrm{Fin}(*X)/(\mu_{\mathrm{w}}(0) \cap \mathrm{Fin}(*X))$, *the quotient space under the seminorm* $^\circ\|\cdot\|$. $\qquad\square$

Using Thm. 2.5, it is easy to show that:

Proposition 2.28. *The mapping* $X \ni x \mapsto \widehat{x} \in \widehat{*X^{\mathrm{w}}}$ *is an isometric embedding.* $\qquad\square$

In general, the above embedding is a proper one.

We have already used biduals in the last section. Now we give yet another representation of the weak nonstandard hull: as a bidual.

Theorem 2.13. *Let* X *be a normed linear space.*

Then there is an isometric isomorphism $\pi : \widehat{*X^{\mathrm{w}}} \to X''$ *such that* $\pi \upharpoonright_X$ *is the canonical embedding given by the evaluation mapping in Prop. 2.23.*

Proof. We first define $\pi : \widehat{*X^{\mathrm{w}}} \to X''$.

For $\widehat{x} \in \widehat{*X^{\mathrm{w}}}$ we let $\pi(\widehat{x})$ to be the function

$$X' \ni \phi \mapsto {}^\circ\big(\,{}^*\phi(x)\big).$$

So $\pi(\widehat{x})$ is a well-defined bounded linear functional on X', i.e. $\pi(\widehat{x}) \in X''$.

Clearly π is linear and injective, and for $x \in X$, $\pi(x) : X' \to \mathbb{F}$ is the evaluation mapping in Prop. 2.23.

We now show that it is surjective and isometric.

Let $F \in X''$. Apply the nonstandard Helly's Theorem (Thm. 2.7, with $I = X'$ and $f_i = \phi$, $\gamma_i = F(\phi)$, for each $i = \phi \in I$, and $r = \|F\|_{X''}$), we see that (2.3) is satisfied. Hence there is some $a \in *X$ such that

$$\|a\|_{*X} \lesssim \|F\|_{X''} \wedge \big(\forall \phi \in X'\,({}^*\phi(a) \approx F(\phi))\big).$$

In particular, $a \in \mathrm{Fin}(*X) \subset \mathrm{Fin}_{\mathrm{w}}(*X)$ and $\forall \phi \in X'\,\big(\pi(\widehat{a})(\phi) = F(\phi)\big)$. The latter implies that $\|\widehat{a}\|_{\widehat{*X^{\mathrm{w}}}} = \|F\|_{X''}$ and $\pi(\widehat{a}) = F$, by Prop. 2.22.

Therefore $\pi : \widehat{*X^{\mathrm{w}}} \to X''$ is an isometric isomorphism. $\qquad\square$

Corollary 2.17. $\mathrm{Fin}_{\mathrm{w}}(*X) = \bigcup \{\mu_{\mathrm{w}}(x) \mid x \in \mathrm{Fin}(*X)\}$.

Proof. This follows from the proof of the above theorem that each $F \in X''$ is $\pi(\widehat{a})$ for some $a \in \mathrm{Fin}(*X)$. $\qquad\square$

2.4.3 *Applications of weak nonstandard hulls*

Recall from Example 2.1 that $c_0' = \ell_1$, $\ell_1' = \ell_\infty$ and $\ell_\infty' = ba(\mathbb{N})$.

Example 2.2. Let $X = c_0$ and $W = \{\, {}^*p_\phi \mid \phi \in \bar{B}_{\ell_1} \,\}$. Then

$$\mathrm{Fin}_w({}^*c_0) = \Big\{ a \in {}^*c_0 \mid \forall x \in \ell_1 \, \big(\, \big| \sum_{n \in {}^*\mathbb{N}} a_n \, {}^*x_n \big| < \infty \big) \Big\} \quad \text{and}$$

$$\mu_w(0) = \Big\{ a \in {}^*c_0 \mid \forall x \in \ell_1 \, \big(\, \sum_{n \in {}^*\mathbb{N}} a_n \, {}^*x_n \approx 0 \big) \Big\}.$$

Observe that elements of $\mu_w(0)$ may have infinite coordinates: Fix any $N \in {}^*\mathbb{N} \setminus \mathbb{N}$. By saturation, there is $\epsilon \in {}^*\mathbb{R}$ such that $0 \approx \epsilon \geq |\, {}^*b_N|$ for all $b \in \ell_1$. Then $\epsilon^{-1/2} \chi_{\{N\}} \in \mu_w(0)$ and has an infinite coordinate.

Note, however, $N^3 \chi_{\{N\}} \notin \mathrm{Fin}_w({}^*c_0)$ for any $N \in {}^*\mathbb{N} \setminus \mathbb{N}$.

Now let $\pi : \widehat{{}^*X^w} \to X''$ be given by Thm 2.13, i.e. $\pi : \widehat{{}^*c_0^w} \to \ell_\infty$.

Let $\hat{a} \in \widehat{{}^*c_0^w}$ and $\pi(\hat{a}) = c \in \ell_\infty$. Then it holds for any $b \in \ell_1$ that

$$\sum_{n \in {}^*\mathbb{N}} a_n \, {}^*b_n \approx \pi(\hat{a})(b) = \sum_{n \in \mathbb{N}} c_n b_n = \sum_{n \in {}^*\mathbb{N}} {}^*c_n \, {}^*b_n.$$

By taking $b = \chi_{\{n\}}$, $n \in \mathbb{N}$, we have $\forall n \in \mathbb{N} \, (a_n \approx c_n)$.

Therefore, $\pi(\hat{a}) = ({}^\circ a_n)_{n \in \mathbb{N}}$. In particular, for $a \in \mathrm{Fin}_w({}^*c_0)$,

$$a \in \mu_w({}^*c_0) \quad \text{iff} \quad \forall x \in \ell_1 \Big(\sum_{n \in {}^*N} a_n \, {}^*x_n \approx 0 \Big) \quad \text{iff} \quad \forall n \in \mathbb{N} \, (a_n \approx 0).$$

Consequently, given any $a \in {}^*c_0$ with $\forall n \in \mathbb{N} \, (a_n \approx 0)$ satisfied,

$$\Big(\exists x \in \ell_1 \, \big(\sum_{n \in {}^*\mathbb{N}} a_n \, {}^*x_n \not\approx 0 \big) \Big) \Rightarrow \Big(\exists x \in \ell_1 \, \big(| \sum_{n \in {}^*\mathbb{N}} a_n \, {}^*x_n | \approx \infty \big) \Big).$$

Also, given $a \in \mathrm{Fin}_w({}^*c_0)$, there is by Thm 2.12 $c \in \mathrm{Fin}_w({}^*c_0)$ so that

$$\forall x \in \ell_1 \Big(\sum_{n \in {}^*\mathbb{N}} a_n \, {}^*x_n \approx \sum_{n \in {}^*\mathbb{N}} c_n \, {}^*x_n \Big) \quad (i.e.\ a \approx_w c) \quad \text{and}$$

$$\sup_{n \in \mathbb{N}} {}^\circ |c_n| = \sup_{x \in \bar{B}_{\ell_1}} \sum_{n \in {}^*\mathbb{N}} a_n \, {}^*x_n \quad (i.e.\ \|c\|_{{}^*c_0} \approx \|a + \mu_w(0)\|_q).$$

$\square$

Similarly, there is a concrete representation of measures in $ba(\mathbb{N})$.

Example 2.3. Let $X = \ell_1$ and $W = \{\, {}^*p_\phi \mid \phi \in \bar{B}_{\ell_\infty} \,\}$. Then

$$\mathrm{Fin}_w({}^*\ell_1) = \Big\{ a \in {}^*\ell_1 \mid \forall x \in \ell_\infty \, \big(\, | \sum_{n \in {}^*\mathbb{N}} a_n \, {}^*x_n | < \infty \big) \Big\} \quad \text{and}$$

$$\mu_w(0) = \Big\{ a \in {}^*\ell_1 \mid \forall x \in \ell_\infty \, \big(\, \sum_{n \in {}^*\mathbb{N}} a_n \, {}^*x_n \approx 0 \big) \Big\}.$$

Again, let $\pi : \widehat{{}^*X^{\mathrm{w}}} \to X''$ be given by Thm 2.13. As we have mentioned (without proof) that $\ell_1'' = ba(\mathbb{N})$, we therefore regard $\pi : \widehat{{}^*\ell_1^{\mathrm{w}}} \to ba(\mathbb{N})$.

So each $\mu \in ba(\mathbb{N})$ corresponds to $\pi(\widehat{a})$ for some $a \in \mathrm{Fin}_{\mathrm{w}}({}^*\ell_1)$.

Let $S \subset \mathbb{N}$, so $\chi_S \in \ell_\infty$ and

$$\mu(S) = \int_S d\mu = \pi(\widehat{a})(\chi_S) \approx \sum_{n \in {}^*\mathbb{N}} a_n {}^*\chi_S(n).$$

That is, by Thm 2.12, for any $\mu \in ba(\mathbb{N})$, there is $a \in \mathrm{Fin}_{\mathrm{w}}({}^*\ell_1)$

$$\forall S \subset \mathbb{N} \left(\mu(S) = {}^\circ \sum_{n \in {}^*S} a_n \right) \quad \text{and} \quad \|\mu\| = {}^\circ \sum_{n \in {}^*\mathbb{N}} |a_n|. \tag{2.15}$$

$\square$

Actually, (2.15) defines a mapping

$$\pi : \ell_1'' \to ba(\mathbb{N})$$

by taking, for each $a \in \mathrm{Fin}_{\mathrm{w}}({}^*\ell_1)$, the $\widehat{a}$ to the μ given by the first equation—it is clear that μ is a finitely additive measure $\mathcal{P}(\mathbb{N}) \to \mathbb{F}$. Also it is not hard to check that π is an isometric isomorphism taking the Banach space ℓ_1'' onto $ba(\mathbb{N})$. (Recall that the finitely additive measures in $ba(\mathbb{N})$ are required to have finite total variation *w.r.t.* finite partitions.) So, by Example 2.1 (and Ex. 9 on p.107), ℓ_∞' is isometrically isomorphic to $ba(\mathbb{N})$ and we identify each $\phi \in \ell_\infty'$ with the unique $\mu \in ba(\mathbb{N})$ that takes each $f \in \ell_\infty$ to $\int_\mathbb{N} f(n) d\mu(n)$.

More generally, let Ω be a compact topological space, we would like to get a similar representation for $C(\Omega)$. For convenience, consider the complex case, *i.e.* $C(\Omega)$ is taken to be a complex Banach space. Recall the definition of $\mathbb{M}(\Omega)$ on p. 88 and we will show that $C(\Omega)'$ can be identified with $\mathbb{M}(\Omega)$.

The Banach space $\mathbb{M}(\Omega)$ consists of finite σ-additive complex Borel measures on the compact space Ω. For each such measure μ, obviously the mapping $C(\Omega) \ni f \mapsto \int_\Omega f(x) d\mu(x)$ is linear and bounded and hence belongs to $C(\Omega)'$. We denote such mapping by ϕ_μ.

Note that $\mathbb{M}(\Omega) \ni \mu \mapsto \phi_\mu \in C(\Omega)'$ is an isometric isomorphism, and we only need to show that it is onto. *i.e.* for each $\psi \in C(\Omega)'$, we need to show that $\psi = \phi_\mu$ for some $\mu \in \mathbb{M}(\Omega)$.

Let's fix some $\psi \in C(\Omega)'$.

By Ω being compact and by the saturation, there is a hyperfinite H such that $\Omega \subset H \subset {}^*\Omega$ and $\mathrm{st}[H] = \Omega$, because $\mathrm{st}[{}^*\Omega] = \Omega$.

We use an internal identification of H with a hyperfinite subset of ${}^*\mathbb{N}$, and naturally regard ${}^*\ell_\infty(H) \subset {}^*\ell_\infty$ as an internal subspace.

For each $f \in C(\Omega)$, let $\tilde{f}$ denote $*f \upharpoonright_H$. So $\tilde{f} \in *\ell_\infty(H)$, hence $\tilde{f} \in *\ell_\infty$. Then for each finite dimensional subspace $X \subset C(\Omega)$ with basis $\{f_1, \ldots, f_n\}$, we let $\tilde{X}$ denote the internal subspace $*\mathrm{Lin}(\tilde{f}_1, \ldots, \tilde{f}_n)$ of $*\ell_\infty$.

Note we have for $f \in C(\Omega)$ that $°\tilde{f}(x) = f(°x)$ by the S-continuity of $*f$ and $\mathrm{st}[H] = \Omega$. Also, if $\{f_1, \ldots, f_n\}$ is linearly independent in $C(\Omega)$, then so is $\{\tilde{f}_1, \ldots, \tilde{f}_n\}$ in $*\ell_\infty$.

Moreover, each finite dimensional subspace $X \subset C(\Omega)$ with basis $\{f_1, \ldots, f_n\}$ gives rise to the internal linear functional in $(\tilde{X})'$ which takes the $\tilde{f}_i$ to $\psi(f_i)$, $i = 1, \ldots, n$. We denote this internal linear functional by δ_X. Hence $\delta_X(\tilde{f}) = \psi(f)$ for all $f \in X$. Clearly δ_X does not depend on a particular choice of basis for X.

Applying the Hahn-Banach Theorem internally, for each finite dimensional subspace $X \subset C(\Omega)$, the δ_X extends to some $\theta_X \in (*\ell_\infty)'$ such that $\|\theta_X\| = \|\delta_X\|$. Note that $\theta_X(\tilde{f}) = \delta_X(\tilde{f}) = \psi(f)$ for every $f \in X$.

Now apply saturation to the δ_X, where the X's range over all finite dimensional subspaces of $C(\Omega)$, we obtain some $\theta \in (*\ell_\infty)'$ such that

$$\|\theta\| \approx \|\psi\| \quad \text{and} \quad \theta(\tilde{f}) = \psi(f) \quad \text{for all} \quad f \in C(\Omega).$$

So by what was shown earlier, $\theta = \phi_\lambda$ for some $\lambda \in *ba(\mathbb{N})$.

Then let ν be the restriction of λ on internal subsets of H. Since the total variation $|\nu|(\Omega) \leq |\lambda|(*\mathbb{N}) = \|\lambda\| = \|\theta\| \approx \|\psi\|$, which is finite, ν is a finite internal complex measure. So we apply Thm. 1.11 to obtain the Loeb measure $L(\nu)$, a finite σ-additive complex measure.

Finally, let μ be $L(\nu) \circ \mathrm{st}^{-1}$, defined on Borel subsets of Ω. Then we have $\mu \in \mathbb{M}(\Omega)$ and, for each $f \in C(\Omega)$,

$$\psi(f) = \theta(\tilde{f}) = \int_{*\mathbb{N}} \tilde{f} d\lambda = \int_H \tilde{f} d\nu = \int_H °\tilde{f} \, dL(\nu) = \int_\Omega f d\mu.$$

i.e. $\psi = \phi_\mu$. Therefore we have proved the following:

Theorem 2.14. (Riesz Representation Theorem) *Let Ω be a compact topological space and consider the complex Banach space $C(\Omega)$.*

For each $\mu \in \mathbb{M}(\Omega)$, let ϕ_μ denote the mapping $C(\Omega) \to \mathbb{C}$ given by $C(\Omega) \ni f \mapsto \int_\Omega f(x) d\mu(x)$. Then $\phi_\mu \in C(\Omega)'$.

Moreover, define $\pi : \mathbb{M}(\Omega) \to C(\Omega)'$ by $\pi(\mu) := \phi_\mu$. Then π is an isometric isomorphism onto the Banach space $C(\Omega)'$. $\qquad\square$

Note that Thm. 2.14 and Thm. 2.28 are namesakes.

According to [Dunford and Schwartz (1988a)], no completely satisfactory representation of $ba(\mathbb{N})'$ seems to be known. However, we have the following representation of $ba(\mathbb{N})'$ as a quotient space:

Example 2.4. Let $X = \ell_\infty$ and take $W = \{\, {}^*p_\mu \mid \mu \in \bar{B}_{ba(\mathbb{N})} \,\}$. Then

$$\mathrm{Fin}_w({}^*\ell_\infty) = \Big\{ a \in {}^*\ell_\infty \mid \forall \nu \in ba(\mathbb{N})\,\big(\,|\sum_{n \in {}^*\mathbb{N}} a_n\, {}^*\nu(\{n\})| < \infty\big) \Big\},$$

$$\mu_w(0) = \Big\{ a \in {}^*\ell_\infty \mid \forall \nu \in ba(\mathbb{N})\,\big(\sum_{n \in {}^*\mathbb{N}} a_n\, {}^*\nu(\{n\}) \approx 0\big) \Big\},$$

$$\mathrm{Fin}_q({}^*\ell_\infty/\mu_w(0)) = \big\{ a + \mu_w(0) \mid a \in {}^*\ell_\infty \wedge \|a + \mu_w(0)\|_q < \infty \big\}, \quad \text{where}$$

$$\|a + \mu_w(0)\|_q := \inf\big\{ {}^\circ \|a + b\|_{{}^*\ell_\infty} \mid b \in \mu_w(0) \big\}.$$

Since $ba(\mathbb{N})'$, is the bidual ℓ_∞'', we have by Thm 2.13 an isometric isomorphism showing $ba(\mathbb{N})' = \widehat{{}^*\ell_\infty^w}$ and we also have by Thm 2.12 that

$$ba(\mathbb{N})' = \mathrm{Fin}_q\big({}^*\ell_\infty/\mu_w(0)\big).$$

$\square$

2.4.4 *Weak compactness and separation*

We are still working with the weak topology on a normed linear space X and $W = \{\, {}^*p_\phi \mid \phi \in \bar{B}_{X'} \,\}$.

By Robinson's characterization of compactness, a subset $K \subset X$ is weakly compact iff $\forall x \in {}^*K\ \exists y \in K\,(x \approx_w y)$.

Immediately the following holds.

Proposition 2.29. *Let X be a normed linear space and $C \subset X$ is weakly compact. Then C is bounded and closed in norm.*

Proof. Let $\phi \in X'$. Suppose $\sup_{x \in C} |\phi(x)|$ is infinite. Then by saturation, there is $c \in {}^*C$ such that ${}^*\phi(c)$ is infinite. But this is impossible, as the above remark shows that $c \approx_w a$ for some $a \in C$, implying $\phi(a)$ is infinite. So it follows from Prop. 2.24 that C is bounded in norm.

Clearly compactness implies closedness in the weak topology, so it follows from Prop. 2.26 that C is closed in the norm topology. $\square$

The converse of the above fails in general. But it does hold in the class of reflexive spaces (Cor. 2.22).

We have the following strong separation property.

Theorem 2.15. *Let X be a normed linear space, K and C be disjoint nonempty subsets of X such that K is weakly compact and C is closed convex. Then $\mathrm{dist}(K, C) > 0$.*

Proof. Suppose $\text{dist}(K, C) = 0$, then for any $\epsilon \in \mathbb{R}^+$, there are $a \in K$ and $c \in C$ such that $\|a - c\| < \epsilon$.

Then, by saturation, let $a \in {}^*K$ and $c \in {}^*C$ such that $a \approx c$.

By assumption and Robinson's characterization of compactness, let $b \in K$ such that $b \approx_w a$.

Since $b \notin C$ and C is closed, $A \cap C = \emptyset$, where $A := B(b, r)$ for some $r \in \mathbb{R}^+$. Now apply the Hahn-Banach Separation Theorem (Thm. 2.11), there is $\phi \in X'$ and $\lambda \in \mathbb{R}$ such that

$$\text{Re}(\phi(b)) < \lambda \quad \text{and} \quad \forall x \in C \left(\text{Re}(\phi(x)) \geq \lambda \right).$$

Transferring the latter, we see that $\text{Re}({}^*\phi(c)) \geq \lambda$, hence $\phi(b) \not\approx {}^*\phi(c)$.

On the other hand, $b \approx_w a$, so ${}^*\phi(a) \not\approx {}^*\phi(c)$, which is impossible, since $a \approx c$. $\square$

2.4.5 Weak* topology and Alaoglu's Theorem

The following shows that the general nonstandard hull could be very small.

Proposition 2.30. *Given a normed linear space X, let $W = \{ {}^*p_a \,|\, a \in \bar{B}_X \}$, where $p_a : X' \to [0, \infty)$ is the seminorm $\theta \mapsto |\theta(a)|$.*

Then $X' = {}^*\widehat{X'^W}$ in the sense that $\pi : X' \to {}^*\widehat{X'^W}$ given by $\phi \mapsto \widehat{{}^*\phi}$ is an isometric isomorphism.

Proof. First note that

$$\text{Fin}_w({}^*X) = \left\{ \theta \in {}^*X' \,\Big|\, \sup_{a \in \bar{B}_X} {}^\circ |\theta(a)| < \infty \right\},$$

$$\mu_w(0) = \left\{ \theta \in {}^*X' \,|\, \forall x \in \bar{B}_X \left(\theta(x) \approx 0 \right) \right\}.$$

Clearly π is linear. To show isometry, let $\phi \in X'$, then

$$\|\pi(\phi)\|_{\widehat{*X'^w}} = \left\| \widehat{{}^*\phi} \right\|_{\widehat{*X'^w}} = \sup_{a \in \bar{B}_X} {}^\circ |\phi(a)| = \|\phi\|_{X'} .$$

To show surjection, let $\widehat{\theta} \in {}^*\widehat{X'^w}$. Define $\phi := {}^\circ\theta \upharpoonright_X$. Then

$$\forall x \in \bar{B}_X \left(|\phi(x)| = {}^\circ |\theta(x)| \leq \|\widehat{\theta}\|_{\widehat{*X^w}} \right),$$

hence $\phi \in X'$. Moreover,

$$\forall x \in \bar{B}_X \left((\theta - {}^*\phi)(x) = \theta(x) - {}^*\phi(x) = \theta(x) - \phi(x) \approx 0, \right)$$

therefore $(\theta - {}^*\phi) \in \mu_W(0)$, i.e. $\widehat{\theta} = \widehat{{}^*\phi}$. $\square$

Given a normed linear space X, the topology on X' given by the above seminorms p_a, $a \in \bar{B}_X$, is called the **weak* topology** on X'. Hence, under the weak* topology, $\theta_n \to 0$ in X' iff $\forall x \in X \left(\theta_n(x) \to 0 \right)$.

One says that a topological vector space X is a **locally convex linear space** if the topology of X is given by a family $\mathcal{F}$ of seminorms. (See [Rudin (1991)].) Moreover, such X is Hausdorff iff

$$\forall x \in X \left((x = 0) \Leftrightarrow \left(\forall p \in \mathcal{F} \left(p(x) = 0 \right) \right) \right).$$

Then we see that, for any normed linear space X, X' forms a Hausdorff locally convex linear space under the weak* topology.

Note that the weak* topology is defined on any normed linear space which is the dual of another space, *i.e.* those having a **predual**.

If $0 \neq \phi \in X'$, then $\|\phi\| \neq 0$, *i.e.* $\phi(a) \neq 0$ for some $a \in X$. Therefore, if $\theta \neq \phi$ in X', then $p_a(\theta) \neq p_a(\phi)$ for some $a \in X$, *i.e.* the p_a's separate points in X', *i.e.* the weak* topology on X' is Hausdorff.

Note that by Prop. 2.28, the weak* topology on X' is weaker (*i.e.* coarser) than the weak topology on X'.

Then the monad of 0 *w.r.t.* the weak* topology is given by the equivalence relation

$$\theta \overset{w^*}{\approx} 0 \quad \text{iff} \quad \forall x \in X \left(\theta(x) \approx 0 \right), \quad \text{where } \theta \in {}^*X'.$$

For $\theta, \phi \in {}^*X'$, we have $\theta \overset{w^*}{\approx} \phi$ iff $(\theta - \phi) \overset{w^*}{\approx} 0$.

By Robinson's characterization (Thm. 1.21), a subset $A \subset X'$ is compact under the weak* topology—**weak* compact**, if

$$\forall \theta \in {}^*A \, \exists \phi \in A \left(\theta \overset{w^*}{\approx} {}^*\phi \right).$$

By Cor. 2.2, unless X is finite dimensional, $\bar{B}_{X'}$ is not compact in the norm topology. However, we have the following.

Theorem 2.16. (Alaoglu's Theorem) *Let X be a normed linear space, then $\bar{B}_{X'}$ is weak* compact.*

Proof. Respectively, we let $\widehat{{}^*X}$ and $\widehat{{}^*X'}$ denote the nonstandard hulls of ${}^*X$ and ${}^*X'$ *w.r.t.* their norms with their elements written as $\widehat{x}$ and $\widehat{\theta}$.

Let $\theta \in {}^*\bar{B}_{X'} = \bar{B}_{{}^*X'}$. So $\widehat{\theta} \in \widehat{{}^*X'}$.

By Prop. 2.16(ii), $\widehat{\theta} \in \left(\widehat{{}^*X} \right)'$. Noting that $X \subset \widehat{{}^*X}$, let $\phi := \left(\widehat{\theta} \restriction_X \right)$. So $\phi \in X'$. It is clear that $\|\phi\|_{X'} \leq \|\widehat{\theta}\|_{\widehat{{}^*X}}$, hence $\phi \in \bar{B}_{X'}$.

Finally, for any $x \in X$, we have $\theta(x) \approx \widehat{\theta}(\widehat{x}) = \phi(x)$, *i.e.* $\theta \overset{w^*}{\approx} {}^*\phi$.

Therefore, by Robinson's characterization, $\bar{B}_{X'}$ is weak* compact. $\quad \square$

Now consider a denseness result concerning the weak* topology on X''.

Theorem 2.17. (Goldstine's Theorem) *Let X be a normed linear space, then $\bar{B}_X$ is weak* dense in $\bar{B}_{X''}$.*

Proof. Let $\widehat{{}^*X^{\mathrm{w}}}$ denote the weak nonstandard hull of X and $\pi : \widehat{{}^*X^{\mathrm{w}}} \to X''$ be the isometric isomorphism given by Thm. 2.13.

Let $a \in \bar{B}_X$. Then $a = \pi(\hat{b}) = \pi(b + \mu_{\mathrm{w}}(0))$ for some $b \in {}^*X$. Moreover, by Cor. 2.15, b can be chosen so that $\|b\|_{{}^*X} \approx \|\pi(\hat{b})\|_{\widehat{{}^*X^{\mathrm{w}}}} = \|a\|_X$. If $b \neq 0$, replace b by $b/\|b\|_{{}^*X}$, so we assume that $b \in \bar{B}_{{}^*X}$. Moreover,

$$\forall \phi \in X' \left(\phi(a) = \pi(\hat{b})(\phi) \approx {}^*\phi(b) \right)$$

i.e. $a \overset{\mathrm{w}^*}{\approx} b$. Therefore, by Prop. 1.28, $\bar{B}_X$ is weak* dense in $\bar{B}_{X''}$. $\quad\square$

2.4.6 Notes and exercises

A more general nonstandard hull construction for uniform spaces was first considered in [Luxemburg (1969)].

In the general nonstandard hull construction considered here, one can alternatively use the fact that $\sup_{p \in \mathrm{W}} {}^\circ p(x)$ forms a seminorm on $\mathrm{Fin}_{\mathrm{w}}(X)$, then by Prop. 2.1(ii), the quotient space $\mathrm{Fin}_{\mathrm{w}}(X)/\mu_{\mathrm{w}}(0)$, *i.e.* the nonstandard hull $\widehat{X}^{\mathrm{w}}$, forms a normed linear space over $\mathbb{F}$.

Let X be a (not necessarily Hausdorff) locally convex linear space with topology given by $\mathcal{F}$, a family of seminorms. Take $\mathrm{W} := \{ {}^*p \mid p \in \mathcal{F} \}$ then $\widehat{{}^*X^{\mathrm{w}}}$, is the nonstandard hull construction of a Banach space from a locally convex linear space.

EXERCISES

(1) Prove Prop. 2.28.
(2) As a consequence of Lem. 2.3 and Thm. 2.12, $\mathrm{Fin}_q\left({}^*X/\mu_{\mathrm{w}}(0)\right)$ is complete. Give a direct proof of this fact.
(3) Prove the Riesz Representation Theorem (Thm. 2.14) where Ω is only assumed to be locally compact.
(4) Find an infinite dimensional Banach space X such that the weak* topology on X is metrizable.
(5) Show that every Banach space is isometrically embeddable into $C(\Omega)$ for some compact space Ω.
(6) Are there Banach spaces X_n, $n \in \mathbb{N}$, such that X_{n+1} is a predual of X_n and nonisomorphic to X_n?

2.5   Reflexive Spaces

To recognize the reflection of a space on a nonstandard mirror, it is
necessary to apply some theorems of James.

A normed linear space X is said to be **reflexive** if $X = X''$ under the
canonical isometric embedding as evaluation mappings given by Prop. 2.23.

Since the dual space is always complete, reflexive normed linear spaces
are necessarily Banach spaces.

This section is mainly about reflexive spaces, including superreflexive
spaces. A collection of characterizations of reflexivity is given and finite
representability is treated.

2.5.1   *Weak compactness and reflexivity*

Example 2.5.   Among the examples in § 2.1.4, Lebesgue spaces ℓ_p and
$L_p(\mu)$, where $p \in (1, \infty)$, are reflexive.

While the others:

$$\ell_1,\ \ell_\infty,\ L_1(\mu), L_\infty(\mu),\ c,\ c_0,\ C_b(\Omega),\ C_0(\Omega), ba(\mathbb{N}), ca(\mathbb{N})$$

are examples of nonreflexive spaces.

Also all finite dimensional normed linear spaces are reflexive.   □

Theorem 2.18.   *Let X be a normed linear space. Then the following are*
equivalent:

(i) *X is reflexive;*
(ii) *$X = \widehat{{}^*X^{\mathrm{w}}}$, the weak nonstandard hull;*
(iii) *$\bar{B}_X$ is weakly compact;*

Proof.   ((i) ⇔ (ii)) : By Thm. 2.13 and the definition.

((ii) ⇒ (iii)) :

This is similar to the proof of Goldstine's Theorem (Thm. 2.17). Let
$a \in {}^*\bar{B}_X$. Then $a \in \mathrm{Fin_w}({}^*X)$, so $\hat{a} \in \bar{B}_{\widehat{{}^*X^{\mathrm{w}}}}$, which is, by (ii), identifiable
with $\bar{B}_X$ under the canonical isometric isomorphism. Hence $\hat{a} = b$ for some
$b \in \bar{B}_X$. Therefore

$$\forall \phi \in X' \left(\phi(b) = \pi(\hat{a})(\phi) \approx {}^*\phi(a) \right),$$

where $\pi : \widehat{{}^*X^{\mathrm{w}}} \to X''$ is the isometric isomorphism given by Thm. 2.13.

That is, $a \approx_{\mathrm{w}} b$ (*i.e.* infinitely close in the weak topology), hence $\bar{B}_X$ is
weakly compact by Robinson's characterization of compactness.

$((\text{iii}) \Rightarrow (\text{ii})):$

Since $\bar{B}_X$ is weakly compact, Robinson's characterization of compactness implies $\forall x \in \text{Fin}(^*X) \exists y \in X \left(x \approx_w y\right)$. Therefore, by Cor. 2.16 that $\widehat{^*X^w} = \text{Fin}(^*X)/\left(\mu_w(0) \cap \text{Fin}(^*X)\right)$ we have $X = \widehat{^*X^w}$. $\square$

In particular, Thm. 2.18 and Lem. 2.3 imply that:

Corollary 2.18. *Reflexive normed linear spaces are Banach spaces.* $\square$

Corollary 2.19. *Let X be a reflexive Banach space, then X' is normed by X, i.e. $\forall \phi \in X' \exists x \in \bar{B}_X \left(\|\phi\| = \phi(x)\right)$.*

Proof. Let $\phi \in X'$ then $\|\phi\| = \sup_{x \in \bar{B}_X} |\phi(x)| = \sup_{x \in \bar{B}_X} \phi(x) \approx {}^*\phi(a)$ for some $a \in {}^*\bar{B}_X$, by saturation. But $\bar{B}_X$ is weakly compact, so there is $b \in \bar{B}_X$ such that $a \approx_w b$, hence $\|\phi\| = \phi(b)$. $\square$

The converse of the above corollary is a deep result due to R.C. James. Since all known proofs are rather involved, the result is stated here without proof. (See [Megginson (1998)] for a presentation of the original proof.)

Theorem 2.19. (James' Theorem) *Let X be a Banach space such that $\forall \phi \in X' \exists x \in \bar{B}_X \left(\|\phi\| = \phi(x)\right)$. Then X is reflexive.* $\square$

Recall in Prop. 2.16 that, for an internal normed linear space X, the norm-nonstandard hull of X' is a closed subspace of the dual of the norm-nonstandard hull $\widehat{X}$.

Here is a characterization of the reflexivity of a (norm-)nonstandard hull via its dual, which simultaneously shows that in general $\widehat{X'} \subsetneq (\widehat{X})'$.

Theorem 2.20. *Let X be an internal normed linear space. Then the norm-nonstandard hull $\widehat{X}$ is reflexive iff $\widehat{X'} = (\widehat{X})'$.*

Proof. $(\Rightarrow)$: Suppose $\widehat{X}$ is reflexive but $\widehat{X'} \subsetneq (\widehat{X})'$. Let $\theta \in \left((\widehat{X})' \setminus \widehat{X'}\right)$.

Since $\widehat{X'}$ is a closed subspace of $(\widehat{X})'$, by Cor. 2.14 and $\widehat{X''} = \widehat{X}$, there is $a \in \text{Fin}(X)$ such that

$$\theta(\widehat{a}) = 1 \wedge \forall \phi \in \text{Fin}(X') \left(\widehat{\phi}(\widehat{a}) = 0\right),$$

i.e.
$$\theta(\widehat{a}) = 1 \wedge \forall \phi \in \text{Fin}(X') \left(\phi(a) \approx 0\right).$$

In particular, $a \not\approx 0$, so by Thm. 2.5, for some $\phi \in S_{X'}$, we have $\phi(a) = \|a\|_X \not\approx 0$, contradicting to the above.

$(\Leftarrow)$: Let $\theta \in (\widehat{X})'$. Then, by assuming $\widehat{X'} = (\widehat{X})'$, we have $\theta = \widehat{\phi}$ for some $\phi \in \text{Fin}(X')$.

By Prop. 2.22, there is $a \in B_X$ such that $\phi(a) \approx \|\phi\|_{X'}$, hence

$$\theta(\widehat{a}) = {}^{\circ}\phi(a) = {}^{\circ}\|\phi\|_{X'} = \|\theta\|_{X'}.$$

That is, every $\theta \in (\widehat{X})'$, is normed by some $\widehat{a} \in \bar{B}_{\widehat{X}}$, therefore $\widehat{X}$ is reflexive by James' Theorem (Thm. 2.19). $\qquad\square$

2.5.2   The Eberlein-Šmulian Theorem

Let W be a set of seminorms on a linear space X. Define an infinite seminorm on ${}^*X$ by $\|x\|_{\mathrm{w}} := \sup_{p \in W} \left({}^{\circ}\,{}^*p(x) \right)$ and let $\mathrm{Fin}_{\mathrm{w}}({}^*X)$ denote the finite part of ${}^*X$ *w.r.t.* this seminorm.

We say that W is **binding** if for any $a \in \mathrm{Fin}_{\mathrm{w}}({}^*X)$, $p_1, \ldots, p_n \in W$, $\epsilon_1, \ldots, \epsilon_n \in \mathbb{R}^+$, where $n \in \mathbb{N}$,

$$\exists b \in X \left((\|b\|_{\mathrm{w}} \leq \|a\|_{\mathrm{w}}) \wedge \bigwedge_{i=1}^{n} \left({}^*p_i(a-b) < \epsilon_i \right) \right).$$

Let X be a normed linear space. By an **endorsement** of the norm $\|\cdot\|_X$ on X, we mean a set W of seminorms on X with the property

$$\forall x \in X \left(\|x\|_X = \sup_{p \in W} p(x) \right).$$

Example 2.6. On a normed linear space X, the following are binding endorsements:

- $\{p_\phi \mid \phi \in \bar{B}_{X'}\}$, where $p_\phi : X \ni x \mapsto |\phi(x)|$, as in §2.4.2.
- $\{p_\phi \mid \phi \in S_{X'}\}$.
- $\{p_{\mathrm{Re}(\phi)} \mid \phi \in S_{X'}\}$, where $\mathrm{Re}(\phi)$ is regarded as a real linear functional.
- $\{p_\phi \mid \phi : X \to \mathbb{R} \text{ is a real linear functional and } \|\phi\| = 1\}$.

If X has a *predual*, *i.e.* there is a normed linear space Y such that $X = Y'$ then $\{p_a \mid a \in \bar{B}_Y\}$ where $p_a : X \ni x \mapsto |x(a)|$, forms a binding endorsement.

Trivially, $\{\|\cdot\|_X\}$ always forms an endorsement for X. It is binding iff X is finite dimensional.

Note that there is $a \in S_{*\ell_1}$ such that $\forall x \in \ell_1 \left({}^*\|a - x\| \approx 1 \right)$.

Another situation is to take W_0 to be any collection of seminorms on a linear space X, with seminorm $\|x\| := \sup_{p \in W_0} p(x)$ and nullspace $Y :=$ $\{x \in X \mid \|x\| = 0\}$. Let $Z := X/Y$ be the quotient space and W the set of seminorms on Z inherited from W_0. Then W is an endorsement on Z. $\qquad\square$

Lemma 2.4. *Let X be a normed linear space and W a binding set of semi-norms. Let $a \in \mathrm{Fin}_w(^*X)$ with $r := \|a\|_w > 0$. Then there are increasing finite subsets $W_n \subset W$, $a_n \in X$, $\|a_n\|_w \le r$, $n \in \mathbb{N}$, such that*

- $\forall p \in W_n \left(|\, ^*p(a_n - a)| \le n^{-1} \right)$;
- $\forall x \in \mathrm{Lin}(\{a, a_1, \ldots, a_n\}) \left(\|x\|_w \le (1 + n^{-1}) \max_{p \in W_{n+1}} \left({}^{\circ}\, ^*p(x) \right) \right).$

Proof. Without loss of generality, we assume that $\|a\|_w = 1$.

Let $W_1 = \{p_1\}$, where $p_1 \in W$ is arbitrary. Then, by W being binding, there is $a_1 \in X$, $\|a_1\|_w \le 1$ such that $|\, ^*p_1(a_1 - a)| \le 1$.

Assume inductively that the W_i, a_i, $i = 1, \ldots, n$, with the above properties are constructed.

CLAIM: Suppose $Y \subset {}^*X$ is a finite dimensional subspace under $\|\cdot\|_w$. Then for any $\epsilon \in \mathbb{R}^+$, we have $\forall y \in Y \left(\|y\|_w \le (1 + \epsilon) \max_{p \in V} \left({}^{\circ}\, ^*p(y) \right) \right)$ for some finite $V \subset W$.

PROOF OF THE CLAIM: It suffices to consider the complement to the nullspace $\{y \in Y \mid \|y\|_w = 0\}$, so we replace Y by such and hence the seminorm $\|\cdot\|_w$ forms a norm on Y denoted by $\|\cdot\|_Y$.

Suppose the conclusion fails for some $\epsilon \in \mathbb{R}^+$. Then

$$\forall^{\mathrm{fin}} V \subset W \, \exists y \in S_Y \left((1 + \epsilon) \max_{p \in V} {}^*p(y) \le 1 \right).$$

Then by saturation, there is $a \in S_{{}^*Y}$ such that

$$\forall p \in W \left({}^*p(a) \le (1 + \epsilon)^{-1} \right).$$

Since Y is finite dimensional, S_Y is compact by Cor. 2.2, so there is $b \in S_Y$ such that $a \approx b$ in the topology given by $\|\cdot\|_Y$, hence by $\|\cdot\|_w$. In particular, $^*p(a) \approx {}^*p(b)$ for all $p \in W$, therefore

$$1 = \|b\|_Y = \sup_{p \in W} \left({}^{\circ}\, ^*p(b) \right) = \sup_{p \in W} \left({}^{\circ}\, ^*p(a) \right) \le (1 + \epsilon)^{-1} < 1,$$

a contradiction, and hence the Claim is proved.

Returning to the construction, since $\mathrm{Lin}(\{a, a_1, \ldots, a_n\}) \subset {}^*X$ is finite dimensional, it follows from the Claim that there is a finite $V \subset W$ such that

$$\forall x \in \mathrm{Lin}(\{a, a_1, \ldots, a_n\}) \left(\|x\|_w \le (1 + n^{-1}) \max_{p \in V} \left({}^{\circ}\, ^*p(x) \right) \right).$$

Then, by letting $W_{n+1} := W_n \cup V$, the requirement for W_{n+1} is satisfied. Moreover, as W is binding, there is $a_{n+1} \in X$, $\|a_{n+1}\|_w \le 1$, such that

$$\forall p \in W_{n+1} \left(|\, ^*p(a_{n+1} - a)| \le (n+1)^{-1} \right)$$

and the requirement for a_{n+1} is satisfied.

Therefore the inductive step of the construction is completed. $\square$

We call $\{(a_n, W_n)\}_{n \in \mathbb{N}} \subset X \times \mathcal{P}(W)$ a **Day sequence** for a.

In the case of a normed linear space X which has a predual Y with $W = \{p_a \mid a \in \bar{B}_Y\}$, we have a trivial example of a Day sequence for $a \in \text{Fin}_w(*X)$ by taking all $a_n = c$ for some $c \in X$ such that $c \approx_w a$ (Alaoglu's Theorem (Thm. 2.16)) and W_n any increasing subsets of W.

For a normed linear space X and $A \subset X$, where $n \in \mathbb{N}$, the **convex hull** of A is denoted by $\text{conv}(A)$ and is defined as

$$\Big\{ \sum_{i=1}^{n} \lambda_i a_i \ \Big| \ a_1, \ldots, a_n \in A, \ n \in \mathbb{N}, \ \lambda_1, \ldots, \lambda_n \in [0,1] \wedge \sum_{i=1}^{n} \lambda_i = 1 \Big\}.$$

Elements of $\text{conv}(A)$ are called (finite) **convex combinations** of elements from A.

The norm closure of $\text{conv}(A)$ is denoted by $\overline{\text{conv}}(A)$.

To apply Thm. 2.18, it is important to have a simple characterization of weak compactness such as the following.

Theorem 2.21. (Eberlein-Šmulian Theorem) *Let X be a normed linear space. Suppose $\bigcap_{n \in \mathbb{N}} C_n \neq \emptyset$ for every decreasing sequence $\{C_n\}_{n \in \mathbb{N}}$ of nonempty closed convex subsets of $\bar{B}_X$. Then $\bar{B}_X$ is weakly compact.*

Proof. Let $W := \{p_\phi \mid \phi \in S_{X'}\}$. Let $a \in \bar{B}_{*X}$.

Apply Lem. 2.4 to $W := \{p_\phi \mid \phi \in S_{X'}\}$ and a, then from the Day sequence, we have increasing finite $S_n \subset \bar{B}_{X'}$, $n \in \mathbb{N}$, and $\{a_n\}_{n \in \mathbb{N}} \subset \bar{B}_X$ such that

$$\forall \phi \in S_n \left(\mid {}^*\phi(a_n - a)\mid \ \leq n^{-1} \right);$$

$$\forall x \in \text{Lin}(\{a, a_1, \ldots, a_n\}) \left(\|x\|_w \leq (1 + n^{-1}) \max_{\phi \in S_{n+1}} \left({}^\circ \mid {}^*\phi(x)\mid \right) \right).$$

Let $C_n := \overline{\text{conv}}(\{a_n, a_{n+1}, \ldots\})$, $n \in \mathbb{N}$, then by assumption, there is some $b \in \bigcap_{n \in \mathbb{N}} C_n$. In particular $b \in \bar{B}_X$.

We now show that $\|a - b\|_w = 0$, i.e. $\forall \phi \in S_{X'} \left(\phi(b) \approx {}^*\phi(a) \right)$, i.e. a and b are infinitely close under the weak topology. Since this holds for any $a \in \bar{B}_{*X}$, it follows from Robinson's characterization of compactness that $\bar{B}_X$ is weakly compact.

CLAIM: $\forall \phi \in \bigcup_{n \in \mathbb{N}} S_n \left(\phi(b) \approx {}^*\phi(a) \right)$.

PROOF OF THE CLAIM: Suppose $\phi \in W_n$. Let $\epsilon \in \mathbb{R}^+$. Let $m \geq n$ be large enough satisfying both $m^{-1} \leq \epsilon$ and $\|b - c\|_X \leq \epsilon$ for some $c \in \text{conv}(\{a_m, a_{m+1}, \ldots\})$.

Write c as a convex combination $\sum_{i=0}^{k} \lambda_i a_{m+i}$, where $\lambda_0, \ldots, \lambda_k \in [0,1]$ are such that $\sum_{i=0}^{k} \lambda_i = 1$.

Since $m \geq n$, we have $|\phi(a_{m+i}) - {}^*\phi(a)| \leq m^{-1}$ by the property of the Day sequence. Hence

$$|\phi(c) - {}^*\phi(a)| = \left| \phi\left(\sum_{i=0}^{k} \lambda_i a_{m+i} \right) - {}^*\phi(a) \right| = \left| \sum_{i=0}^{k} \lambda_i \left(\phi(a_{m+i}) - {}^*\phi(a) \right) \right|$$

$$\leq \sum_{i=0}^{k} \lambda_i |\phi(a_{m+i}) - {}^*\phi(a)| \leq m^{-1} \leq \epsilon.$$

On the other hand, since $\|b - c\|_X \leq \epsilon$ and $\phi \in S_{X'}$, we have $|\phi(b) - \phi(c)| \leq \epsilon$. Therefore, $|\phi(b) - {}^*\phi(a)| \leq 2\epsilon$. But $\epsilon \in \mathbb{R}^+$ is arbitrary, so $\phi(b) \approx {}^*\phi(a)$ and the Claim is proved.

Now for any $n \in \mathbb{N}$ and $c \in \mathrm{conv}(\{a_1, \ldots, a_n\})$, we have by the property of the Day sequence and the Claim that

$$\|a - c\|_w \leq (1 + n^{-1}) \max_{\phi \in S_{n+1}} {}^\circ {}^*\phi(a - c)$$

$$= (1 + n^{-1}) \max_{\phi \in S_{n+1}} {}^\circ {}^*\phi(b - c) \leq (1 + n^{-1}) \|b - c\|_w.$$

Hence

$$\|a - b\|_w \leq \|a - c\|_w + \|b - c\|_w \leq 3 \|b - c\|_w.$$

So for any $\epsilon \in \mathbb{R}^+$, choose $n \in \mathbb{N}$ large enough so that $\|b - c\|_w \leq \epsilon$ for some $n \in \mathbb{N}$ and $c \in \mathrm{conv}(\{a_1, \ldots, a_n\})$, then $\|a - b\|_w \leq 3\epsilon$, therefore $\|a - b\|_w = 0$. □

What happens in the above proof is essentially that, in the weak topology, the Day sequence gives a Cauchy sequence from the $\mathrm{conv}(\{a_1, \ldots, a_n\})$, $n \in \mathbb{N}$, which potentially may have a limit in $\bigcap_{n \in \mathbb{N}} C_n$ infinitely close to the $a \in {}^*X$. Also, in general, $\bigcap_{n \in \mathbb{N}} C_n$ is either empty or a singleton.

Conversely, if $\{a_n\}_{n \in \mathbb{N}}$ is any Cauchy sequence in the weak topology, then with W corresponding to the weak topology and any increasing finite subsets of W, $\{(a_n, W_n)\}_{n \in \mathbb{N}}$ forms a Day sequence.

In a seminormed linear space X a subset $S \subset X$ is said to satisfy the *Šmulian condition* if for every decreasing sequence $\{C_n\}_{n \in \mathbb{N}}$ of nonempty closed convex subsets of S, we have $\bigcap_{n \in \mathbb{N}} C_n \neq \emptyset$.

By Prop. 2.27, closed convex subset of a normed linear space is weakly closed, hence a closed convex subset of a weakly compact set is weakly compact, therefore weakly compact sets satisfy the Šmulian condition.

In the proof of Thm. 2.21, $\bar{B}_X$ can be replaced by any bounded closed convex subset. Therefore we have the following.

Corollary 2.20. *Let X be a normed linear space and $A \subset X$ be bounded closed convex. Then A is weakly compact iff the Šmulian condition is satisfied.*

Moreover, for a weakly compact $C \subset X$, $S \cap C$ is weakly compact for any closed convex $S \subset X$.   □

In particular, together with Thm. 2.18, we have

Corollary 2.21. *Let X be a normed linear space. Then the following are equivalent:*

- *X is reflexive;*
- *$\overline{B_X(a,r)}$ is weakly compact for some $a \in X$ and $r \in \mathbb{R}^+$;*
- *$\overline{B_X(a,r)}$ satisfies the Šmulian condition for some $a \in X$ and $r \in \mathbb{R}^+$.*

□

Immediately we also have the following converse of Prop. 2.29 for reflexive spaces. In the setting of normed linear spaces, this infinite dimensional analogue of the classical Heine-Borel Theorem is perhaps more satisfactory than Thm. 1.22.

Corollary 2.22. *Let X be a reflexive Banach space and $C \subset X$.*

Then C is weakly compact iff it is bounded and closed in norm.   □

Notice that for a normed linear space X and a closed subspace $Y \subset X$, a subset $C \subset Y$ is closed convex in Y iff it is closed convex in X. It follows then from Thm. 2.21 that X is reflexive iff every closed subspace is. Moreover, suppose X is nonreflexive, let $\{C_n\}_{n\in\mathbb{N}} \subset \bar{B}_X$ be a strictly decreasing sequence of closed convex sets such that $\bigcap_{n\in\mathbb{N}} C_n = \emptyset$. Choose $c_n \in (C_n \setminus C_{n+1})$ and $Y := \overline{\mathrm{Lin}}(\{c_n \,|\, n \in \mathbb{N}\})$, then Y is a separable closed subspace of X failing the Šmulian condition for $\bar{B}_Y$. Hence we have the following:

Corollary 2.23. *Let X be a normed linear space. Then the following are equivalent:*

- *X is reflexive;*
- *every closed subspace of X is reflexive;*
- *every closed separable subspace of X is reflexive.*   □

Corollary 2.24. *Let X be a reflexive Banach space. Then for every nonempty closed convex $C \subset X$ and $a \in X$, there is $c \in C$ such that* $\text{dist}(a, C) = \|a - c\|$.

Proof.   Write $r := \text{dist}(a, C)$.

For each $n \in \mathbb{N}$, let $C_n := C \cap \overline{B_X(a, r + n^{-1})}$. Then the C_n's are decreasing nonempty bounded convex subsets of X hence, by the Šmulian condition and Cor. 2.21, we let $b \in \cap_{n \in \mathbb{N}} C_n$, then the required property is satisfied. $\square$

Actually the above can be re-stated in a more general setting:

Corollary 2.25. *Let X be a Banach space. Then for every nonempty weakly compact convex $C \subset X$ and $a \in X$, there is $c \in C$ such that* $\text{dist}(a, C) = \|c - a\|$. $\square$

We leave it as an exercise to prove the following generalization.

Theorem 2.22.   (Eberlein-Šmulian Theorem—general version) *Let X be a linear space and W a binding set of seminorms.*

Then, w.r.t. $\|\cdot\|_w$, *if $\bar{B}_X$ satisfies the Šmulian condition, then $\bar{B}_X$ is compact.* $\square$

2.5.3 James' characterization of reflexivity

Although we skip the proof of James' Theorem (Thm. 2.19), we now give a complete treatment of another important characterization of reflexivity due to James. It is more convenient to state this result in a negative form.

Theorem 2.23.   (James' Characterization) *Let X be a Banach space. Then the following are equivalent:*

(i) X *is nonreflexive.*

(ii) *There are $r \in (0, 1)$, $\{a_n\}_{n \in \mathbb{N}} \subset \bar{B}_X$ and $\{\phi_n\}_{n \in \mathbb{N}} \subset S_{X'}$ such that*
$$\forall n, m \in \mathbb{N}, \quad n = m \Rightarrow \phi_n(a_n) \in [r, \infty),$$
$$n < m \Rightarrow \text{Re}(\phi_n)(a_m) \in [r, \infty),$$
$$n > m \Rightarrow \phi_n(a_m) = 0.$$

Proof.   ((i) $\Rightarrow$ (ii)) : By Prop. 2.23, we identify $X \subset X''$ as a closed subspace. Since X is nonreflexive, $X \neq X''$, so $\left(\bar{B}_{X''} \setminus X\right) \neq \emptyset$.

Let $\pi : \widehat{^*X^w} \to X''$ be the isometric isomorphism given by Thm. 2.13. Consequently, for some $\hat{a} \in \bar{B}_{\widehat{^*X^w}}$ we obtain $\text{dist}_{X''}(\pi(\hat{a}), X) \in \mathbb{R}^+$, as X is a closed subspace of X''.

Let $s := \mathrm{dist}_{X''}(\pi(\widehat{a}), X)$.

By Cor. 2.15 and scaling, we can assume that $a \in \bar{B}_{*X}$.

Choose $r, \epsilon \in \mathbb{R}^+$ such that $s > (1 + \epsilon)r$.

The required sequences $\{a_n\}_{n \in \mathbb{N}} \subset \bar{B}_X$ and $\{\phi_n\}_{n \in \mathbb{N}} \subset S_{X'}$ will be constructed inductively. For each $n \in \mathbb{N}$, we define the property

$$P_n : \begin{cases} \phi_k(a_k) \in [r, \infty), & \text{if } 1 \leq k \leq n, \\ \mathrm{Re}(\phi_k)(a_m) \in [r, \infty), & \text{if } 1 \leq k < m \leq n, \\ \phi_k(a_m) = 0, & \text{if } 1 \leq m < k \leq n, \\ \pi(\widehat{a})(\phi_k) \in [s(1+\epsilon)^{-1}, \infty), & \text{if } 1 \leq k \leq n, \end{cases}$$

for $\{a_k\}_{1 \leq k \leq n} \subset \bar{B}_X$ and $\{\phi_k\}_{1 \leq k \leq n} \subset S_{X'}$.

First, $\mathrm{dist}_{X''}(\pi(\widehat{a}), X) = s > s(1+\epsilon)^{-1}$ implies that

$$\|\pi(\widehat{a})\|_{X''} \geq \pi(\widehat{a})(\phi_1) = {}^{\circ} {}^*\phi_1(a) \in [s(1+\epsilon)^{-1}, \infty)$$

for some $\phi_1 \in S_{X'}$. (Recall that we can require $\pi(\widehat{a})(\phi_1)$ to be a positive real number by replacing ϕ_1 by an appropriate rotation of ϕ_1.) Then, by a rotation of a if needed and by $s(1+\epsilon)^{-1} > r$, we obtain the following:

$$\exists x \in \bar{B}_{*X} \left(\phi_1(x) \in {}^*[r, \infty) \right).$$

By transfer, there is $a_1 \in \bar{B}_X$ satisfying the above and, together with the $\phi_1 \in S_{X'}$, P_1 holds.

Now, assume inductively that $\{a_k\}_{1 \leq k \leq n} \subset \bar{B}_X$ and $\{\phi_k\}_{1 \leq k \leq n} \subset S_{X'}$ have been constructed so that P_n holds.

For any $\{\alpha_k\}_{1 \leq k \leq n+1} \subset \mathbb{F}$, we have

$$\left| \sum_{k=1}^{n} \alpha_k \cdot 0 + \alpha_{n+1} s \right| = |\alpha_{n+1}| s = |\alpha_{n+1}| \, \mathrm{dist}_{X''}(\pi(\widehat{a}), X)$$

$$\leq \left\| \sum_{k=1}^{n} \alpha_k \, a_k + \alpha_{n+1} \, \pi(\widehat{a}) \right\|_{X''}.$$

Therefore, by Helly's Theorem (Cor. 2.10 applied to X''), there is $\theta \in X'$ with $\|\theta\|_{X'} \leq 1 + \epsilon$ such that

$$\begin{cases} a_k(\theta) = 0, & \text{if } 1 \leq k \leq n, \\ \pi(\widehat{a})(\theta) = s, \end{cases} \quad i.e. \quad \begin{cases} \theta(a_k) = 0, & \text{if } 1 \leq k \leq n, \\ {}^*\theta(a) \approx s. \end{cases}$$

Clearly, such $\theta \neq 0$. Let $\phi_{n+1} := \theta \|\theta\|_{X'}^{-1}$, then we have $\phi_{n+1} \in S_{X'}$ and

$$\phi_{n+1}(a_k) = 0, \quad \text{if } 1 \leq k \leq n,$$

$$\pi(\widehat{a})(\phi_{n+1}) \approx {}^*\phi_{n+1}(a) = \frac{{}^*\theta(a)}{\|\theta\|_{X'}} \gtrsim s(1+\epsilon)^{-1} > r.$$

Hence, by a rotation of a if needed and $\pi(\widehat{a})(\phi_k) \in [s(1+\epsilon)^{-1}, \infty)$, for $1 \le k \le n$, the following holds:

$$\exists x \in \bar{B}_{*X} \left({}^*\phi_{n+1}(x) \in {}^*[r, \infty) \wedge \bigwedge_{k=1}^{n} \left({}^*\mathrm{Re}(\phi_k)(x) \in {}^*[r, \infty) \right) \right).$$

By transfer, let $a_{n+1} \in \bar{B}_X$ satisfy the above, then P_{n+1} is satisfied by $\{a_k\}_{1 \le k \le n+1}$ and $\{\phi_k\}_{1 \le k \le n+1}$, so the inductive step of the construction is complete.

$((ii) \Rightarrow (i))$: Let r, $\{a_n\}_{n \in \mathbb{N}}$ and $\{\phi_n\}_{n \in \mathbb{N}}$ be given by (ii).

Define $C_n := \overline{\mathrm{conv}}(\{a_n, a_{n+1}, \dots\})$, $n \in \mathbb{N}$. So the C_n's are decreasing nonempty closed convex subsets of $\bar{B}_X$.

Note that $\forall x \in C_n \left(\mathrm{Re}(\phi_n)(x) \in [r, \infty) \right)$.

Suppose X is reflexive, then as a consequence of the Eberlein-Šmulian Theorem (Cor.. 2.21), $\bigcap_{n \in \mathbb{N}} C_n \ne \emptyset$.

Let $a \in \bigcap_{n \in \mathbb{N}} C_n \ne \emptyset$ and choose $n \in \mathbb{N}$ large enough so that for some $\lambda_k \in [0, 1]$, $1 \le k < n$, with $\sum_{1 \le k < n} \lambda_k = 1$ we have

$$\left\| a - \sum_{1 \le k < n} \lambda_k a_k \right\|_X < r.$$

Since $\phi_n \in S_{X'}$ and $\phi_n(a_k) = 0$ for all $1 \le k < n$, we have $|\phi_n(a)| < r$.

On the other hand, $a \in C_n$, so $\mathrm{Re}(\phi_n)(a) \in [r, \infty)$, a contradiction.

Therefore X is nonreflexive. $\square$

We call the sequence in Thm. 2.23 a **James sequence**.

2.5.4   Finite representability and superreflexivity

We say that a normed linear space Y is **finitely representable** in a normed linear space X, if for every $\epsilon \in \mathbb{R}^+$ and every finite dimensional subspace $Y_0 \subset Y$, there is a linear mapping $\pi : Y_0 \to X$ such that

$$\forall y \in Y_0 \left((1+\epsilon)^{-1} \|y\|_Y \le \|\pi(y)\|_X \le (1+\epsilon) \|y\|_Y \right).$$

So for such π we have $\pi \in \mathcal{B}(Y_0, X)$.

Note that finite representability is a generalization of isometric embedding. But as a consequence of saturation, as embeddings into a nonstandard hull, the two notions are the same.

Proposition 2.31. *Let X be an internal normed linear space and $\widehat{X}$ its norm-nonstandard hull. Let Y be a normed linear space which is finitely representable in $\widehat{X}$, then Y is isometrically embeddable in $\widehat{X}$.*

Proof.   Assume that Y is finitely representable in $\widehat{X}$.

Let $e_1, \ldots, e_n \in Y$, $n \in \mathbb{N}$, be linearly independent.

Write $Y_0 := \mathrm{Lin}(\{e_1, \ldots, e_n\})$.

Suppose $\epsilon \in \mathbb{R}^+$ and $\pi_0 \in \mathcal{B}(Y_0, \widehat{X})$ so that

$$\forall y \in Y_0 \left((1+\epsilon)^{-1} \|y\|_Y \le \|\pi_0(y)\|_{\widehat{X}} \le (1+\epsilon) \|y\|_Y \right).$$

Write $\pi_0(e_1) = \widehat{a}_1, \ldots, \pi_0(e_n) = \widehat{a}_n$ for some $a_1, \ldots, a_n \in \mathrm{Fin}(X)$. So $\widehat{a}_1, \ldots, \widehat{a}_n$ are linearly independent. By Prop. 2.8, $a_1, \ldots, a_n$ are *linearly independent.

Let $\pi \in {}^*\mathcal{B}({}^*Y_0, X)$ be given by

$$\pi\left(\sum_{k=1}^n \alpha_k e_k \right) := \sum_{k=1}^n \alpha_k a_k, \qquad \alpha_1, \ldots, \alpha_n \in {}^*\mathbb{F}.$$

Then we have

$$\forall y \in \bar{B}_{{}^*Y_0} \left((1+2\epsilon)^{-1} \|y\|_{{}^*Y} \le \|\pi(y)\|_X \le (1+2\epsilon) \|y\|_{{}^*Y} \right).$$

Therefore we also have

$$\forall y \in {}^*Y_0 \left((1+2\epsilon)^{-1} \|y\|_{{}^*Y} \le \|\pi(y)\|_X \le (1+2\epsilon) \|y\|_{{}^*Y} \right).$$

The above π can be identified with an element in ${}^*\mathcal{B}({}^*Y, X)$, which maps elements in ${}^*Y \setminus {}^*Y_0$ to 0.

Therefore, if we let $\mathcal{F}_{Y_0, \epsilon}$, $\epsilon \in \mathbb{R}^+$, denote the internal subset of ${}^*\mathcal{B}({}^*Y, X)$ consisting of π such that

$$\forall y \in {}^*Y_0 \left((1+\epsilon)^{-1} \|y\|_{{}^*Y} \le \|\pi(y)\|_X \le (1+\epsilon) \|y\|_{{}^*Y} \right),$$

then $\mathcal{F}_{Y_0, \epsilon} \ne \emptyset$.

Now, over all finite dimensional subspaces $Y_0 \subset Y$ and $\epsilon \in \mathbb{R}^+$, the families $\mathcal{F}_{Y_0, \epsilon}$ are directed under the inclusion relation, so they satisfy the finite intersection property.

By saturation, let π be an element in the common intersection of the families $\mathcal{F}_{Y_0, \epsilon}$.

Then $\pi \in {}^*\mathcal{B}({}^*Y, X)$ and $\forall y \in \mathrm{Fin}({}^*Y) \left(\|y\|_{{}^*Y} \approx \|\pi(y)\|_X \right)$.

Define $\tilde{\pi} : Y \to \widehat{X}$ by $y \mapsto \widehat{\pi(y)}$.

It is easy to check that $\tilde{\pi}$ is well-defined.

Moreover, $\tilde{\pi} \in \mathcal{B}(Y, \widehat{X})$ and $\forall y \in Y \left(\|y\|_Y = \|\pi(y)\|_{\widehat{X}} \right)$ as required. $\square$

On the other hand, the following exhibits the close relation between a normed linear space and its nonstandard hull.

Proposition 2.32.   *Let X be a normed linear space, then ${}^*X$ is finitely representable in X.*

Proof.   Let $\epsilon \in \mathbb{R}^+$.

Let $Y := \mathrm{Lin}(\{\widehat{e}_1, \ldots, \widehat{e}_n\}) \subset \widehat{{}^*X}$, where $n \in \mathbb{N}$ and $\widehat{e}_1, \ldots, \widehat{e}_n \in \widehat{{}^*X}$ are linearly independent. Then by Prop. 2.8, $e_1, \ldots, e_n \in {}^*X$ are *linearly independent.

Define $\mathbb{F}_0 := \{\alpha \in \mathbb{F} \mid |\alpha| \le 1\}$. In particular,

$$\forall \alpha_1, \ldots, \alpha_n \in \mathbb{F}_0^n \left(\left\| \sum_{i=1}^{n} \alpha_i \widehat{e}_i \right\|_{\widehat{{}^*X}} \approx \left\| \sum_{i=1}^{n} \alpha_i e_i \right\|_{{}^*X} \right).$$

Define $\theta : \mathbb{F}_0^n \to [0, \infty)$ by $\theta(\alpha_1, \ldots, \alpha_n) := \left\| \sum_{i=1}^{n} \alpha_i \widehat{e}_i \right\|_{\widehat{{}^*X}}$ and similarly $\Theta : {}^*\mathbb{F}_0^n \to {}^*[0, \infty)$ by $\Theta(\alpha_1, \ldots, \alpha_n) := \left\| \sum_{i=1}^{n} \alpha_i e_i \right\|_{{}^*X}$.

Note that θ is continuous and

$$\forall \alpha_1, \ldots, \alpha_n \in {}^*\mathbb{F}_0^n \left({}^*\theta(\alpha_1, \ldots, \alpha_n) \approx \Theta(\alpha_1, \ldots, \alpha_n) \right).$$

So there are linearly independent $x_1, \ldots, x_n \in {}^*X$, $\forall \alpha_1, \ldots, \alpha_n \in {}^*\mathbb{F}_0^n$

$$(1 - \epsilon)^{-1}\, {}^*\theta(\alpha_1, \ldots, \alpha_n) \le \left\| \sum_{i=1}^{n} \alpha_i x_i \right\|_{{}^*X} \le (1 + \epsilon)\, {}^*\theta(\alpha_1, \ldots, \alpha_n).$$

Now, by transfer, let $a_1, \ldots, a_n \in X$ be linearly independent so that $\forall \alpha_1, \ldots, \alpha_n \in \mathbb{F}_0^n$

$$(1 - \epsilon)^{-1}\theta(\alpha_1, \ldots, \alpha_n) \le \left\| \sum_{i=1}^{n} \alpha_i a_i \right\|_X \le (1 + \epsilon)\theta(\alpha_1, \ldots, \alpha_n).$$

Hence $\forall \alpha_1, \ldots, \alpha_n \in \mathbb{F}_0^n$

$$(1 - \epsilon)^{-1} \left\| \sum_{i=1}^{n} \alpha_i \widehat{e}_i \right\|_{\widehat{{}^*X}} \le \left\| \sum_{i=1}^{n} \alpha_i a_i \right\|_X \le (1 + \epsilon) \left\| \sum_{i=1}^{n} \alpha_i \widehat{e}_i \right\|_{\widehat{{}^*X}}.$$

By letting $\pi \in \mathcal{B}(Y, X)$ be given by

$$\forall \alpha_1, \ldots, \alpha_n \in \mathbb{F}_0^n \left(\pi\Big(\sum_{i=1}^{n} \alpha_i \widehat{e}_i \Big) := \sum_{i=1}^{n} \alpha_i a_i \right),$$

we have a bounded linear mapping witnessing the finite representability requirement for ϵ. $\square$

A normed linear space is called **superreflexive** if every normed linear space which is finitely representable in it is reflexive.

Trivially, superreflexivity implies reflexivity. Due to saturation, both notions coincide for nonstandard hulls.

Theorem 2.24. *Let X be an internal normed linear space.*
Then $\widehat{X}$ is reflexive iff it is superreflexive.

146   *Nonstandard Methods in Functional Analysis*

Proof.   For the nontrivial direction, assume that $\widehat{X}$ is reflexive.

Let Y be a normed linear space which is finitely representable in $\widehat{X}$. Then by Prop. 2.31, Y is isometrically embeddable into $\widehat{X}$.

Then by Cor. 2.23, Y is reflexive.

Hence $\widehat{X}$ is superreflexive.   □

Corollary 2.26. *Let X be a normed linear space. Then X is superreflexive iff $*\widehat{X}$ is.*

Proof.   ($\Rightarrow$) : By Prop. 2.32, $\widehat{*X}$ is finitely representable in X. So if X is superreflexive, then $*\widehat{X}$ is reflexive, hence also superreflexive by Thm. 2.24.

($\Leftarrow$) : It is clear that a closed subspace of a superreflexive space is superreflexive.   □

2.5.5   Notes and exercises

What we call Day sequence here is constructed on the basis of M. M. Day's idea. (See [Day (1973)].)

A similar but more direct proof of the Eberlein-Šmulian Theorem (Thm. 2.21) can be found in [Baratella and Ng (2003)].

For a geometric characterization of superreflexivity in terms of supporting a uniform rotund norm, see [Megginson (1998)].

Thm. 2.20 was given in [Henson and Moore (1983)]. Prop. 2.32 and Thm. 2.24 were proved in [Heinrich (1980)].

EXERCISES

(1) Use Alaoglu's Theorem to prove the implication ((i) $\Rightarrow$ (iii)) in Thm. 2.18.
(2) Verify the claims in Example 2.5.
(3) Verify the statements in Example 2.6.
(4) Prove Thm. 2.22. Generalize the notion of binding and generalize Thm. 2.22 for arbitrary compact sets w.r.t. $\|\cdot\|_{\mathrm{w}}$.
(5) Show that $\aleph_1$-saturation is sufficient for proving Thm. 2.24.
(6) Show that in a normed linear space X, $\bar{B}_X$ is weakly compact iff for every finite dimensional normed linear space Y over the same field, for every $b \in \bar{B}_{*X}$, there is $a \in \bar{B}_X$ such that $*f(b) \approx f(a)$ for every $f \in \mathcal{B}(X, Y)$.

Show that this equivalence does not hold in general when arbitrary normed linear spaces Y are used.

(7) Show that there are nonreflexive Banach spaces X such that X and X'' are isometrically isomorphic.

(8) Prove that a Banach space X is reflexive iff X' is reflexive.

(9) Continuing with the last problem, is it true for any Banach space X that X is superreflexive iff X' is superreflexive?

(10) Let X be an internal normed linear space such that $\widehat{X}$ is reflexive. Then for internal $C \subset \text{Fin}(X)$, we have C is *convex iff $\widehat{C}$ is convex.

(11) Find an example of a reflexive space X which is not superreflexive. (So, by Cor. 2.26, $*\widehat{X}$ is not reflexive.)

(12) Prove the ($\Leftarrow$) direction in Thm. 2.20 using James Characterization (Thm. 2.23).

2.6   Hilbert Spaces

> *The parallelogram law produces superreflexivity, with Hilbertness*
> *emerging from Banachness.*

Before we consider Hilbert spaces, the main topic of this section, we need some general notion.

A **pre-inner product** on a linear space X is a mapping $\langle \cdot, \cdot \rangle : X^2 \to \mathbb{F}$ such that

(i) $\forall x \in X \left(\langle x, x \rangle \in [0, \infty) \right)$;
(ii) $\forall x, y \in X \left(\langle x, y \rangle = \overline{\langle y, x \rangle} \right)$;
(iii) $\forall \alpha \in \mathbb{F} \, \forall x, y \in X \left(\langle \alpha x, y \rangle = \alpha \langle x, y \rangle \right)$;
(iv) $\forall x_1, x_2, y \in X \left(\langle x_1 + x_2, y \rangle = \langle x_1, y \rangle + \langle x_2, y \rangle \right)$.

In the case $\mathbb{F} = \mathbb{R}$ a pre-inner product is symmetric, hence it is a bilinear form.

If $\mathbb{F} = \mathbb{C}$, then we have linearity in the first variable and antilinearity in the second variable, *i.e.*

$$\forall \alpha \in \mathbb{F} \, \forall x, y_1, y_2 \in X \left(\langle x, \alpha y_1 + y_2 \rangle = \bar{\alpha} \langle x, y_1 \rangle + \langle x, y_2 \rangle \right).$$

Such mapping is called a **sesquilinear form**.

Observe that if $\mathbb{F} = \mathbb{C}$, $\text{Im}(\langle x, y \rangle) = \text{Re}(\langle x, iy \rangle)$ holds for every x, y in the pre-inner product space.

A linear space equipped with a pre-inner product is called a **pre-inner product space**.

A pre-inner product $\langle \cdot, \cdot \rangle$ on a linear space X is called an **inner-product** if it further satisfies

- $\forall x \in X \left(\left(\langle x, x \rangle = 0 \right) \Leftrightarrow \left(x = 0 \right) \right)$.

Such a space is called an **inner-product space**.

According to $\mathbb{F} = \mathbb{R}$ or $\mathbb{C}$, we call such space a **real inner-product space** or a **complex inner-product space**.

2.6.1   *Basic properties*

First some elementary properties about pre-inner product spaces whose proofs are left as exercises.

Theorem 2.25. *Let X be a pre-inner product space. Define for all $x \in X$ a function $\|x\| := \sqrt{\langle x, x \rangle}$, Then we have*

(i) (Cauchy-Schwarz Inequality) $\forall x, y \in X \left(|\langle x, y \rangle| \leq \|x\| \, \|y\| \right)$.

(ii) *The function* $\|\cdot\|$ *is a seminorm on* X.

  Moreover, if $\langle \cdot, \cdot \rangle$ *is an inner-product, then* $\|\cdot\|$ *is a norm.*

(iii) (Polar Identity) $\forall x, y \in X \left(\|x + y\|^2 = \|x\|^2 + 2\mathrm{Re}(\langle x, y \rangle) + \|y\|^2 \right)$.

(iv) (Parallelogram Law)

$$\forall x, y \in X \left(\|x + y\|^2 + \|x - y\|^2 = 2 \|x\|^2 + 2 \|y\|^2 \right).$$

□

In particular, pre-inner product spaces are seminormed linear spaces and inner-product spaces are normed linear spaces. Unless specified otherwise, given a pre-inner product space, $\|\cdot\|$ always denotes the seminorm defined as above.

Proposition 2.33. *Let* X *be a pre-inner product space and let* I *be the closed subspace* $\{x \in X \mid \|x\| = 0\}$. *Then the quotient normed linear space* X/I *forms an inner product space.*

Proof. It is clear that I is a closed subspace and, by Prop. 2.1(ii), X/I forms a normed linear space. Moreover, the mapping $\langle x+I, y+I \rangle := \langle x, y \rangle_X$ where $\langle \cdot, \cdot \rangle_X$ denotes the pre-inner product on X, defines an inner product on X/I and gives the quotient norm. □

Here are two properties equivalent to the parallelogram law.

(Although, by Prop. 2.34 below, the parallelogram law in a seminormed linear space implies the existence of a pre-inner product that gives the seminorm, so the proof of the following can be done more straightforwardly by calculations using the seminorm, we prefer to produce a proof free of any reference to the pre-inner product.)

Theorem 2.26. *Let* X *be a seminormed linear space. Then the following are equivalent:*

(i) *The parallelogram law holds in* X.

(ii) *The following holds for any* $x_1, \ldots, x_n \in X$, $1 < n \in \mathbb{N}$:

$$\sum_{1 \leq i < j \leq n} \|x_i + x_j\|^2 = (n-2) \sum_{i=1}^{n} \|x_i\|^2 + \left\| \sum_{i=1}^{n} x_i \right\|^2. \qquad (2.16)$$

(iii) *The following holds for any* $x_1, \ldots, x_n \in X$, $1 < n \in \mathbb{N}$:

$$\sum_{1 \leq i < j \leq n} \|x_i - x_j\|^2 = n \sum_{i=1}^{n} \|x_i\|^2 - \left\| \sum_{i=1}^{n} x_i \right\|^2. \qquad (2.17)$$

Proof. $((ii) \Rightarrow (i))$: Let $x, y \in X$. Apply (2.16) to $x, y, -y$, we have:

$$\|x+y\|^2 + \|x-y\|^2 + \|y-y\|^2 = \|x\|^2 + \|y\|^2 + \|-y\|^2 + \|x+y-y\|^2,$$

hence the parallelogram law holds.

$((iii) \Rightarrow (i))$: Similarly, apply (2.17) to $x, y, -y$, and get the parallelogram law from:

$$\|x-y\|^2 + \|x+y\|^2 + \|y-(-y)\|^2 = 3\left(\|x\|^2 + \|y\|^2 + \|-y\|^2\right) - \|x+y-y\|^2.$$

$((i) \Rightarrow (ii))$: When $n = 2$, (2.16) is trivial.

Let $x, y, z \in X$. Apply the parallelogram law to $(x+y)$, $(y+z)$, we have $\|x+2y+z\|^2 = 2\|x+y\|^2 + 2\|y+z\|^2 - \|x-z\|^2$. Another application of the parallelogram law then gives

$$\|x+2y+z\|^2 = 2\|x+y\|^2 + 2\|y+z\|^2 + \|x+z\|^2 - 2\|x\|^2 - 2\|z\|^2. \quad (2.18)$$

Yet another application of the parallelogram law gives

$$\|(x+2y+z)+(x+z)\|^2 + \|(x+2y+z)-(x+z)\|^2$$
$$= 2\|x+2y+z\|^2 + 2\|x+z\|^2,$$

i.e. $2\|x+y+z\|^2 + 2\|y\|^2 = \|x+2y+z\|^2 + \|x+z\|^2$,

which yields the following by combining with (2.18):

$$\|x+y\|^2 + \|x+z\|^2 + \|y+z\|^2 = \|x\|^2 + \|y\|^2 + \|z\|^2 + \|x+y+z\|^2.$$

Hence (2.16) for the case $n = 3$ is proved.

Now fix $n > 3$, assume inductively that (2.16) holds for all sequence of length at least 2 but $< n$.

Let $x_i \in X$, $1 \le i \le n$. Then by (2.16) for the case $n = 3$, we have

$$\|x_1 + x_2\|^2 + \left\|x_1 + \sum_{3 \le i \le n} x_i\right\|^2 + \left\|x_2 + \sum_{3 \le i \le n} x_i\right\|^2$$
$$= \|x_1\|^2 + \|x_2\|^2 + \left\|\sum_{3 \le i \le n} x_i\right\|^2 + \left\|\sum_{i=1}^{n} x_i\right\|^2. \quad (2.19)$$

By the inductive assumptions, we have

$$\left\| \sum_{3 \le i \le n} x_i \right\|^2 = \sum_{3 \le i < j \le n} \|x_i + x_j\|^2 - (n-4) \sum_{3 \le i \le n} \|x_i\|^2,$$

$$\left\| x_1 + \sum_{3 \le i \le n} x_i \right\|^2 = \sum_{3 \le i \le n} \|x_1 + x_i\|^2 + \sum_{3 \le i < j \le n} \|x_i + x_j\|^2$$
$$- (n-3)\|x_1\|^2 - (n-3) \sum_{3 \le i \le n} \|x_i\|^2,$$

$$\left\| x_2 + \sum_{3 \le i \le n} x_i \right\|^2 = \sum_{3 \le i \le n} \|x_2 + x_i\|^2 + \sum_{3 \le i < j \le n} \|x_i + x_j\|^2$$
$$- (n-3)\|x_2\|^2 - (n-3) \sum_{3 \le i \le n} \|x_i\|^2.$$

Substituting the left sides of the above into (2.19), we see that (2.16) holds for the $\{x_i\}_{1 \le i \le n}$.

$((ii) \Rightarrow (iii))$: We apply (2.16) to the sequence

$$x_1, x_2, \ldots, x_n,\ x_{n+1} = -x_1,\ x_{n+2} = -x_2,\ \ldots,\ x_{2n} = -x_n \in X.$$

Then

$$\sum_{1 \le i < j \le 2n} \|x_i + x_j\|^2 = (2n-2) \sum_{i=1}^{2n} \|x_i\|^2 + \left\| \sum_{i=1}^{2n} x_i \right\|^2. \qquad (2.20)$$

We have the following for the left side of (2.20)

$$\sum_{1 \le i < j \le n} \|x_i + x_j\|^2 + \sum_{1 \le i,j \le n} \|x_i - x_j\|^2 + \sum_{1 \le i < j \le n} \|-x_i - x_j\|^2$$

$$= 2 \sum_{1 \le i < j \le n} \|x_i + x_j\|^2 + 2 \sum_{1 \le i < j \le n} \|x_i - x_j\|^2$$

$$= 2(n-2) \sum_{i=1}^{n} \|x_i\|^2 + 2 \left\| \sum_{i=1}^{n} x_i \right\|^2 + 2 \sum_{1 \le i < j \le n} \|x_i - x_j\|^2.$$

Meanwhile the right side of (2.20) equals

$$(2n-2) \sum_{i=1}^{n} \|x_i\|^2 + (2n-2) \sum_{i=1}^{n} \|-x_i\|^2 + 0 = 2(2n-2) \sum_{i=1}^{n} \|x_i\|^2.$$

Therefore the equality in (2.17) holds. $\qquad \square$

So in particular, both (2.16) and (2.17) hold in any pre-inner product space.

Note also by adding the two equalities together we have the following equality which is just the result of adding the associated $(n-1)$ parallelogram law equations:

$$\sum_{1 \leq i < j \leq n} \|x_i + x_j\|^2 + \sum_{1 \leq i < j \leq n} \|x_i - x_j\|^2 = 2(n-1) \sum_{i=1}^{n} \|x_i\|^2.$$

The following shows that the sum of the squares of the distances between n unit vectors in a pre-inner product space is always bounded by n^2.

Corollary 2.27. *Let X be a pre-inner product space. For any given $a_1, \ldots, a_n \in S_X$, $n \in \mathbb{N}$, we have*

$$\sum_{1 \leq i < j \leq n} \|a_i - a_j\|^2 \leq n^2.$$

Moreover, the maximum n^2 is attained precisely when $\sum_{i=1}^{n} a_i = 0$.

Proof. Since $\|a_i\| = 1$, $n \in \mathbb{N}$, (2.17) implies that

$$\sum_{1 \leq i < j \leq n} \|a_i - a_j\|^2 = n^2 - \left\| \sum_{i=1}^{n} a_i \right\|^2.$$
$\square$

Note that in particular, the maximum in the above is attained when the a_i's are vertices of a regular polygon or a regular polyhedron.

Proposition 2.34. *Let X be a seminormed linear space. Then X is a pre-inner product space with its seminorm obtained as in Thm. 2.25 iff the parallelogram law is satisfied.*

Note that if X is a normed linear space, the above statement holds with "pre-inner product" replaced by "inner product".

Proof. One direction is in Thm. 2.25 (iv).

For the other implication, in the case $\mathbb{F} = \mathbb{R}$, take

$$\langle x, y \rangle := \frac{1}{2} \Big(\|x + y\|^2 - \|x\|^2 - \|y\|^2 \Big),$$

and in the case $\mathbb{F} = \mathbb{C}$, an imaginary part with similar form is needed:

$$\langle x, y \rangle := \frac{1}{2} \Big(\|x + y\|^2 - \|x\|^2 - \|y\|^2 \Big) + \frac{i}{2} \Big(\|x + iy\|^2 - \|x\|^2 - \|y\|^2 \Big).$$

Then it can be verified that $\langle \cdot, \cdot \rangle$ forms the required pre-inner product: Properties (i) and (ii) of the definition of pre-inner product are straightforward. (iv) can be shown from (2.16) in Thm. 2.26 for the case $n = 3$. (iii) can be proved first by using (iv) for integer α, then for rational α followed by using the transfer (or by a limit argument) for general $\alpha \in \mathbb{F}$.) $\square$

One can view the pre-inner product $\langle x, y \rangle$ in Prop. 2.34 as being defined from some given $\langle x + y, x + y \rangle, \langle x + iy, x + iy \rangle, \langle x, x \rangle, \langle y, y \rangle$. This is usually referred to as **definition by polarization**.

A complete inner-product space is called a **Hilbert space**. So Hilbert spaces are Banach spaces. In fact Hilbert spaces are reflexive Banach spaces, as we shall see in a moment.

Note also that a closed subspace of a Hilbert space is a Hilbert space.

An **extreme point** of a convex subset C of a linear space is defined to be an element $c \in C$ such that whenever $a, b \in C$ satisfying $(a + b)/2 = c$, then $a = b = c$.

The following says that in a Hilbert space, the surface of a closed ball has no flat region.

Proposition 2.35. *Let X be a Hilbert space, then every $c \in S_X$ is an extreme point of $\bar{B}_X$.*

Similarly, for every $a \in X$ and $r \in \mathbb{R}^+$, every $c \in S_X(a, r)$ is an extreme point of $\overline{B_X(a, r)}$.

Proof.   Let $a, b \in S_X$ such that $(a + b)/2 = c \in S_X$. Then $\|a + b\| = 2$, so the parallelogram law in Thm. 2.25 (iv) gives that $\|a - b\| = 0$, *i.e.* $a = b = c$.   □

Given an internal pre-inner product space X, the seminorm-nonstandard hull $\widehat{X}$ is of course a Banach space (Thm. 2.1). But $\widehat{X}$ is also a Hilbert space, because of Prop. 2.34 and the fact that the parallelogram law is preserved from X to $\widehat{X}$.

More directly, we can see that the corresponding inner product on $\widehat{X}$ is given by $\langle \widehat{x}, \widehat{y} \rangle := {}^\circ\langle x, y \rangle$, where $x, y \in \text{Fin}(X)$.

We say that the **approximate parallelogram law** holds in an internal seminormed linear space X if $\|x + y\|^2 + \|x - y\|^2 \approx 2\|x\|^2 + 2\|y\|^2$ holds for any $x, y \in \text{Fin}(X)$.

The following is an easy way to construct Hilbert spaces, generalizing what we have just remarked slightly.

Proposition 2.36. *Let X be an internal seminormed linear space satisfying the approximate parallelogram law.*

Then $\widehat{X}$, the seminorm-nonstandard hull, is a Hilbert space.

Proof.   By Thm. 2.1, $\widehat{X}$ forms a Banach space. By a saturation argument, the parallelogram law holds in $\widehat{X}$, so it is a Hilbert space by Prop. 2.34. □

Elements x, y in an inner-product space are said to be **orthogonal**, if $\langle x, y \rangle = 0$. We write $x \perp y$ when x, y are orthogonal.

A subset A of a Hilbert space H is called **orthonormal** if

$$A \subset S_H \wedge \forall x, y \in A \, (x \neq y \Rightarrow x \perp y).$$

By a **basis** for H we mean a maximal orthonormal subset of H.

Clearly, all bases of a Hilbert space are of the same cardinality called the **dimension** of the Hilbert space and written as $\dim(H)$.

Note that for infinite dimensional spaces, this is not the same as the dimension of H as a linear space, although the same symbol is used in both cases.

Theorem 2.27.

 (i) (The Gram-Schmidt Process) *Every Hilbert space has a basis.*
 (ii) *S is a basis for a Hilbert space X iff $\forall x \in X \, (x \perp S \Leftrightarrow x = 0)$.*
 (iii) (Parseval's Identity) *If S is a basis for a Hilbert space X, then*

$$\forall x \in X \left(\left(\|x\|^2 = \sum_{y \in S} |\langle x, y \rangle|^2 \right) \wedge \left(x = \sum_{y \in S} \langle x, y \rangle y \right) \right).$$

 (iv) *Hilbert spaces of the same dimension are isometrically isomorphic.* $\square$

If Y is a subspace of a Hilbert space X, the **orthogonal complement** of Y is denoted by $Y^\perp$ and defined as $\{x \in X \mid \forall y \in Y \, (x \perp y)\}$.

Note that $Y^\perp$ is a closed subspace of X.

A mapping $f : X \to Y$ between normed linear spaces X, Y is said to be **antilinear** if

$$\forall x_1, x_2 \in X \left(f(x_1 + x_2) = f(x_1) + f(x_2) \right) \wedge \forall x \in X \, \forall \alpha \in \mathbb{F} \left(f(\alpha x) = \bar{\alpha} f(x) \right).$$

So for $\mathbb{F} = \mathbb{R}$ linear is the same as antilinear.

If there is an antilinear isometric isomorphism between X and X', we say that X is **self-dual**.

Theorem 2.28.   (Riesz Representation Theorem) *Let X be a Hilbert space. For each $x \in X$ let ϕ_x denote the mapping $X \ni y \mapsto \langle y, x \rangle$.*

Then $\forall x \in X \, (\phi_x \in X')$.

Moreover, the mapping $\pi : X \to X'$ given by $\pi(x) = \phi_x$ is an antilinear isometric isomorphism, i.e. X is self-dual.

Proof.   Clearly, for $x \in X$, the ϕ_x is linear. Also, by the Cauchy-Schwarz inequality, for $x \neq 0$,

$$\|x\| = \left| \left\langle \frac{x}{\|x\|}, x \right\rangle \right| \leq \sup_{\|y\|=1} |\langle y, x \rangle| \leq \sup_{\|y\|=1} \|y\| \, \|x\| = \|x\|,$$

hence $\forall x \in X \left(\phi_x \in X' \right)$ and $\forall x \in X \left(\|\phi_x\|_{X'} = \|x\| \right)$, *i.e.* π is an isometry.

Clearly π is antilinear.

Moreover, $\forall x \in X \left((\phi_x = 0) \Leftrightarrow (x = 0) \right)$, *i.e.* π is injective.

To show that π is surjective, let $\phi \in X'$. Since $\phi_0 = 0$, we assume that $\phi \neq 0$. Then there is $b \in \mathrm{Ker}(\phi)^{\perp}$ such that $\phi(b) \in \mathbb{R}$.

Set $a := \phi(b) \|b\|^{-2} b$. Then $\langle b, a \rangle = \phi(b)$ and

$$\forall x \in X \left(\left(x - \frac{\phi(x)}{\phi(b)} b \right) \in \mathrm{Ker}(\phi) \right), \text{ so } \forall x \in X \left(\left\langle x - \frac{\phi(x)}{\phi(b)} b, a \right\rangle = 0 \right).$$

Therefore, for all $x \in X$,

$$\phi_a(x) = \langle x, a \rangle = \left\langle x - \frac{\phi(x)}{\phi(b)} b, a \right\rangle + \left\langle \frac{\phi(x)}{\phi(b)} b, a \right\rangle = 0 + \frac{\phi(x)}{\phi(b)} \phi(b) = \phi(x).$$
$\square$

(Note that Thm. 2.14 is another theorem by the same name.)

Corollary 2.28. *Hilbert spaces are superreflexive.*

Proof. Let X be a Hilbert space. So $\widehat{^*X}$ is also a Hilbert space by Prop 2.36. From Thm. 2.28, every $\phi \in \left(\widehat{^*X} \right)'$ is normed, so $\widehat{^*X}$ is reflexive by James' Theorem (Thm. 2.19), hence $\widehat{^*X}$ is superreflexive by Thm. 2.24. But X is a closed subspace of $\widehat{^*X}$, so X is superreflexive.

Or more directly, iterate Thm. 2.28 to get $\left(\widehat{^*X} \right)' = \left(\widehat{^*X} \right)''$, and by composing the antilinear isometric isomorphisms, one sees that $\widehat{^*X} = \left(\widehat{^*X} \right)''$ via the identity embedding, then argue as above. $\square$

Corollary 2.29. *By regarding a Hilbert space X as its dual X', a net $\{a_i\}_{i \in I} \subset X$ converges to $a \in X$ in the weak topology iff $\lim_{i \in I} \langle a_i, b \rangle = \langle a, b \rangle$ for every $b \in X$.* $\square$

By combining the Riesz Representation Theorem and the Banach-Steinhaus Theorem (Cor. 2.9), we have the following.

Corollary 2.30. *Given sequence $\{a_n\}_{n \in \mathbb{N}}$ in a Hilbert space X, suppose $\lim_{n \to \infty} \langle x, a_n \rangle$ exists for every $x \in X$, then $\lim_{n \to \infty} a_n$ exists.* $\square$

Proposition 2.37. *Let X be a Hilbert space. Suppose a net $\{a_i\}_{i \in I} \subset X$ converges to $a \in X$ in the weak topology.*

Then $\lim_{i \in I} a_i = a$ in norm iff $\lim_{i \in I} \|a_i\| = \|a\|$.

Moreover $\|a\| \leq \liminf_{i \in I} \|a_i\|$.

Proof.   For each $i \in I$ we have

$$\|a - a_i\|^2 = \langle a - a_i, a - a_i \rangle = \langle a, a \rangle + \langle a_i, a_i \rangle - \langle a, a_i \rangle - \langle a_i, a \rangle,$$

therefore, by Cor. 2.29,

$$\lim_{i \in I} \|a - a_i\|^2 = \|a\|^2 + \lim_{i \in I} \|a_i\|^2 - \lim_{i \in I} \overline{\langle a_i, a \rangle} - \lim_{i \in I} \langle a_i, a \rangle$$

$$= \|a\|^2 + \lim_{i \in I} \|a_i\|^2 - \overline{\langle a, a \rangle} - \langle a, a \rangle = \lim_{i \in I} \|a_i\|^2 - \|a\|^2,$$

hence $\lim_{i \in I} a_i = a$ in norm iff $\lim_{i \in I} \|a_i\| = \|a\|$.

Moreover, by the Cauchy-Schwarz inequality,

$$\|a\|^2 = \langle a, a \rangle = \lim_{i \in I} \langle a_i, a \rangle = \lim_{i \in I} |\langle a_i, a \rangle| \le \lim_{i \in I} \|a_i\| \, \|a\| \le \|a\| \liminf_{i \in I} \|a_i\|.$$

$\square$

For Hilbert spaces, we can strengthen the conclusion of Cor. 2.24

Theorem 2.29.   (The Projection Theorem) *Let X be a Hilbert space. Let $a \in X$ and $\emptyset \neq C \subset X$ be closed convex.*

Then there is a unique $c \in C$ such that $\mathrm{dist}(a, C) = \|c - a\|$.

*Such c is called the **projection** of a onto C.*

Moreover, $\mathrm{dist}(\cdot, C)$ is a uniformly continuous function.

Let $p : X \to C$ denote the projection onto C. Then p is continuous if C is also compact.

Proof.   By Cor. 2.28, X is reflexive. So it follows from Cor. 2.24 that such $c \in C$ exists.

Let $r := \mathrm{dist}(a, C)$. If there were distinct $c_1, c_2 \in C$ such that $\|c_1 - a\| = \|c_2 - a\| = r$, then since $2^{-1}(c_1 + c_2) \in C$, we have $\|2^{-1}(c_1 + c_2) - a\| \ge r$, and since $c_1, c_2 \in S_X(a, r)$ it follows that $2^{-1}(c_1 + c_2) \in B_X(a, r)$. Therefore it must be the case that $\|2^{-1}(c_1 + c_2) - a\| = r$, *i.e.* $2^{-1}(c_1 + c_2) \in S_X(a, r)$. But, as $c_1 \neq c_2$, it is not an extreme point of $\overline{B_X(a, r)}$, contradicting to Prop. 2.35.

For any $x, y \in {}^*X$, if $x \approx y$, we have

$${}^*\mathrm{dist}(x, {}^*C) \le {}^*\|x - {}^*p(y)\| \le {}^*\|x - y\| + {}^*\|y - {}^*p(y)\| \approx {}^*\mathrm{dist}(y, {}^*C).$$

By interchanging x, y we have ${}^*\mathrm{dist}(y, {}^*C) \lesssim {}^*\mathrm{dist}(x, {}^*C)$, hence ${}^*\mathrm{dist}(y, {}^*C) = {}^*\mathrm{dist}(x, {}^*C)$. Therefore $\mathrm{dist}(\cdot, C)$ is uniformly continuous.

Now suppose C is also compact. Let $a \in X$ and $x \in {}^*X$ such that $x \approx a$. Since C is compact, ${}^*p(x) \approx c$ for some $c \in C$. But

$$\|a - c\| = {}^\circ\left({}^*x - {}^*p(x) \right) = {}^\circ\left({}^*\mathrm{dist}(x, {}^*C) \right) = \mathrm{dist}(a, C) = \|a - p(a)\|,$$

so $c = p(a)$ by uniqueness, hence ${}^*p(x) \approx p(a)$. Therefore p is continuous at every $a \in X$.

$\square$

In particular, in a Euclidean space $\mathbb{R}^n$, $n \in \mathbb{N}$, every point has a unique projection onto a nonempty closed convex set.

In the Projection Theorem above, if C is a closed subspace $Y \subset X$, the mapping $X \to Y$ that takes each $a \in X$ to such the unique $c \in Y$ is denoted by π_Y and called the *orthogonal projection*.

Here orthogonality is taken in the following sense: By Thm. 2.27, let S_0 be a basis of Y extending to a basis S of X. Then, *w.r.t.* the basis S, each $a \in X$ is represented uniquely by the Parseval Identity as $c + b$ where $c = \sum_{y \in S_0} \langle a, y \rangle y$ and $b = \sum_{y \in S \setminus S_0} \langle a, y \rangle y$. Then it is not hard to see that $c = \pi_Y(a)$ and $b = (a - \pi_Y(a)) \in Y^\perp$.

Furthermore, we have the following observations which are left as an exercise. Recall the definition of a projection given on p.103.

Proposition 2.38. *Let X be a Hilbert space and $p \in \mathcal{B}(X)$ be a projection such that $p[X] = Y$, then Y be a closed subspace of X and $p = \pi_Y$.* $\square$

Proposition 2.39. *Let Y be a closed subspace of a Hilbert space X. Then*

 (i) *$\pi_Y : X \to Y$ is linear and $\|\pi_Y\| = 1$, hence $\pi_Y \in \mathcal{B}(X)$.*
 (ii) *$(\pi_Y) \circ (\pi_Y) = \pi_Y$, i.e. π_Y is a projection.*
 (iii) *$Y^\perp = \mathrm{Ker}(\pi_Y)$.*
 (iv) *$\pi_{Y^\perp} = 1 - \pi_Y$ (where 1 is the identity mapping).*
 (v) *Let $f \in \mathcal{B}(X)$. If $\forall x \in X \left((x - f(x)) \in Y^\perp \right)$ then $f = \pi_Y$.*
 (vi) *Y is complemented.* $\square$

We have seen on p.104 that projections have norm at least one, but could be arbitrarily large for projections on general Banach spaces. However, by Prop. 2.20 and 2.39, we have the following.

Corollary 2.31. *Let X be a Hilbert space and $p \in \mathcal{B}(X)$ be a projection. Then $\|p\| = 1$.* $\square$

2.6.2 *Examples*

All finite-dimensional inner-product spaces are Hilbert spaces.

Example 2.7.

- For each $n \in \mathbb{N}$, the inner product space with $\langle x, y \rangle := \sum_{k=1}^{n} x_k \bar{y}_k$ on $\mathbb{F}^n$ forms a Hilbert space. So every finite dimensional inner-product space is a Hilbert space.

- The Lebesgue space $L_2(\mu)$ forms a Hilbert space with inner-product given by

$$\langle f, g \rangle := \int f(\omega)\bar{g}(\omega)\, d\mu(\omega).$$

So by the Riesz Representation Theorem (Thm. 2.28), bounded linear functionals on $L_2(\mu)$ are precisely mappings of the form

$$L_2(\mu) \ni f \mapsto \int f(\omega)\bar{g}(\omega)\, d\mu(\omega), \text{ for some } g \in L_2(\mu).$$

Note that the other Lebesgue spaces $L_p(\mu)$ for $p \neq 2$ fail to be inner-product spaces. In particular, the parallelogram law fails in these space.

- The set of absolutely continuous functions $f : [0, 1] \to \mathbb{C}$ with $f(0) = 0$ and $f' \in L_2(\mu)$, where μ is the Lebesgue measure forms a Hilbert space under the inner-product

$$\langle f, g \rangle := \int_0^1 f'(t)\overline{g'}(t)\, d\mu(t).$$

- The Sobolev spaces $W^{k,2}(\Omega)$, where $k \in \mathbb{N}$ and Ω is open in some $\mathbb{R}^n$, are Hilbert spaces.

- The Hardy space H_2 forms a Hilbert space. H_2 consists of complex functions analytic on the open unit disc having a finite norm given by

$$\|f\| := \lim_{r \to 1^-} \left(\frac{1}{2\pi} \int_{-\pi}^{\pi} \left| f(re^{i\theta}) \right|^2 d\theta \right)^{1/2}, \quad f \in H_2$$

and inner product given by

$$\langle f, g \rangle := \lim_{r \to 1^-} \left(\frac{1}{2\pi} \int_{-\pi}^{\pi} f(re^{i\theta})\overline{g(re^{i\theta})}\, d\theta \right), \quad f, g \in H_2.$$

$\square$

Note that by Thm. 2.27 (iv), all separable Hilbert spaces are isometrically isomorphic to ℓ_2.

Moreover, for any infinite dimensional Hilbert space X, we have the following isometric embeddings:

$$X \subset \widehat{\ell_2} \subset \widehat{{}^*X}.$$

(Here X and ℓ_2 are over the same $\mathbb{F}$.)

We also remark that by Thm. 2.27 (iv) again, because of κ-saturation, all infinite dimensional Hilbert space (over the same $\mathbb{F}$) which are nonstandard hulls have a dimension at least κ.

Similar to those described on p.91, a Hilbert space can be constructed by a **direct sum** of a family of Hilbert spaces as follows.

Let X_i, $i \in I$, be a family of Hilbert spaces over the same field. Define

$$\bigoplus_{i \in I} X_i := \left\{ x \mid x \in \prod_{i \in I} X_i \wedge \sum_{i \in I} \|x_i\|^2 < \infty \right\}$$

with coordinatewise linear operations. It can be verified that $\bigoplus_{i \in I} X_i$ forms a Hilbert space under the inner product given by

$$\langle x, y \rangle := \sum_{i \in I} \langle x_i, y_i \rangle, \qquad x, y \in \bigoplus_{i \in I} X_i.$$

Note that each element in $\bigoplus_{i \in I} X_i$ contains only countably many nonzero coordinates.

2.6.3 Notes and exercises

A pre-inner product is also called a **semi-inner product** and is the same as a positive semi-definite Hermitian sesquilinear form.

Note that it is common for physicists to adopt a definition of pre-inner product where the linearity and antilinearity are switched, *i.e.* antilinear in the first variable and linear in the second.

Hilbert spaces can be viewed as strong infinite dimensional analogue of Euclidean spaces. Such study was first taken by David Hilbert in early 20th century.

Types of Hilbert spaces are not as varied as Banach spaces in general— as Thm. 2.27 indicates that there is only one Hilbert space over $\mathbb{F}$ for each dimension. In particular, in the setting of this book, all nonstandard hulls of Hilbert spaces over $\mathbb{F}$ are the same. This is basically due to the simple geometry features that Hilbert spaces inherited from the Euclidean spaces. However, operators on Hilbert spaces turn out to be rather complex and rich in structure which we will investigate in the following section and in the next chapter.

On the other hand, although Hilbert spaces over $\mathbb{F}$ of a given dimension are unique up to isometric isomorphism, some are more convenient to work with than the other, depending on the problems on hand. For example, the complex Hilbert spaces $L_2([0, 2\pi])$ and $\ell_2(\mathbb{Z})$ are isometrically isomorphic through the Fourier transform and the Fourier transform is an important tool to help people to jump between the two representations and solve problems, including those from sciences.

The unique separable Hilbert space turns out to be fundamental in quantum physics as $\mathbb{C}$ or $\mathbb{R}$ in classical physics. Von Neumann was the one who initiated the use of this Hilbert space to represent states in quantum mechanics.

See §4.1 for more important applications of the identities (2.16) and (2.17) in Thm. 2.26.

Cor. 2.27 can be verified more straightforwardly by using the pre-inner product. A special case for $\mathbb{R}^3$ appeared as a Putnam Mathematical Competition problem in 1968.

EXERCISES

(1) Verify Thm. 2.25 and supply details for the proof of Prop. 2.34.
(2) (The Pythagorean Theorem) Show that if $a_1, \ldots, a_n$, $n \in \mathbb{N}$ are pairwise orthogonal in a Hilbert space, then $\left\| \sum_{i=1}^{n} a_i \right\|^2 = \sum_{i=1}^{n} \|a_i\|^2$.
(3) Prove Thm. 2.27.
(4) Show that isometries between Hilbert spaces need not be isomorphisms.
(5) Show that for a Hilbert space, a basis is a Hamel basis iff the space is finite dimensional.
(6) Give a direct proof of Thm. 2.29 using the parallelogram law instead of quoting Cor. 2.24.
(7) Show that in Thm. 2.29 the uniqueness fails for reflexive spaces in general.
(8) Complete the proof of Prop. 2.39.

2.7   Miscellaneous Topics

The dessert menu includes compact operators, extreme points, bases
and fixed points.

2.7.1   *Compact operators*

Let X, Y be normed linear spaces and $A \subset X$. A function $f : A \to Y$ is
called **compact** if $\overline{f[B]}$ is compact whenever $B \subset A$ is bounded.

A **compact operator** $f : X \to Y$ is a compact function which is also
linear. Note immediately that a linear function $f : X \to Y$ is a compact
operator iff $\overline{f[B_X]}$ is compact.

We mainly deal with compact operators on Hilbert spaces.

Proposition 2.40. *Let X, Y be normed linear spaces and $f : X \to Y$ be
linear. Then the following are equivalent:*

(i) f *is compact*
(ii) $\forall x \in \mathrm{Fin}(^*X)\ \exists y \in Y\ \left({}^*f(x) \approx y \right)$, *i.e.* ${}^*f[\mathrm{Fin}(^*X)] \subset \mathrm{ns}[^*Y]$.
(iii) $\widehat{{}^*f[^*X]} \subset Y$.

Proof.   ((ii) $\Leftrightarrow$ (iii)) is clear from the nonstandard hull construction.
((i) $\Rightarrow$ (ii)) : Without loss of generality, consider $x \in B_{*X}$. So ${}^*f(x) \in$
${}^*f[B_X]$. Apply Robinson's compactness characterization to the compact set
$\overline{f[B_X]}$, we see that there is $y \in \overline{f[B_X]}$ such that ${}^*f(x) \approx y$.
((ii) $\Rightarrow$ (i)) : To show that $\overline{f[B_X]}$ is compact by Robinson's characterization, let $a \in {}^*\overline{f[B_X]}$, and find $c \in \overline{f[B_X]}$ such that $a \approx c$. Note that
$a \approx {}^*f(b)$ for some $b \in B_{*X}$. In particular, $b \in \mathrm{Fin}(^*X)$, so ${}^*f(b) \approx c$ for
some $c \in Y$. Such c must belong to $\overline{f[B_X]}$, for otherwise $\mathrm{dist}(c, \overline{f[B_X]}) > 0$
and, by transfer, we have $a \not\approx c$, a contradiction.   □

We remark that by Prop. 2.13 (x), another consequence of Prop. 2.40
is that a compact operator $f : X \to Y$ must be an element in $\mathcal{B}(X, Y)$.
The collection of compact operators in $\mathcal{B}(X, Y)$ is denoted by $\mathcal{B}_c(X, Y)$.
Likewise we write $\mathcal{B}_c(X)$ when $X = Y$.

Corollary 2.32. *For a normed linear space X and a Banach space Y over
the same field, $\mathcal{B}_c(X, Y)$ is a closed subspace of $\mathcal{B}(X, Y)$.*

Proof. $\mathcal{B}_c(X, Y)$ is clearly a linear space, we only need to show that it is
closed.

Let $f_n \in \mathcal{B}_c(X, Y)$, $n \in \mathbb{N}$, such that $f_n \to f$ in $\mathcal{B}(X, Y)$.

Let $x \in B_{*X}$. For each $n \in \mathbb{N}$, let $m \in \mathbb{N}$ so that $\|f - f_m\| < (2n)^{-1}$, hence $\|{}^*f(x) - {}^*f_m(x)\| < (2n)^{-1}$. By $f_m \in \mathcal{B}_c(X, Y)$ and Prop. 2.40, choose $y_n \in Y$ such that $\|y_n - {}^*f_m(x)\| < (2n)^{-1}$. Hence

$$\|{}^*f(x) - y_n\| < n^{-1}$$

which implies that $\{y_n\}_{n \in \mathbb{N}}$ is Cauchy and, since Y is complete, $y_n \to y$ for some $y \in Y$. Then we have ${}^*f(x) \approx y \in Y$.

The same conclusion holds for all $x \in \mathrm{Fin}({}^*X)$, so another application of Prop. 2.40 shows that $f \in \mathcal{B}_c(X, Y)$. $\qquad \square$

Corollary 2.33. *Let X be a normed linear space. Then $1 \in \mathcal{B}_c(X)$ (i.e. the identity mapping is compact) iff* $\dim(X) < \infty$.

Proof. By Cor. 2.1, $\dim(X) < \infty$ iff $\mathrm{ns}({}^*X) = \mathrm{Fin}({}^*X)$. Now note that the latter is equivalent to $1 \in \mathcal{B}_c(X)$ by Prop. 2.40. $\qquad \square$

Proposition 2.41. *Let X, Y be normed linear spaces and $f \in \mathcal{B}(X, Y)$. Then $f \in \mathcal{B}_c(X, Y)$ iff $\widehat{{}^*f} \in \mathcal{B}_c(\widehat{{}^*X}, \widehat{{}^*Y})$.*

Proof. ($\Rightarrow$) : We have the following:

$$\widehat{{}^*f}[B_{\widehat{{}^*X}}] = \widehat{{}^*f[{}^*B_X]} = \left(\overline{{}^*f[{}^*B_X]}\right)^{\wedge} = \left({}^*(\overline{f[B_X]})\right)^{\wedge} = \widehat{\overline{f[B_X]}} = \overline{f[B_X]},$$

where the second last equality comes from the the Robinson's characterization of compactness for $\overline{f[B_X]}$.

In particular, $\widehat{{}^*f}[B_{\widehat{{}^*X}}]$ is compact in $\widehat{{}^*Y}$.

($\Leftarrow$) is clear, as $\overline{f[B_X]}$ is a closed subset of $\widehat{{}^*f}[B_{\widehat{{}^*X}}]$ which is compact in $\widehat{{}^*Y}$. $\qquad \square$

A related result is the following.

Proposition 2.42. *Let X, Y be normed linear spaces and $f \in {}^*\mathcal{B}({}^*X, {}^*Y)$. Suppose $f[\mathrm{Fin}({}^*X)] \subset \mathrm{ns}({}^*Y)$. Then $(\widehat{f} \restriction_X) \in \mathcal{B}_c(X, Y)$.*

Proof. Let $g = \widehat{f} \restriction_X$. Then

$$g[B_X] = \widehat{f}[B_X] \subset \widehat{f}[B_{\widehat{{}^*X}}] = \mathrm{st}[f[B_{*X}]] \subset Y.$$

By Prop. 1.23, $\mathrm{st}[f[B_{*X}]]$ is compact, therefore $\overline{g[B_X]}$ is compact, *i.e.* g is a compact operator. $\qquad \square$

Note that given normed linear spaces X, Y, Z with $f \in \mathcal{B}_c(Y, Z)$ and $g \in \mathcal{B}(X, Y)$ then $fg \in \mathcal{B}_c(X, Z)$. Likewise, as a consequence of Prop. 2.42, we have:

Corollary 2.34. *Let X, Y, Z be normed linear spaces.*
Suppose that $f \in {}^*\mathcal{B}({}^*Y, {}^*Z)$ satisfies $f[\mathrm{Fin}({}^*Y)] \subset \mathrm{ns}({}^*Z)$. Then for any given $g \in \mathrm{Fin}({}^*\mathcal{B}({}^*X, {}^*Y))$, we have $(\widehat{fg} \restriction_X) \in \mathcal{B}_c(X, Z)$. $\square$

As an application, we can produce compact operators from internal powers of a compact operator.

Corollary 2.35. *Let X be a normed linear space and $f \in \mathcal{B}_c(X)$. Let $N \in {}^*\mathbb{N}$. Suppose ${}^*f^N \neq 0$, define $g := \left\| {}^*f^N \right\|^{-1} {}^*f^{N+1}$ and $h := \widehat{g} \restriction_X$. Then $h \in \mathcal{B}_c(X)$.*

Proof. The conclusion follows from an application of Cor. 2.34 with $X = Y = Z$ to the operators ${}^*f$ and $\left\| {}^*f^N \right\|^{-1} {}^*f^N$. $\square$

Given a normed linear space and $f \in \mathcal{B}(X)$, we define $f^{[0]} := f^0 = 1$, the identity operator, and for each $0 < n \in {}^*N$ we define

$$f^{[n]} := \begin{cases} \left(\left\| {}^*f^n \right\|^{-1} {}^*f^{n+1} \right)^{\wedge} \restriction_X & \text{if } {}^*f^n \neq 0 \\ 0 & \text{otherwise.} \end{cases}$$

So, whenever $f \in \mathcal{B}_c(X)$, the $f^{[n]}$ are not just elements of $\mathcal{B}(X, \widehat{X})$, they are elements of $\mathcal{B}(X)$.

Given $f \in \mathcal{B}(X)$, the symbol $\{f\}'$ denotes the **commutant** of f, *i.e.*

$$\{f\}' := \left\{ g \in \mathcal{B}(X) \,\middle|\, fg = gf \right\}.$$

Then note that $\overline{\mathrm{Lin}}(\{f^{[n]} \mid n \in {}^*\mathbb{N}\}) \subset \{f\}'$ for every $f \in \mathcal{B}_c(X)$.
Moreover, Cor. 2.35 implies that $\overline{\mathrm{Lin}}(\{f^{[n]} \mid n \in {}^*\mathbb{N}\}) \subset \mathcal{B}_c(X)$ whenever $f \in \mathcal{B}_c(X)$.

Note also that for $f \in \mathcal{B}(X)$, if ${}^*f^n = 0$ for some $n \in {}^*\mathbb{N}$, then, by transfer, we must have $f^n = 0$ for some $n \in \mathbb{N}$, *i.e.* f is **nilpotent**.

Non-nilpotent compact operators do exist.

Proposition 2.43. *Let X be an infinite dimensional Banach space and $f \in \mathcal{B}_c(X)$. Then $\widehat{{}^*f}[\widehat{X}] \not\supset X$.*

Proof. By Prop. 2.40, we have $\widehat{{}^*f}[\widehat{{}^*X}] \subset X$. Suppose $\widehat{{}^*f}[\widehat{{}^*X}] \supset X$, then we have $\widehat{{}^*f}[\widehat{{}^*X}] = X$. Hence by the Open Mapping Theorem (Thm. 2.3), $\widehat{{}^*f} \in \mathcal{B}(\widehat{{}^*X}, X)$ is an open mapping. By Prop. 2.17, there is $r \in \mathbb{R}^+$ so that $\widehat{{}^*f}[B_{\widehat{{}^*X}}] \supset r\bar{B}_X$. Since f is compact, the proof of Prop. 2.41 shows that $\widehat{{}^*f}[B_{\widehat{{}^*X}}]$ is compact, therefore $\bar{B}_X$ is compact. Then X must be finite dimensional (Cor. 2.2), a contradiction. □

We leave as an exercise to show the following

Proposition 2.44. *Let X, Y be Banach spaces, where X is reflexive. Let $f \in \mathcal{B}(X)$. Then $f \in \mathcal{B}_c(X)$ iff it holds for every net $\{a_i\}_{i \in I} \subset X$ and $a \in X$ that $a_i \to_w a$ implies $\lim_{i \in I} f(a_i) = f(a)$.* □

In general, given Banach spaces X, Y, $f \in \mathcal{B}(X, Y)$ is said to be **completely continuous** if the weak convergence condition in the above proposition is satisfied. It is easy to check that compact operators are always completely continuous, but the converse may fail if the target space is not reflexive.

Example 2.8. Recall the Loeb measure theory in §1.5.2 and §1.5.3.

Let Ω be a hyperfinite set and μ an internal probability measure on Ω. Consider real Hilbert spaces

$$X = L_2(\Omega, \mathcal{B}_1, L(\mu)) \quad \text{and} \quad Y = L_2(\Omega^2, \mathcal{B}_2, L(\mu^2)),$$

where $\mathcal{B}_1 = L({}^*\mathcal{P}(\Omega))$ and $\mathcal{B}_2 = L({}^*\mathcal{P}(\Omega^2))$ are the Loeb algebras.

We remark that in general $\left(L(\mu)\right)^2$-measurability implies $L(\mu^2)$-measurability but not the other way around.

Fix any SL^2-function $g : \Omega^2 \to {}^*\mathbb{R}$.

For each $f \in X$, let $F : \Omega \to {}^*\mathbb{R}$ be given by

$$F(t) = \sum_{s \in \Omega} g(s, t) \, {}^*f(s) \, \mu(\{s\}), \quad t \in \Omega.$$

It is easy to see that F is SL^2 hence $°F \in X$.

Now define $T : X \to X$ by $T(f) := °F$ for each f and F as above.

It can be verified that $T \in \mathcal{B}(X)$ and T is compact. □

Examples from L^2-spaces with ordinary probability measure can be obtained from the above example via liftings. For example, it follows that the Volterra operator $T : L^2([0,1]) \to L^2([0,1])$, with $T(f)(t) = \int_0^t f(s) d\text{Leb}(s)$, $f \in L^2([0,1])$, is compact.

For $f, g \in \mathcal{B}(X)$, we write fg for the composition $f \circ g$ and note that $fg \in \mathcal{B}(X)$. With this product, $\mathcal{B}(X)$ turns into an algebra.

A linear operator $f : X \to Y$ between normed linear spaces is said to have **finite rank** if $\dim(f[X]) < \infty$. *i.e.* the range is a finite dimensional subspace. Note also that operators in $\mathcal{B}(X,Y)$ of finite rank are compact, but there are unbounded linear operators of finite rank.

The following shows that compact operator on a Hilbert space behaves like a *finite rank operator.

Proposition 2.45. *Let X be a Hilbert space and $f \in \mathcal{B}(X)$. Then the following are equivalent:*

(i) $f \in \mathcal{B}_c(X)$.

(ii) *For some $g \in {}^*\mathcal{B}_c(X)$ of *finite rank, we have $^*f \approx g$.*

(iii) *For any $\epsilon \in \mathbb{R}^+$, there is $g \in \mathcal{B}(X)$ of finite rank with $\|f - g\| < \epsilon$.*

Proof. ((i) $\Rightarrow$ (ii)) : Since $\overline{f[B_X]}$ is compact, it is separable (Ex. 13 on p.75). By Thm. 2.27 (iv), separable Hilbert spaces are isometrically isomorphic to ℓ_2, so we assume with loss of generality that $f[X] \subset \ell_2 \subset X$.

Let $\pi_n : \ell_2 \to \ell_2$ be the projection of the first n coordinates, *i.e.* for $a = \{a_n\}_{n \in \mathbb{N}} \in \ell_2$, $\pi_n(a)_m = a_m$ if $m \leq n$ and $\pi_n(a)_m = 0$ otherwise.

Let $N \in {}^*\mathbb{N} \setminus \mathbb{N}$.

For any $b \in B_{^*X}$, by Prop. 2.40, there is $a = \{a_n\}_{n \in \mathbb{N}} \in \ell_2$, such that $^*f(b) \approx {}^*a$. Hence also $\pi_N {}^*f(b) \approx \pi_N({}^*a)$ and

$$\|({}^*f - \pi_N {}^*f)(b)\|^2 \approx \|{}^*a - \pi_N({}^*a)\|^2 \approx \sum_{m > N} a_m^2 \approx 0.$$

Therefore $\|{}^*f - \pi_N {}^*f\| \approx 0$.

Take $g = \pi_N {}^*f$, then $g \in {}^*\mathcal{B}_c(X)$ is of *finite rank and $^*f \approx g$.

((ii) $\Rightarrow$ (iii)) : Given *finite rank $h \in {}^*\mathcal{B}_c(X)$ with $^*f \approx h$, for any $\epsilon \in \mathbb{R}^+$ we have $\|{}^*f - h\| < \epsilon$. Transfer this, we conclude that for some $g \in \mathcal{B}_c(X)$ of finite rank, $\|f - g\| < \epsilon$.

((iii) $\Rightarrow$ (i)) : We already noted that finite rank operators in $\mathcal{B}(X)$ are compact, so it follows from Cor. 2.32 that if there are finite rank $f_n \in \mathcal{B}_c(X)$, such that $f_n \to f$, then $f \in \mathcal{B}_c(X)$. $\square$

We now define the ***adjoint*** of a linear bounded linear operator between Hilbert spaces.

Proposition 2.46. *Let X, Y be Hilbert spaces. Then for each $f \in \mathcal{B}(X,Y)$ there is a unique element in $\mathcal{B}(Y,X)$, denoted by f^*, such that the mapping $* : \mathcal{B}(X,Y) \to \mathcal{B}(Y,X)$ given by $f \mapsto f^*$ is continuous and satisfies*

(i) $\forall f \in \mathcal{B}(X,Y) \ \forall x \in X \ \forall y \in Y \ (\langle f(x), y \rangle_Y = \langle x, f^*(y) \rangle_X$.

(ii) $\forall f, g \in \mathcal{B}(X,Y) \ \forall \alpha \in \mathbb{F} \ ((f + \alpha g)^* = (f^* + \bar{\alpha} g^*))$.

(iii) $\forall f \in \mathcal{B}(X,Y) \left((f^*)^* = f \right)$.

(iv) $\forall f \in \mathcal{B}(X,Y) \left(\|f\| = \|f^*\| = \sqrt{\|f^* \circ f\|} \right)$.

    Such unique f^* is called the adjoint of f.

    Moreover, $\forall f,g \in \mathcal{B}(X,Y) \left((fg)^* = g^* f^* \right)$.

Proof.   Let $y \in Y$. Let $\theta : X \to \mathbb{F}$ be given by $X \ni x \mapsto \langle f(x), y \rangle_Y$. Then θ is linear and $\sup_{x \in B_X} |\langle f(x), y \rangle_Y| \le \|f\| \, \|y\|$, therefore $\theta \in X'$.

So $\theta = \phi_z$ for some unique $z \in Y$ as given in the Riesz Representation Theorem (Thm. 2.28). We define $f^*(y)$ to be such $z \in Y$.

Then for any $x \in X$ and $y \in Y$ we have

$$\langle f(x), y \rangle_Y = \phi_{f^*(y)}(x) = \langle x, f^*(y) \rangle_X$$

and thus (i) holds.

(ii) follows from $\phi_{x+\alpha y} = \phi_x + \bar{\alpha} \phi_y$ for any $x, y \in X$ and $\alpha \in \mathbb{F}$.

For (iii), let $x \in X$, then we have for any $z \in X$ that

$$\phi_{f^{**}(x)}(z) = \langle z, (f^*)^*(x) \rangle = \overline{\langle f^*(z), x \rangle}$$
$$= \overline{\langle x, f^*(z) \rangle} = \overline{\langle f(x), z \rangle} = \langle z, f(x) \rangle = \phi_{f(x)}(z),$$

so, by applying the Riesz Representation Theorem, $(f^*)^* = f$.

For (iv), let $f \in \mathcal{B}(X,Y)$ and consider the following:

$$\|f\|^2 = \sup_{x \in B_X} \|f(x)\|^2 = \sup_{x \in B_X} |\langle f(x), f(x) \rangle| = \sup_{x \in B_X} |\langle x, f^*(f(x)) \rangle|$$
$$\le \sup_{x \in B_X} \|f^*(f(x))\| \le \|f^*\| \sup_{x \in B_X} \|f(x)\|$$
$$= \|f^*\| \, \|f\|. \tag{2.21}$$

So $\|f\| \le \|f^*\|$ and hence also $\|f^*\| \le \|(f^*)^*\| = \|f\|$, by (iii). *i.e.* $\|f\| = \|f^*\|$. Then (2.21) gives

$$\|f\|^2 \le \sup_{x \in B_X} \|f^*(f(x))\| = \|f^* \circ f\| \le \|f\|^2.$$

To see the continuity of the adjoint mapping, let $f \in \mathcal{B}(X,Y)$, then

$$^*f \approx 0 \;\Leftrightarrow\; \forall x \in B_{*X} \left({}^*f(x) \approx 0 \right)$$
$$\Leftrightarrow\; \forall x \in B_{*X} \, \forall y \in B_{*Y} \left(\langle {}^*f(x), y \rangle_Y \approx 0 \right)$$
$$\Leftrightarrow\; \forall x \in B_{*X} \, \forall y \in B_{*Y} \left(\langle x, {}^*f^*(y) \rangle_X \approx 0 \right)$$
$$\Leftrightarrow\; \forall y \in B_{*Y} \left({}^*f^*(y) \approx 0 \right)$$
$$\Leftrightarrow\; {}^*f^* \approx 0.$$

Finally, let $f, g \in \mathcal{B}(X)$ and $x, y \in X$ then

$$\langle x, (fg)^*(y) \rangle = \langle (fg)(x), y \rangle = \langle g(x), f^*(y) \rangle = \langle x, (g^* f^*)(y) \rangle,$$

i.e. $(fg)^* = g^* f^*$. $\qquad\qquad\qquad\qquad\qquad\qquad\qquad\qquad\qquad\qquad\square$

Despite the unsightliness, we trust that the various possibly ways of placing the $*$ sign would not cause much confusion.

Note that for finite dimensional Hilbert spaces, the adjoint of a linear operator can be identified with the **Hermitian transpose** (*i.e. conjugate transpose*) of a square matrix representing the operator.

Now we add another property to Prop. 2.39 with the following characterization of the orthogonal projection operator.

Proposition 2.47. *Let X be a Hilbert space and $p \in \mathcal{B}(X)$. Then p is an orthogonal projection iff $p = p^* = p^2$.*

Proof. ($\Rightarrow$) : If $p = \pi_Y$ for some closed subspace $Y \subset X$, then clearly $p^2 = p$. The fact that $p = p^*$ follows from the Parseval's Identity (Thm. 2.27).

($\Leftarrow$) : Suppose $p = p^* = p^2$. Let $Y = p[X]$. One sees that Y is closed because whenever $a \in X$ is such that $a \approx {}^*p(b) \in {}^*Y$ for some $b \in {}^*X$, then $p(a) \approx {}^*p^2(b) = {}^*p(b) \approx a$, implying $a = p(a)$, *i.e.* $a \in p[X] = Y$.

Now one can verify that $p = \pi_Y$: Let $a \in X$ and $b \in Y$. Write $b = p(c)$. Then $\langle p(a), p(c) \rangle = \langle a, p^* \circ p(c) \rangle = \langle a, p^2(c) \rangle = \langle a, p(c) \rangle$, hence

$$\langle a - p(a), b \rangle = \langle a - p(a), p(c) \rangle = \langle a, p(c) \rangle - \langle p(a), p(c) \rangle = 0,$$

i.e. $(a - p(a)) \in Y^\perp$ for every $a \in X$, therefore $p = \pi_Y$. $\square$

When dealing with $\mathcal{B}(X)$ for a Hilbert space X, we often call an orthogonal projection simply a **projection**, so we require such p not only satisfies $p^2 = p$ as on p.103 but also $p = p^*$.

Being a compact operator is a dual property.

Theorem 2.30. *Let X, Y be Hilbert spaces. Let $f \in \mathcal{B}(X, Y)$. Then f is compact iff f^* is.*

Proof. By $(f^*)^* = f$, we only need to prove one direction.

First note that for any $h \in \mathcal{B}(X, Y)$ and $y \in Y$,

$$y \in \operatorname{Ker}(h^*) \Leftrightarrow \forall x \in X \left(\langle x, h^*(y) \rangle_X = 0 \right)$$
$$\Leftrightarrow \forall x \in X \left(\langle h(x), y \rangle_Y = 0 \right) \Leftrightarrow y \in (h[X])^\perp,$$

i.e. $\operatorname{Ker}(h^*) = (h[X])^\perp$. In particular, for any h of finite rank, $h^*[Y]$ is finite dimensional, so h^* is also of finite rank.

Suppose $f \in \mathcal{B}_c(X, Y)$. is compact. Then by Prop. 2.45, for some $g \in {}^*\mathcal{B}_c(X)$ of *finite rank we have ${}^*f \approx g$.

Then ${}^*f^* \approx g^*$. Since g^* is of *finite rank, $f^* \in \mathcal{B}_c(Y, X)$, by Prop. 2.45 again. $\square$

When X, Y are Banach spaces, the **adjoint** of $f \in \mathcal{B}(X, Y)$ is defined as some $f^* \in \mathcal{B}(Y', X')$ such that for $\phi \in Y'$, $f^*(\phi)$ is the functional in X' given by $X \ni x \mapsto \phi(f(x))$. It is straightforward to check that this is well-defined and generalizes the previous definition in the case of Hilbert spaces. We leave it as an exercise to verify that Thm. 2.30 generalizes for the adjoint in Banach space setting.

For a Banach space X and $f \in \mathcal{B}(X)$, a subspace $Y \subset X$ is call an **invariant subspace** for f if $f[Y] \subset Y$. It is called **nontrivial**, if it is different from the zero space $\{0\}$ and X. Usually ones is more interested in closed invariant subspaces, as non-closed ones are easy to get.

Note that Y is invariant for f iff $(\pi_Y f \pi_Y) = f \pi_Y$.

Theorem 2.31. *Let X be a complex Banach space of dimension > 1. Then each $f \in \mathcal{B}_c(X)$ has a nontrivial closed invariant subspace.*

Proof. We assume that $f \neq 0$; for otherwise, trivially, any closed subspace of X would be an invariant subspace.

By saturation, let Y be a hyperfinite dimensional subspace of $^*X$ such that $X \subset Y$.

By Prop. 2.40, since $\widehat{^*f}[\,^*\widehat{X}] \subset X$, we have $\widehat{^*f}[\widehat{Y}] \subset \widehat{Y}$.

Let $g := (\pi_Y \circ {}^*f) \restriction_Y$, then $g \in {}^*\mathcal{B}(Y)$ and $g \approx {}^*f$ on $\mathrm{Fin}(Y)$. Therefore, $\widehat{g} = \widehat{^*f} \restriction_{\widehat{Y}}$ and extends f.

Let $N = {}^*\dim(Y)$. Since $g \in {}^*\mathcal{B}(Y)$, by a transfer of the Jordan Canonical Form Theorem, there is a basis $\{e_1, \ldots, e_N\}$ w.r.t. which g can be represented by an upper triangular matrix.

Therefore, if we let $Y_n := \mathrm{Lin}(\{e_1, \cdots, e_n\})$, then

- $\{0\} = Y_0 \subset \cdots \subset Y_n \subset Y_{n+1} \subset \cdots \subset Y_N = Y$;
- $^*\dim(Y_{n+1}) = {}^*\dim(Y_n) + 1$;
- by the triangular matrix, for $n \leq N$, $g[Y_n] \subset Y_n$, hence $\widehat{g}[\widehat{Y_n}] \subset \widehat{Y_n}$.

Define closed subspaces $Z_n := \widehat{Y_n} \cap X$, $n \leq N$.
Since $\widehat{^*f}[\widehat{X}] \subset X$, we have $\widehat{g}[\widehat{Y}] \subset X$.
Then $f[Z_n] \subset \widehat{^*f}[\widehat{Y_n}] = \widehat{g}[\widehat{Y_n}] \subset \widehat{Y_n}$ and of course $f[Z_n] \subset X$, therefore

$$\forall n \leq N \, (f[Z_n] \subset Z_n).$$

i.e. the Z_n's are closed invariant subspaces for f.

Suppose f has no nontrivial invariant subspace. Then we must have

$$Z_n = \{0\} \quad \text{or} \quad Z_n = X \quad \text{for all } n \leq N. \tag{2.22}$$

Note it follows that $Z_N = X$, because $f \neq 0$ implies

$$\{0\} \neq f[X] = \widehat{g}[X] \subset \widehat{Y} \cap X = Z_N.$$

Also $f \neq 0$ implies that $f(a) \neq 0$ for some $a \in X = Z_N$. Let $r \in \mathbb{R}^+$ such that $\|f(a)\| > r > 0$. By $\widehat{g}(a) = f(a)$, we have $^\circ \|g(^*a)\| > r$.

Now define $\pi_n := \pi_{Y_n}$. Then it holds that

$$\|(g\pi_0)(^*a)\| = 0 < r < {}^\circ \|g(^*a)\| = {}^\circ \|(g\pi_N)(^*a)\|.$$

By $r \in \mathbb{R}^+$, this gives $\|(g\pi_0)(^*a)\| < r < \|(g\pi_N)(^*a)\|$.

So, by N being hyperfinite, there is some $M \leq N$ such that

$$\|(g\pi_{M-1})(^*a)\| \leq r < \|(g\pi_M)(^*a)\|. \tag{2.23}$$

Since $\|\pi_M(^*a)\| \leq \|^*a\| = \|a\| < \infty$, by Prop. 2.40, for some $b \in X$,

$$g\big(\pi_M(^*a)\big) \approx {}^*f\big(\pi_M(^*a)\big) \approx b.$$

As $\|(g\pi_M)(^*a)\| \not\approx 0$, we have $b \neq 0$. Then $Z_M \neq \{0\}$, since

$$b = \widehat{g}\big(\widehat{\pi_M(^*a)}\big) \in \big(\widehat{g}\,[\widehat{Y}_M] \cap X\big) \subset \big(\widehat{Y}_M \cap X\big) = Z_M.$$

Hence $Z_M = X$ according to (2.22).

On the other hand, it cannot be the case that $Z_{M-1} = X$. For otherwise, $a \in Z_{M-1}$, so $\pi_{M-1}(^*a) \approx {}^*a$, then $\|g\pi_{M-1}(^*a)\| \approx \|g(^*a)\| \gtrsim r$, contradicting to (2.23).

So we conclude that $Z_{M-1} = \{0\}$. But $\{0\} = Z_{M-1} \subset Z_M = X$ is an impossibility, since, from $^*\dim(Y_M) = {}^*\dim(Y_{M-1}) + 1$, we would then have $\dim(\widehat{Y}_M) \leq \dim(\widehat{Y}_{M-1}) + 1$ by Prop 2.8. Consequently, $\dim(Z_M) \leq \dim(Z_{M-1}) + 1$, making $\dim(X) \leq 1$, contrary to the assumption.

Therefore we have shown that (2.22) is false. $\qquad\qquad\square$

2.7.2 The Krein-Milman Theorem

In Euclidean spaces, any polyhedron is a closed convex hull of its vertices. By Prop. 2.35, in a Hilbert space, the closed unit ball has the same property as it is the closed convex hull of extreme points, namely points on the sphere.

The Krein-Milman Theorem provides a simple and useful extension of these results.

Recall the definition of a locally convex linear space on p.132. Given a locally convex linear space X, we extend the notation for the dual space of a normed linear space by letting X' to denote the space of continuous linear functions $X \to \mathbb{F}$.

By an application of the Extended Hahn-Banach Theorem (Thm. 2.10) and by following the same line of argument for proving Cor. 2.13, the following general separation result can be shown.

Lemma 2.5. *Let X be a Hausdorff locally convex linear space. Let $a \in X$ and $C \subset X$ be a nonempty and convex subset.*

Suppose $a \notin C$, then $\exists \phi \in X' \left(\operatorname{Re}(\phi)(a) > \sup_{x \in C} \operatorname{Re}(\phi(x)) \right)$.   □

Theorem 2.32. (Krein-Milman Theorem) *Let X be a Hausdorff locally convex linear space. Let $C \subset X$ be nonempty, compact and convex. Let $E \subset C$ be the set of extreme points in C.*

Then $E \neq \emptyset$ and $C = \overline{\operatorname{conv}}(E)$.

Proof.   We first need the following:

CLAIM: Let $K \subset C$ be nonempty, compact and convex and $\phi \in X'$. Define $K^\phi := \{x \in K \mid \operatorname{Re}(\phi(x)) = r\}$, where $r = \max_{x \in K} \operatorname{Re}(\phi(x))$. Then K^ϕ is nonempty, compact and convex.

PROOF OF THE CLAIM: By saturation, let A be a hyperfinite set so that $K \subset A \subset {}^*K$. Let $c \in A$ with ${}^*\operatorname{Re}({}^*\phi(c)) = {}^*\max_{x \in A} {}^*\operatorname{Re}({}^*\phi(x))$. Then

$$\sup_{x \in K} \operatorname{Re}(\phi(x)) \lesssim {}^*\operatorname{Re}({}^*\phi(c)) \leq {}^*\left(\sup_{x \in K} \operatorname{Re}(\phi(x)) \right).$$

By Robinson's characterization of compactness, let $a \in K$ such that $a \approx c$. Hence

$$\operatorname{Re}(\phi(a)) = {}^\circ\left({}^*\operatorname{Re}({}^*\phi(c)) \right) = \sup_{x \in K} \operatorname{Re}(\phi(x)) = \max_{x \in K} \operatorname{Re}(\phi(x))$$

is realized and $r = \max_{x \in K} \operatorname{Re}(\phi(x))$ is defined. In particular $K^\phi \neq \emptyset$.

Let $a \in X$ such that $a \approx c \in {}^*(K^\phi)$ for some c. Then by continuity, $\phi(a) \approx {}^*\phi(c) \approx r$, so $\phi(a) = r$, i.e. $a \in K^\phi$. Therefore K^ϕ is closed. As a subset of K, K^ϕ is compact as well.

Now let $a, b \in K^\phi$ and $t \in [0,1]$. Then $(ta + (1-t)b) \in K$. Since

$$\operatorname{Re}(\phi(ta + (1-t)b)) = t\operatorname{Re}(\phi(a)) + (1-t)\operatorname{Re}(\phi(b)) = r,$$

we have $(ta + (1-t)b) \in K^\phi$. i.e. K^ϕ is convex and the Claim is proved.

By saturation, let H be an internal set of hyperfinite cardinality $N \in {}^*\mathbb{N}$ such that $X' \subset H \subset {}^*X'$. Let $\mathbb{L}$ denote the internal set of all internal bijections $\{1, \cdots, N\} \to H$

For each $\lambda \in \mathbb{L}$, we let $K_0^\lambda := {}^*C$ and for $n < N$, define recursively $K_{n+1}^\lambda := \left(K_n^\lambda\right)^{\lambda(n)}$ as given by transferring the Claim. Let $K^\lambda := \bigcap_{n \leq N} K_n^\lambda$.

Then the Claim implies that $\emptyset \neq K^\lambda \subset {}^*C$ is *compact and *convex. Let $\lambda \in \mathbb{L}$. Since ${}^*C \subset \operatorname{ns}({}^*X)$, $\operatorname{st}[K^\lambda]$ is defined.

Moreover, $\mathrm{st}[K^\lambda]$ is a singleton: For otherwise, there would be $a, b \in K^\lambda$ with $^\circ a \neq {}^\circ b$. By Lem. 2.5, there is $\phi \in X'$ such that $\phi(^\circ a - {}^\circ b) \neq 0$. Rotate, if necessary, we can assume that $\phi(^\circ a - {}^\circ b) \in \mathbb{R}^+$. But $^*\phi = \lambda(n)$ for some $n \leq N$, so a, b could not be both in the same K_n^λ, a contradiction.

Furthermore, the unique element in $\mathrm{st}[K^\lambda]$ is an extreme point of C:

Let $c \in \mathrm{st}[K^\lambda]$, with $c \approx c_0 \in K^\lambda$. Suppose $a, b \in C$ are such that $c = (a + b)/2$. Let $a_0 := {}^*a$ and $b_0 := 2c_0 - {}^*a$.

Then $^\circ a_0 = a$, $^\circ b_0 = b$ and $c_0 = (a_0 + b_0)/2$.

Clearly, $a_0, b_0 \in {}^*C = K_0^\lambda$. Let $0 < n \leq N$. Suppose $a_0, b_0 \in K_{n-1}^\lambda$. Since $c_0 \in K_n^\lambda = \left(K_{n-1}^\lambda\right)^{\lambda(n)}$, $\lambda(n)(c_0) = {}^*\max_{x \in K_{n-1}^\lambda} {}^*\mathrm{Re}\big(\lambda(n)(x)\big)$. Hence, by $c_0 = (a_0 + b_0)/2$, we have $\lambda(n)(a_0) = \lambda(n)(c_0) = \lambda(n)(b_0)$, i.e. $a_0, b_0 \in K_n^\lambda$. Therefore $a_0, b_0 \in K^\lambda$ and thus both $a = {}^*a_0$ and $b = {}^*b_0$ equal to the unique $c \in \mathrm{st}[K^\lambda]$, showing that c is an extreme point of C.

Finally, let $E_0 := \bigcup_{\lambda \in \mathbb{L}} \mathrm{st}[K^\lambda]$, a set of extreme points in C. i.e. $E_0 \subset E$, and so, $\overline{\mathrm{conv}}(E_0) \subset \overline{\mathrm{conv}}(E) \subset C$.

Suppose there is some $a \in \overline{\mathrm{conv}}(E_0) \setminus C$. Then by Lem. 2.5, let $\phi \in X'$ be such that $\mathrm{Re}\big(\phi(a)\big) > \mathrm{Re}\big(\phi(x)\big)$ for all $x \in \overline{\mathrm{conv}}(E_0)$. Let $\lambda \in \mathbb{L}$ be such that $\lambda(1) = {}^*\phi$. Let $c \in \mathrm{st}[K^\lambda]$. So $c \in E_0$, therefore $\mathrm{Re}\big(\phi(a)\big) > \mathrm{Re}\big(\phi(c)\big)$. But this is impossible, as $c \in \mathrm{st}[K_1^\lambda] = \mathrm{st}[{}^*C^{*\phi}]$. It follows that $\mathrm{Re}\big(\phi(c)\big) = \max_{x \in C} \mathrm{Re}\big(\phi(x)\big)$. So such a does not exist. i.e. $C = \overline{\mathrm{conv}}(E_0)$.

Hence we conclude that $C = \overline{\mathrm{conv}}(E)$. $\qquad\square$

By the remark on p.132, we obtain the following consequence.

Corollary 2.36. *Let X be a normed linear space. Let $\emptyset \neq C \subset X'$ be convex and weak* compact. Then C equals the closed convex hull of its extreme points.* $\qquad\square$

2.7.3   Schauder bases

In a Banach space X, a sequence $\{e_n\}_{n \in \mathbb{N}}$ of X is called a **Schauder Basis**, if for every $a \in X$, there is a *unique* sequence $\{\alpha_n\}_{n \in \mathbb{N}}$ of $\mathbb{F}$ such that

$$\sum_{n \in \mathbb{N}} \alpha_n e_n \quad \text{converges and} \quad \sum_{n \in \mathbb{N}} \alpha_n e_n = \lim_{m \to \infty} \sum_{n \leq m} \alpha_n e_n = a.$$

In particular, a Banach space having a Schauder basis is necessarily separable. But an example given by Enfo shows that there are separable Banach spaces with no Schauder basis.

It is clear that the only case where Hamel bases form Schauder bases is when the Banach space is finite dimensional.

Many classical sequence spaces, such as c_0, ℓ_p, where $p \in [1, \infty)$ do have Schauder basis. In fact their standard bases are examples of such. But ℓ_∞ does not have a Schauder basis as the space is not separable.

In a separable Hilbert space, any basis (*i.e.* a maximal orthonormal subset) forms a Schauder basis.

It is easily seen that a Banach space X having a Schauder basis $\{e_n\}_{n \in \mathbb{N}}$ is isometrically isomorphic to the sequence space where an equivalence class of a sequence $\{\alpha_n\}_{n \in \mathbb{N}}$ is given the induced norm $\left\| \sum_{n \in \mathbb{N}} \alpha_n e_n \right\|_X$.

Theorem 2.33.   *Let $\{e_n\}_{n \in \mathbb{N}}$ be a Schauder basis of a Banach space $(X, \|\cdot\|)$. Define $\pi_m : X \to X$, $m \in \mathbb{N}$, by $\sum_{n \in \mathbb{N}} \alpha_n e_n \mapsto \sum_{n \leq m} \alpha_n e_n$. (So $x = \lim_{m \to \infty} \pi_m(x)$ holds for all $x \in X$.)*

Let $\|\cdot\|$ be the mapping on X given by

$$\| \cdot \| := \sup_{n \in \mathbb{N}} \|\pi_n(\cdot)\|.$$

Then the following hold:

(i) $\| \cdot \| : X \to [0, \infty)$ *and forms a norm on X.*

(ii) $\|\cdot\|$ *and* $\| \cdot \|$ *are equivalent norms.*

(iii) $\forall x \in {}^*X \left(({}^*\|x\| \approx 0) \Leftrightarrow ({}^*\|x\| \approx 0) \right)$.

(iv) π_m, $m \in \mathbb{N}$, *are projections in* $\mathcal{B}(X)$.

Proof.   (i) : First of all, $\| \cdot \|$ is finite: for if there is $a \in X$ such that $\|a\| = \infty$, there would be some $m_k \in \mathbb{N}$ such that $\pi_{m_k}(a) \to \infty$ as $k \to \infty$, and hence the representation of a is some divergent series $\sum_{n \in \mathbb{N}} \alpha_n e_n$, a contradiction.

Also, $\|a\| = \lim_{m \to \infty} \|\pi_m(a)\| \leq \sup_{m \in \mathbb{N}} \|\pi_m(a)\|$, therefore $\|\cdot\| \leq \| \cdot \|$.

In particular, $\forall x \in X \left(\|x\| = 0 \Leftrightarrow (x = 0) \right)$.

It is straightforward to check that $\| \cdot \|$ is subadditive and homogeneous, therefore it forms a norm on X.

(ii) : We first show that X is complete *w.r.t.* $\| \cdot \|$.

Let $\{c_n\}_{n \in \mathbb{N}} \subset X$ be Cauchy *w.r.t.* $\| \cdot \|$. Then for each $\epsilon \in \mathbb{R}^+$, there is $n_\epsilon \in \mathbb{N}$, such that $\|c_{n_1} - c_{n_2}\| < \epsilon$ holds for all $n_\epsilon \leq n_1, n_2 \in \mathbb{N}$. Write

$$a_{nm} := \pi_m(c_n), \quad n, m \in \mathbb{N}.$$

Then for any $\epsilon \in \mathbb{R}^+$ and $n_\epsilon \leq n_1, n_2 \in \mathbb{N}$, we have for any $m \in \mathbb{N}$ that

$$\|a_{n_1 m} - a_{n_2 m}\| = \|\pi_m(c_{n_1} - c_{n_2})\| \leq \|c_{n_1} - c_{n_2}\| < \epsilon. \tag{2.24}$$

Therefore, (1.5) in Thm. 1.18 is satisfied. Hence, by Thm. 1.18, for any $N, M \in ({}^*\mathbb{N} \setminus \mathbb{N})$,

$$\lim_{m \to \infty} \lim_{n \to \infty} a_{nm} = \lim_{m \to \infty} {}^\circ({}^*a_{Nm}) = {}^\circ({}^*a_{NM}) \in X,$$

where the limits and standard parts are taken $w.r.t.$ $\|\cdot\|$.

Fix any $N, M \in ({}^*\mathbb{N} \setminus \mathbb{N})$ and let $c := {}^\circ({}^*a_{NM})$.

Then, for all $m \in \mathbb{N}$, we have $\pi_m(c) = {}^\circ({}^*a_{Nm})$, because

$$c = {}^\circ({}^*a_{Nm}) + \underbrace{\lim_{k\to\infty}\left({}^\circ({}^*a_{Nk}) - {}^\circ({}^*a_{Nm})\right)}_{\in\,\overline{\mathrm{Lin}}(\{e_k \mid m<k\in\mathbb{N}\})}.$$

Therefore, for any $m \in \mathbb{N}$, $\epsilon \in \mathbb{R}^+$ and $n_\epsilon \leq n \in \mathbb{N}$, we have by transferring (2.24) that

$$\|\pi_m(c - c_n)\| = \|{}^\circ({}^*a_{Nm}) - a_{nm}\| = {}^\circ\left({}^*\|{}^*a_{Nm} - {}^*a_{nm}\|\right) < \epsilon.$$

In other words, $\|c - c_n\| \to 0$ as $n \to \infty$.

Hence $\{c_n\}_{n\in\mathbb{N}} \subset X$ has a limit in X $w.r.t.$ $\|\cdot\|$.

Therefore X is complete $w.r.t.$ $\|\cdot\|$.

Let $f : (X, \|\cdot\|) \to (X, \|\cdot\|)$ be the identity mapping. This linear mapping f is bounded because of $\|\cdot\| \leq \|\cdot\|$ from the proof of (i). $i.e.$ $f \in \mathcal{B}\left((X, \|\cdot\|), (X, \|\cdot\|)\right)$.

Therefore, by Cor. 2.5, $\|\cdot\|$ and $\|\cdot\|$ are equivalent.

(iii) : By Prop. 2.3 and (ii).

(iv) : By (ii), let $k \in \mathbb{R}^+$ so that $\|\cdot\| \leq k\|\cdot\|$. (By Prop. 2.4.)

Let $m \in \mathbb{N}$. For each $a \in X$ with $\|a\| = 1$,

$$\|\pi_m(a)\| \leq \|a\| \leq k\|a\| \leq k.$$

In particular, $\pi_m \in \mathcal{B}(X)$.

Moreover, π_m is clearly a projection onto the finite dimensional subspace $\mathrm{Lin}(\{e_0, \cdots, e_m\})$, a complemented closed subspace of X. $\qquad\square$

The following shows that Banach spaces with a Schauder basis behaves like ℓ_1.

Corollary 2.37. *Let $\{e_n\}_{n\in\mathbb{N}}$ be a Schauder basis of a Banach space X. Then $\mathrm{ns}({}^*X) =$*

$$\mathrm{Fin}({}^*X) \cap \left\{ \sum_{n\in\,{}^*\mathbb{N}} \alpha_n\,{}^*e_n \;\middle|\; (\{\alpha_n\}_{n\in\,{}^*\mathbb{N}} \subset {}^*\mathbb{F}) \wedge \forall N \in ({}^*\mathbb{N}\setminus\mathbb{N})\left(\sum_{n>N} \alpha_n\,{}^*e_n \approx 0\right) \right\}.$$

Proof. We continue to use the notations in Thm. 2.33.

($\subset$): Let $c \in \mathrm{ns}({}^*X)$, with $c \approx {}^*a$, where $a \in X$. Then by Thm. 2.33, ${}^*\|c - {}^*a\| \approx 0$.

Let $N \in ({}^*\mathbb{N} \setminus \mathbb{N})$. Then ${}^*\pi_N(c - {}^*a) \approx (c - {}^*a)$. But, as $\{\pi_n(a)\}_{n\in\mathbb{N}}$ converges, ${}^*\pi_N({}^*a) \approx {}^*a$, therefore ${}^*\pi_N(c) \approx c$, as required.

($\supset$): Let $c \in \text{Fin}(^*X)$ so that $^*\pi_N(c) \approx c$ for any $N \in (^*\mathbb{N} \setminus \mathbb{N})$.

Since $^*\|c\|$ is finite, for each $n \in \mathbb{N}$, $^*\| \, ^*\pi_n(c)\|$ is finite; moreover, $\pi_n[X]$ is finite dimensional, so $^*\pi_n(c) \in \text{ns}(^*\pi_n[^*X]) \subset \text{ns}(^*X)$ by Cor. 2.1. *i.e.* $\{ \, ^\circ(\, ^*\pi_n(c))\}_{n \in \mathbb{N}} \subset X$.

Since $\forall N \in (^*\mathbb{N} \setminus \mathbb{N})(\, ^*\pi_N(c) \approx c)$, we note that $\{ \, ^\circ(\, ^*\pi_n(c))\}_{n \in \mathbb{N}}$ is Cauchy and hence converges to some $a \in X$.

By saturation, $^*a \approx \, ^\circ(\, ^*\pi_N(c))$ for any $N \in (^*\mathbb{N} \setminus \mathbb{N})$. Let $N \in (^*\mathbb{N} \setminus \mathbb{N})$, then $^*\pi_N(c) \approx c$, therefore $^*a \approx c$, making $c \in \text{ns}(^*X)$. $\square$

2.7.4 *Schauder's Fixed Point Theorem*

Typically fixed point theorems are results about fixed points of a ***self-mapping***, *i.e.* the existence of some x such that $f(x) = x$ for some $f : C \to C$, where C is a set satisfying some compactness or convexity conditions. We have already mention two essential ones, namely the Banach Contraction Principle (Thm. 1.19) and Brouwer's Fixed Point Theorem (Thm. 1.20). Many important fixed point theorems are motivated by re-formulating solutions to certain system of equations as fixed points.

In a normed linear space, given $f : C \to C$, in order to find a fixed point, the next best possibility is to have a fixed point for $^*f : \widehat{^*C} \to \widehat{^*C}$. But first we need to know $^*f$ can be defined, which was possible in earlier sections when we were dealing with bounded linear operators. We need $^*C$ to be in some finite part *i.e.* some sort of boundedness condition on C, and $^*f$ such that $^*f(x) \approx \, ^*f(y)$ in order to define $\widehat{^*f}(\widehat{x})$, *i.e.* some sort of continuity condition on f. If we want to have a fixed point in the original C we need to make sure that $^*f(x)$ is infinitely close to some element in C, *i.e.* some compactness condition on the image $f[C]$. Moreover, it also turns out that with convexity condition on C, we may define a kind of projection which makes it possible to apply Brouwer's Fixed Point Theorem. Hence, all these conditions are very natural ones. Here we present some examples along this line in the case of normed linear spaces.

Motivated by the notion of totally boundedness (see Prop. 1.25 and the Generalized Heine-Borel Theorem (Thm. 1.22)), we consider the following notion.

Given a normed linear space X and $\epsilon \in \mathbb{R}^+$, we say that a subset $A \subset X$ is ϵ-***bounded*** (*w.r.t.* the norm on X) if there is a finite $H \subset A$ such that $A \subset H^\epsilon$, where

$$H^\epsilon := \big\{ x \in X \, | \, \exists y \in H \, (\, \|x - y\| < \epsilon) \big\}.$$

Lemma 2.6. *Let A be an ϵ-bounded subset in a normed linear space X for some $\epsilon \in \mathbb{R}^+$. So let $A \subset H^\epsilon$ for some finite $H \subset A$.*

Then there is a continuous function $\pi : A \to \operatorname{conv}(H)$, the convex hull of A, such that $\forall x \in A \left(\|x - \pi(x)\| < \epsilon \right)$.

Proof. For each $y \in Y$, define a continuous function $\tau_y : X \to [0, \infty)$ by

$$X \ni x \mapsto \tau_y(x) := \max\left(\epsilon - \|x - y\|, 0 \right).$$

Then $\forall x \in A \,\exists y \in H \left(\tau_y(x) > 0 \right)$. Now define $\pi : A \to \operatorname{conv}(H)$ by

$$\pi(x) = \sum_{y \in H} \lambda_y(x)\, y \quad \text{where } \lambda_y(x) := \tau_y(x) \Big(\sum_{y \in H} \tau_y(x) \Big)^{-1} \in [0, 1].$$

It is clear that π is continuous.

Moreover, for $x \in A$ we have

$$\|x - \pi(x)\| \leq \sum_{y \in H} \lambda_y(x)\, \|x - y\| < \epsilon,$$

by noticing that $\sum_{y \in H} \lambda_y(x) = 1$ and $\lambda_y(x) = 0$ if $\|x - y\| \geq \epsilon$. $\square$

Given an internal metric space X, a (possibly external) subset $A \subset X$ is called ***infinitesimally bounded*** if for some $0 \approx \epsilon \in {}^*[0, \infty)$, there is a hyperfinite internal subset $H \subset A$ such that $A \subset H^\epsilon$.

Note that the definition remains the same even we only require $H \subset X$.

In an internal metric space X and $A \subset X$, we say that an internal function $f : A \to X$ is ***S-continuous*** on some $Y \subset A$ if

$$\forall x_1, x_2 \in Y \left((x_1 \approx x_2) \Rightarrow \left(f(x_1) \approx f(x_2) \right) \right).$$

We simply say that f is S-continuous if it is S-continuous on its own domain.

So, in a metric space X, given $A \subset X$ and $f : A \to X$, f is continuous iff ${}^*f$ is S-continuous on A and f is uniformly continuous iff ${}^*f$ is S-continuous, *i.e.* S-continuous on ${}^*A$. If X is an internal normed linear space and $f \in \operatorname{Fin}\big(\mathcal{B}(X)\big)$, then f is S-continuous. For general internal functions, there is no implication between *continuity and S-continuity.

The following result is a consequence of Brouwer's Fixed Point Theorem.

Theorem 2.34. (Schauder's Fixed Point Theorem—nonstandard version) *Let X be an internal normed linear space. Let $\emptyset \neq C \subset X$ be *convex and $f : C \to C$ be an internal *continuous function such that $f[C]$ is infinitesimally bounded. Then*

(i) $f(c) \approx c$ *for some $c \in C$.*

(ii) *If* $C \subset \text{Fin}(X)$ *and* f *is S-continuous, then* $\widehat{f} : \widehat{C} \to \widehat{C}$ *given by* $\widehat{x} \mapsto \widehat{f(x)}$, $x \in C$, *is well-defined and has a fixed point.*

Proof. We first note the following implication from the Brouwer's Fixed Point Theorem.

CLAIM: Let Y be a finite dimensional normed linear space, $\emptyset \neq K \subset Y$ be bounded closed convex and $g : K \to K$ be continuous. Then g has a fixed point.

PROOF OF THE CLAIM: By Cor. 2.3, let $\theta : Y \to \mathbb{R}^n$ for some $n \in \mathbb{N}$, be a homeomorphism. By scaling if needed, we may assume that $\theta[K] \subset \bar{B}_{\mathbb{R}^n}$.

Then $\theta[K]$ is closed convex and $(\theta \circ g \circ \theta^{-1})$ is a continuous self-mapping on $\theta[K]$. Moreover, g has a fixed point iff $(\theta \circ g \circ \theta^{-1})$ has a fixed point.

So we may assume without loss of generality that $Y = \mathbb{R}^n$, a Euclidean space. Note that, by the Heine-Borel Theorem (Thm. 1.22), K is compact.

Let $p : \mathbb{R}^n \to K$ be the projection onto K given by the Projection Theorem (Thm. 2.29). So by Thm. 2.29, p is continuous, hence $g \circ p : \bar{B}_{\mathbb{R}^n} \to \bar{B}_{\mathbb{R}^n}$ is continuous.

By Brouwer's Fixed Point Theorem (Thm. 1.20), let $c \in \bar{B}_{\mathbb{R}^n}$ be such that $g(p(c)) = c$. In particular, $c \in g[K] \subset K$. Then we have an element $c \in K$ such that $g(c) = g(p(c)) = c$ and the Claim is proved.

To prove (i), we let $\epsilon \approx 0$ such that $f[C] \subset H^\epsilon$ for some hyperfinite $H \subset f[C]$. Let $K = {}^*\text{conv}(H)$.

Since H is hyperfinite, K is *bounded, *closed and *convex. (In fact it is also *compact.)

Moreover, because $H \subset f[C] \subset C$ and C is *convex, $K \subset C$.

Let $\pi : f[C] \to K$ be the *continuous function given by transferring Lem. 2.6.

Then $(\pi \circ f) : K \to K$ and is *continuous.

Since H is hyperfinite, $K = {}^*\text{conv}(H)$ is a subset of some hyperfinite dimensional subspace of X. By transferring the Claim, we see for some $c \in K$ that $\pi(f(c)) = c$.

By Lem. 2.6, $\|\pi(f(c)) - f(c)\| < \epsilon \approx 0$, therefore $f(c) \approx c$.

For (ii), clearly, $\widehat{x} \mapsto \widehat{f(x)}$, $x \in C$ is well-defined under the additional conditions on C and f. Then for the $c \in C$ from (i) such that $f(c) \approx c$. we have $\widehat{f}(\widehat{c}) = \widehat{c}$. $\square$

Corollary 2.38. *Let* X *be an internal normed linear space. Suppose for some* $f \in \text{Fin}(\mathcal{B}(X))$, *there is a nonempty* **convex* $C \subset \text{Fin}(X)$ *such that*

$f[C] \subset C$ and $f[C]$ *is infinitesimally bounded.*
  Then $\widehat{f}$ *has a fixed point in* $\widehat{C}$.

Proof. As remarked earlier, such f is S-continuous, so the conclusion follows form Thm. 2.34(ii). □

In Cor. 2.38, the interesting case is of course when $0 \notin \widehat{C}$.

Corollary 2.39. (Schauder's Fixed Point Theorem) *Let* X *be a normed linear space and* $\emptyset \neq C \subset X$ *be convex.*
  Suppose $f : C \to C$ *is continuous,* $\overline{f[C]} \subset C$ *and* $\overline{f[C]}$ *is compact. Then* f *has a fixed point.*

Proof. Clearly, $^*C$ is *convex, $^*f$ is *continous. Note, by Thm. 1.22, Prop. 1.25 and $^*f[^*C] \subset {}^*(\overline{f[C]})$, that $^*f[^*C]$ is infinitesimally bounded.
  So it follows from Thm. 2.34(i) that for some $c \in {}^*C$ we have $^*f(c) \approx c$. But by the compactness of $\overline{f[C]}$ and $\overline{f[C]} \subset C$, for some $a \in C$ we have $^*f(c) \approx a$. Since f is continuous at a it follows that $f(a) \approx {}^*f(c)$, therefore $f(a) \approx {}^*f(c) \approx a$, *i.e.* $f(a) = a$, because both are in X. □

Recall the definition on p.161 of a compact function. If the C above is also closed, we have immediately that $\overline{f[C]} \subset C$. So we have:

- *Any compact self-mapping on a nonempty bounded closed convex set has a fixed point.*

In Cor.2.39, $\overline{f[C]}$ is bounded closed convex and the restriction of f on it is a compact self-mapping, so the above statement has the same strength as Schauder's Fixed Point Theorem.
  In a metric space (X, d), if $Y \subset X$, a function $f : Y \to X$, of Lipschitz constant 1, *i.e.* $\forall y_1, y_2 \in Y \left(\|f(y_1) - f(y_2)\| \leq \|y_1 - y_2\| \right)$, is called **nonexpansive**.
  So all nonexpansive functions are continuous and, in the case of internal metric spaces, nonexpansive internal functions are S-continuous. Also if X is a normed linear space, then operators in S_X are nonexpansive.
  Thm. 2.34(i) is about the existence of an almost fixed point. Another situation of the existence of such point is the following Cor. 2.40. But we first prove a result which basically says that in a nonstandard hull, nonexpansive self-mapping that can be lifted to an internal function always possesses a fixed point. The proof relies on an application of the Banach Contraction Principle.

Theorem 2.35. *In an internal normed linear space X, consider some internal $A \subset \mathrm{Fin}(X)$ and an internal S-continuous function $f : A \to X$.*

Let $\widehat{f} : \widehat{A} \to \widehat{X}$ be given by $\widehat{x} \mapsto \widehat{f(x)}$, $x \in A$.

Suppose $\widehat{f}[C] \subset C$ for some nonempty convex $C \subset \widehat{A}$, and $\widehat{f} \restriction_C$ is nonexpansive, then $\widehat{f}$ has a fixed point.

Proof.   First note that by $A \subset \mathrm{Fin}(X)$ and the S-continuity of f, we also have $f[A] \subset \mathrm{Fin}(X)$, so $\widehat{f} : \widehat{A} \to \widehat{X}$ is well-defined.

Moreover, by f being S-continuous, $\widehat{f}$ is uniformly continuous. By Prop. 2.5, $\widehat{A}$ is closed, so the closure $\bar{C} \subset \widehat{A}$. Note that $\bar{C}$ is convex. Let $g := \widehat{f} \restriction_{\bar{C}}$. Then by continuity, $g : \bar{C} \to \bar{C}$ and is nonexpansive.

Fix any $\widehat{a} \in C$, where a is chosen to be an element of A, by $C \subset \widehat{A}$. For any $n \in \mathbb{N}$, the function given by

$$\bar{C} \ni x \mapsto \left(n^{-1}\widehat{a} + (1 - n^{-1})g(x) \right) \in \bar{C}$$

is of Lipschitz constant $(1 - n^{-1}) < 1$, *i.e.* a contraction. Hence, by the Banach Contraction Principle (Thm. 1.19), for some $\widehat{a}_n \in \bar{C} \subset \widehat{A}$, we have

$$n^{-1}\widehat{a} + (1 - n^{-1})\widehat{f}(\widehat{a}_n) = \widehat{a}_n.$$

Again, the choice an be made so that $a_n \in A$.

Note that

$$\|f(a_n) - a_n\|_X \approx \left\| \widehat{f}(\widehat{a}_n) - \widehat{a}_n \right\|_{\widehat{X}} \leq n^{-1}\left(\|\widehat{f}(\widehat{a}_n)\|_{\widehat{X}} + \|\widehat{a}_n\|_{\widehat{X}} \right) \to 0$$

as $n \to \infty$ in $\mathbb{N}$.

Extend $\{a_n\}_{n \in \mathbb{N}}$ to an internal sequence in A, let $N \in {}^*\mathbb{N} \setminus \mathbb{N}$ be small enough, then

$$\|f(a_N) - a_N\|_X \approx 0, \quad \text{hence} \quad f(a_N) \approx a_N,$$

therefore $\widehat{f}(\widehat{a_N}) = \widehat{a_N}$. □

Consequently we have the following.

Corollary 2.40. *Let X be an internal normed linear space with an internal nonempty and *convex $C \subset \mathrm{Fin}(X)$ and an internal S-continuous function $f : C \to C$.*

Suppose $\widehat{f} : \widehat{C} \to \widehat{C}$ is nonexpansive, then $\widehat{f}$ has a fixed point. □

Note that the $\widehat{C}$ above is necessarily bounded closed convex.

Corollary 2.41. *Let X be a normed linear space, $C \subset X$ be bounded closed convex and $f : C \to C$ be nonexpansive. Then $\exists x \in {}^*C \left({}^*f(x) \approx x \right)$.*

Proof. Note that $^*C \subset \mathrm{Fin}(^*X)$ is nonempty *convex and $^*f$ is *nonexpansive, hence S-continuous.

It is also easy to check that $\widehat{^*f} : \widehat{^*C} \to \widehat{^*C}$ is nonexpansive. So it follows from Cor. 2.40 that for some $c \in {}^*C$ that $\widehat{^*f}(\widehat{c}) = \widehat{c}$, hence $^*f(c) \approx c$. $\qquad\square$

See [Benyamini and Lindenstrauss (2000)] for more discussion about nonexpansive mappings and the approximation of fixed points.

2.7.5 Notes and exercises

Thm. 2.31 was first proved by von Neumann, Aronszajn and Smith ([Aronszajn and Smith (1954)]). It was generalized to the case for polynomial compact operators by Robinson and with improvement by Bernstein ([Bernstein and Robinson (1966)])—a result hailed as a success in the early days of nonstandard analysis. It was subsequently converted by Halmos to a proof which avoids the use of nonstandard analysis. The strongest extension of results in this direction is the Lomonosov's Theorem, in which the assumption in Thm. 2.31 is weakened to that $\{f\}'$ contains a compact operator, but a stronger conclusion that there is a nontrivial closed subspace which is invariant for any element of $\{f\}'$ was produced. It can be viewed as an infinite dimensional analogue of Burnside's Theorem—*i.e.* the *Fundamental Theorem of Noncommutative Algebra*.

Lomonosov's Theorem was first proved using Schauder's Fixed Point Theorem. See [Aupetit (1991)] for an elegant presentation. A proof for the compact operator case using the Schauder's Fixed Point Theorem can be found in [Fabian *et al.* (2001)]. Despite simplified proofs of the Lomonosov's Theorem are available nowadays, the proof of Thm. 2.31 given here is more concrete and, in a sense, more constructive as well.

An open problem called the *Invariant Subspace Problem* asks whether every $f \in \mathcal{B}(X)$, where X is a complex Hilbert space of dimension > 1, necessarily possesses a nontrivial closed invariant subspace. This statement is known to fail for complex Banach spaces in general. The problem is essentially about infinite dimensional separable complex Hilbert spaces. For the statement holds for finite dimensional ones (just consider the subspace generated by a nonzero eigenvector) and it holds for nonseparable Hilbert spaces (take $\mathrm{Lin}(\{f^n(a) \mid n \in \mathbb{N}\})$ for any nonzero $a \in X$).

There is an abundance of fixed point theorems in the literature, including too many artificial ones. At least when metric spaces are involved, the Brouwer's Fixed Point Theorem is by far the most fundamental result.

180   *Nonstandard Methods in Functional Analysis*

The proof of Thm. 2.34 is based on the same idea used in the usual proof of Schauder's Fixed Point Theorem (see [Conway (1990)], for example) and they are really corollaries of Brouwer's Fixed Point Theorem. See [Goebel and Kirk (1990)] for more study of the fixed point properties.

A major open problem is whether every nonexpansive self-mapping on a nonempty weakly compact convex subset of a reflexive space has a fixed point. See [Aksoy and Khamsi (1990)] for some partial solutions to this problem using the ultraproduct construction.

EXERCISES

(1) Complete the proofs of Prop. 2.47.
(2) Show that for $f \in \mathcal{B}(X)$, where X is a Hilbert space, if $f^2 = f$ (*i.e.* idempotent) and $\|f\| = 1$, then f is a projection.
(3) Verify the compactness of the operator given in Example 2.8.
(4) Are the characterizations of compact operators in Prop. 2.45 valid for an arbitrary Banach space?
(5) Let X be a Banach space. Show that $\mathcal{B}_c(X)$ is a closed ideal of $\mathcal{B}_X$, *i.e.* a closed subspace so that $f\,\mathcal{B}_c(X) \subset \mathcal{B}_c(X)$ and $\mathcal{B}_c(X)\,f \subset \mathcal{B}_c(X)$ for any $f \in B_c(X)$.
(6) Prove Prop. 2.44.
(7) Generalize Thm. 2.30: Let X, Y be Banach spaces, show that $f \in \mathcal{B}(X, Y)$ is compact iff f^* is compact. (This result is called Schauder's Theorem.)
(8) Give an example of $f \in \mathcal{B}(X)$, where X is a Hilbert space, so that $f[X]$ is not closed.
(9) Let X, Y be Banach spaces and $f \in \mathcal{B}(X, Y)$. Suppose for some $r \in \mathbb{R}^+$ that $\forall x \in X \left(\|f(x)\|_Y \geq r \|x\|_X \right)$. Show that $f[X]$ is closed.
(10) In a normed linear space X, find necessary conditions on $f \in \mathcal{B}(X)$ so that $\overline{\mathrm{Lin}}\big(\{f^{[n]} \mid n \in {}^*\mathbb{N}\}\big) = \{f\}'$.
(11) Find example of a compact operator which is not nilpotent.
(12) Show that the statement in Thm. 2.31 fails for some real Hilbert spaces.
(13) Show that for a Hilbert space X, if $f^* = f \in \mathcal{B}(X)$ (*i.e. self-adjoint*), then for every invariant closed subspace $Y \subset X$, $Y^\perp$ is invariant.
(14) Prove Lem. 2.5.
(15) Give an elementary and direct proof of the Krein-Milman Theorem (Thm. 2.32) for the finite dimensional case and use it to prove the general case.
(16) Find a Schauder basis of $C([0, 1])$.

Chapter 3

Banach Algebras

3.1   Normed Algebras and Nonstandard Hulls

In the land of Banach algebras, it is natural for algebra and topology to form a strategic partnership.

In this section, the focus is on unital Banach algebras, invertibility and properties about the spectrum. The nonstandard hull construction extends to this context.

A **normed algebra** $\mathcal{M}$ over $\mathbb{F}$ is a normed linear space over $\mathbb{F}$ that forms a ring so that

- $\forall \alpha \in \mathbb{F} \ \forall x, y \in \mathcal{M} \ \big((\alpha x)y = \alpha(xy) = x(\alpha y)\big)$;
- $\forall x, y \in \mathcal{M} \ \big(\|xy\| \leq \|x\| \, \|y\| \big)$.

where the ring product of two elements x, y from x is written as xy. It follows immediately that the multiplication is continuous *w.r.t.* the norm.

The **commutativity** of the normed algebra refers to that of the underlying ring product. Most of the normed algebras considered will be noncommutative.

Depending on whether $\mathbb{F} = \mathbb{R}$ or $\mathbb{F} = \mathbb{C}$, the normed algebra is called a **real normed algebra** or **complex normed algebra**.

When a normed algebra $\mathcal{M}$ is complete *w.r.t.* its norm, $\mathcal{M}$ is called a **Banach algebra**.

3.1.1   *Examples and basic properties*

We first continue with some more definitions.

If the ring structure of a normed algebra $\mathcal{M}$ has a unit, *i.e.* a multiplicative identity, we call $\mathcal{M}$ a **unital normed algebra**. This necessarily unique identity is denoted by 1, not to be confused with the unit element in $\mathbb{F}$ under the same symbol.

In a unital normed algebra, for $\alpha \in \mathbb{F}$, we write α instead of $\alpha 1$. Again, this should not cause any confusions.

Note from the definition of normed algebras that $\|1\| \geq 1$ holds for any unital normed algebra. For most normed algebras we considered, $\|1\| = 1$ holds naturally. In any case, this can be achieved with the norm $\|\cdot\|$ replaced by the equivalent norm $\|\cdot\| \, / \, \|1\|$.

For convenience, we assume hereafter that

$$\boxed{\|1\| = 1}$$

holds in all unital normed algebras under consideration.

If x is an element of a unital normed algebra, we define x^0 as 1 and $x^n := xx^{n-1}$ for $1 \leq n \in \mathbb{N}$.

Note that $\forall n \in \mathbb{N} \left(\|x^n\| \leq \|x\|^n \right)$.

A **subalgebra** of a normed algebra $\mathcal{M}$ is a linear subspace which is also a subring of $\mathcal{M}$. if $\mathcal{M}$ is unital, a unital subalgebra is one that contains 1. A subalgebra is closed if it is closed *w.r.t.* the norm.

Every normed algebra $\mathcal{M}$ is embedded in a unital normed algebra in the following canonical way. Regard $\mathbb{F}$ as a normed algebra over itself. On the L_1-direct sum of $\mathcal{M} \oplus \mathbb{F}$ (as a direct sum of normed linear spaces), we define a product by

$$(x, \alpha)(y, \beta) := (xy + \alpha y + \beta x, \alpha \beta), \quad \text{where } (x, \alpha), (y, \beta) \in \mathcal{M} \oplus \mathbb{F}.$$

Then it is clear that $\mathcal{M} \oplus \mathbb{F}$ forms a normed algebra under this product and the L_1-norm.

For example, for $(x, \alpha), (y, \beta) \in \mathcal{M} \oplus \mathbb{F}$,

$$\|(x, \alpha)(y, \beta)\| = \|xy + \alpha y + \beta x\| + |\alpha\beta| \leq \|x\| \|y\| + |\alpha| \|y\| + |\beta| \|x\| + |\alpha\beta|$$
$$= (\|x\| + |\alpha|)(\|y\| + |\beta|) = \|(x, \alpha)\| \, \|(y, \beta)\|.$$

It is easy to check that $(0, 1)$ acts as the identity in $\mathcal{M} \oplus \mathbb{F}$.

Moreover, the mapping $\mathcal{M} \ni x \mapsto (x, 0)$ is an isometric algebra isomorphism of $\mathcal{M}$ into a subalgebra of $\mathcal{M} \oplus \mathbb{F}$. (Hereafter, when dealing with normed algebras, isomorphism is taken *w.r.t.* the algebra structure.) Note that even if $\mathcal{M}$ is unital, $(1, 0)$ is not the identity in $\mathcal{M} \oplus \mathbb{F}$ and under the above embedding, $\mathcal{M}$ always has codimension 1 in $\mathcal{M} \oplus \mathbb{F}$.

The normed algebra $\mathcal{M} \oplus \mathbb{F}$ is referred to as the **unitization** of $\mathcal{M}$.

We remark that the unitization of a Banach algebra is a Banach algebra.

Here are some examples of normed algebras and Banach algebras.

Example 3.1.

- Given a normed linear space X over $\mathbb{F}$, the normed linear space $\mathcal{B}(X)$ forms a normed algebra over $\mathbb{F}$ with the the product given by the composition. Since $\mathcal{B}(X)$ contains the identity mapping, it is unital. Unless $\dim X = 1$, $\mathcal{B}(X)$ is noncommutative. If X is a Banach space, $\mathcal{B}(X)$ is a Banach algebra. Later we will pay particular attention to the case when X is Hilbert space.

- If the normed linear space X is finite dimensional, the normed algebra $\mathcal{B}(X)$ is isomorphic to $\mathcal{B}(\mathbb{F}^n)$ for some $n \in \mathbb{N}$, hence forms a Banach algebra. In fact $\mathcal{B}(\mathbb{F}^n)$ can be represented as the Banach algebra of $n \times n$ matrices over $\mathbb{F}$, the *matrix algebra*.

- Let X be a normed linear space. By Prop. 2.40, the composition of compact operators in $\mathcal{B}(X)$ is compact. So $\mathcal{B}_c(X)$ forms a subalgebra of $\mathcal{B}(X)$. By Cor. 2.32, $\mathcal{B}_c(X)$ is actually a closed subalgebra of $\mathcal{B}(X)$.

- If X is an infinite dimensional normed linear space, the identity mapping is not compact (Cor. 2.33), and hence $\mathcal{B}_c(X)$ is nonunital.

- Given a topological space Ω, the Banach space $C_b(\Omega)$ forms a Banach algebra under the pointwise multiplication of functions, *i.e.* for $f, g \in C_b(\Omega)$, $fg(x)$ is defined to be $f(x)g(x), x \in \Omega$, and it is easy to check that $(fg) \in C_b(\Omega)$. Moreover, $C_b(\Omega)$ is commutative and is unital with the unit given by the constant unit function. Note that unless Ω is finite, $C_b(\Omega)$ is not reflexive as a Banach space.

- If Ω is locally compact, $C_0(\Omega)$ still forms a commutative Banach algebra in the same way as above. Of course $C_0(\Omega)$ is the same as $C_b(\Omega)$ when Ω is compact. If Ω is locally compact but not compact, $C_0(\Omega)$ is nonunital. In particular, c_0 is a commutative nonunital Banach algebra.

- Let $n \in \mathbb{N}$ and $|\Omega| = n$, then $C_b(\Omega)$ is isomorphic to $\mathbb{F}^n$. In particular, the Euclidean space $\mathbb{F}^n$ forms a Banach algebra under the pointwise multiplication.

- The Lebesgue space $L_\infty(\Omega, \mathcal{B}, \mu)$ forms a commutative Banach algebra under the pointwise multiplication when μ is a probability measure, or even a σ-finite complex measure.

- The quaternions forms a noncommutative unital Banach algebra. The quaternions can be identified with 2×2 complex matrices of the form $\begin{bmatrix} x + yi & u + vi \\ -u + vi & x - yi \end{bmatrix}$, $x, y, u, v \in \mathbb{R}$, with the usual matrix operations and the norm given by the square root of the determinant. $\quad\square$

Unless in the trivial case when Ω is finite, the Lebesgue space $L_1(\Omega, \mathcal{B}, \mu)$

would not form a normed algebra under the pointwise product. However, if Ω is a locally compact topological group (*i.e.* a topological space which also forms a group under a continuous binary operation $+$), the $L_\infty(\Omega, \mathcal{B}, \mu)$ does form a Banach algebra.

The following is such an example of a **group algebra**.

Example 3.2. Let Ω be a locally compact topological Abelian group, then by [Hewitt and Ross (1963)] for example, Ω has a Haar measure which is unique up to multiplication by $r \in \mathbb{R}^+$. Here, by a **Haar measure** μ we mean a a σ-finite positive measure which coincides with the one generated by its restriction on the Borel sets of Ω and such that it is **invariant**, *i.e.* $\mu(X) = \mu(a + X)$ (equivalently, $\mu(X) = \mu(X + a)$, since Ω is Abelian) for all Borel $X \subset \Omega$ and $a \in X$, and is a Radon measure.

Now consider the complex Banach space $L_1(\mu)$. Let $f, g \in L_1(\mu)$, then the **convolution product** of f, g is defined as follows:

$$(f \star g)(x) := \int_\Omega f(x - y)g(y)d\mu(y), \quad \text{for all } x \in \Omega.$$

First note that by Lebesgue integration, $f \star g$ is μ-measurable. To show that $L_1(\mu)$ forms a Banach algebra under the convolution, just note by Fubini's Theorem and the invariance of μ that:

$$\int_\Omega \left|(f \star g)(x)\right|d\mu(x) = \lim_{n \to \infty} \int_{\Omega_n} \left|\int_\Omega f(x - y)g(y)d\mu(y)\right|d\mu(x)$$

$$\leq \lim_{n \to \infty} \int_{\Omega_n} \int_\Omega |f(x - y)|\,|g(y)|\,d\mu(y)d\mu(x)$$

$$= \lim_{n \to \infty} \int_\Omega \int_{\Omega_n} |f(x - y)|\,d\mu(x)\,|g(y)|\,d\mu(y) \leq \|f\|\,\|g\|,$$

where $\{\Omega_n\}_{n \in \mathbb{N}}$ is a family of Borel sets, or even compact sets, such that $\mu(\Omega_n) < \infty$ and $\Omega = \bigcup_{n \in \mathbb{N}} \Omega_n$.

In the non-Abelian case of Ω, the above still work by a corresponding modification using left- or right-Haar measures.

It can be shown that $L_1(\mu)$ is unital precisely when Ω is a discrete space and $L_1(\mu)$ is commutative precisely when Ω is Abelian.

When Ω is $\mathbb{Z}$ under the usual addition and μ is the counting measure, we have the commutative unital Banach algebra ℓ_1, where, for $a, b \in \ell_1$,

$$\forall n \in \mathbb{N} \left((a \star b)_n = \sum_{m \in \mathbb{Z}} a_{n-m}b_m\right).$$

Observe that the identity is the sequence a given by $a_0 = 1$ and $a_n = 0$ whenever $n \neq 0$.

Moreover, it can be shown that ℓ_1 is isomorphic to the **Wiener algebra** consisting of continuous functions $[0, 2\pi] \to \mathbb{C}$ having absolutely convergent Fourier sequence. $\square$

Similar to p.90, given a family of normed algebras $\mathcal{M}_i$, $i \in I$, and $p \in [1, \infty]$, the L_p-**direct sum** of the family is

$$\bigoplus_p \mathcal{M}_i := \left\{ x \in \prod_{i \in I} \mathcal{M}_i \mid \|x\|_p < \infty \right\},$$

where the norm is given by $\|x\|_p := \left(\sum_{i \in I} \|x_i\|_{\mathcal{M}_i}^p \right)^{1/p}$, if $p \neq \infty$ and $\|x\|_\infty := \sup_{i \in I} \|x_i\|_{\mathcal{M}_i}$ otherwise.

For $x, y \in \bigoplus_p \mathcal{M}_i$, the product xy is defined by pointwise product, *i.e.* $(xy)_i = x_i y_i$ for every $i \in I$.

To show that $\bigoplus_p \mathcal{M}_i$ forms a normed algebra, let $x, y \in \bigoplus_p \mathcal{M}_i$. If $p \neq \infty$, we have $\|xy\|_p^p =$

$$\sum_{i \in I} \|x_i y_i\|_{\mathcal{M}_i}^p \leq \sum_{i \in I} \|x_i\|_{\mathcal{M}_i}^p \|y_i\|_{\mathcal{M}_i}^p \leq \sum_{i \in I} \|x_i\|_{\mathcal{M}_i}^p \sum_{i \in I} \|y_i\|_{\mathcal{M}_i}^p = \|x\|_p^p \|y\|_p^p.$$

Similarly, if $p = \infty$, we have $\|xy\|_\infty \leq \|x\|_\infty \|x\|_\infty$.

Moreover, if all factors $\mathcal{M}_i$ are Banach algebras, $\bigoplus_p \mathcal{M}_i$ is also a Banach algebra.

Note that for each $k \in I$, the mapping $\rho : \mathcal{M}_k \to \bigoplus_p \mathcal{M}_i$ given by for each $a \in \mathcal{M}_k$ such that $\rho(a)_i = a$, if $i = k$, and $\rho(a)_i = 0$ otherwise, is an isometric isomorphism taking $\mathcal{M}_k$ into $\bigoplus \mathcal{M}_i$.

Observe that if $\mathcal{M}$ is a unital normed algebra, then its unitization $\mathcal{M} \oplus \mathbb{F}$ is isomorphic to the normed algebra L_1-direct sum $\mathcal{M} \oplus_1 \mathbb{F}$ via the mapping $\mathcal{M} \oplus \mathbb{F} \ni (x, \alpha) \mapsto (x + \alpha, \alpha) \in \mathcal{M} \oplus_1 \mathbb{F}$.

A subalgebra $I \subset \mathcal{M}$ is called a **left ideal** of the normed algebra $\mathcal{M}$ if $\forall x \in \mathcal{M} \ \forall y \in I \ (xy \in I)$, *i.e.* $\forall x \in \mathcal{M} (xI \subset I)$.

Similarly, it is called a **right ideal** if $\forall x \in \mathcal{M} (Ix \subset I)$.

If I is both a left and a right ideal, it is called a **two-sided ideal**, or simply an **ideal**.

For example, if $a \in \mathcal{M}$, then $\mathcal{M}a$ is a left ideal and $a\mathcal{M}$ is a right ideal. If $\mathcal{M}$ is commutative, then $a\mathcal{M} = \mathcal{M}a$ forms a two-sided ideal.

Note that even if $\mathcal{M}$ is unital, by an ideal or a left or right ideal $I \subset \mathcal{M}$, it is not required that I be a unital subalgebra. In fact it is easy to see that if such I is unital in a unital $\mathcal{M}$, then $I = M$.

We call an ideal $I \subsetneq \mathcal{M}$ a **proper ideal**. So there is no proper unital ideal in a unital normed algebra. Note that the trivial ideal $\{0\}$ is regarded as a proper ideal.

In a commutative normed algebra $\mathcal{M}$, for every $a \in \mathcal{M} \setminus \{0\}$, the ideal $a\mathcal{M}$ is a proper and nontrivial ideal.

Example 3.3. Let X be a normed linear space. By Example 3.1, $\mathcal{B}_c(X)$ is a subalgebra of $\mathcal{B}(X)$. Since the composition of a compact operator with a bounded linear operator is compact, we have $\mathcal{B}_c(X)\,T \subset B_c(X)$ holds for every $T \in \mathcal{B}(X)$, *i.e.* $\mathcal{B}_c(X)$ is a right ideal of $\mathcal{B}(X)$.

On the other hand, if $T \in \mathcal{B}(X), S \in \mathcal{B}_c(X)$ and $Y \subset X$ is bounded, then $\overline{S[Y]}$ is compact, so, as a continuous image of a compact set, $T\left[\overline{S[Y]}\right]$ is compact (Prop. 1.26). Since $TS[Y] \subset T\left[\overline{S[Y]}\right]$, $\overline{TS[Y]}$ is compact. Therefore $TS \in \mathcal{B}_c(X)$. Hence $\mathcal{B}_c(X)$ is a left ideal of $\mathcal{B}(X)$.

i.e. $\mathcal{B}_c(X)$ is a two-sided ideal of $\mathcal{B}(X)$. Also, $\mathcal{B}_c(X)$ is nonunital when X is infinite dimensional.

Moreover, $\mathcal{B}_c(X)$ is a dense but not closed ideal of $\mathcal{B}(X)$. □

The following defines the **quotient algebra** *w.r.t.* a closed proper ideal.

Proposition 3.1. *Let $\mathcal{M}$ be a normed algebra and $I \subset \mathcal{M}$ a closed proper ideal. Then $\mathcal{M}/I$ forms a normed algebra under the product*

$$(x + I)(y + I) := (xy + I), \ x, y \in \mathcal{M}.$$

Moreover, if $\mathcal{M}$ is a Banach algebra, $\mathcal{M}/I$ is a Banach algebra; and if $\mathcal{M}$ is unital, $\mathcal{M}/I$ is unital.

Proof.   By Prop. 2.1, $\mathcal{M}/I$ forms a Banach space. By I being a two-sided ideal, the product is well-defined.

Let $a, b \in \mathcal{M}$. Then

$$\|ab + I\| = \inf_{z \in I} \|ab + z\| \leq \inf_{z_1, z_2 \in I} \|(a + z_1)(b + z_2)\|$$

$$\leq \inf_{z \in I} \|a + z\| \inf_{z \in I} \|b + z\| = \|a + I\| \ \|b + I\|,$$

So $\mathcal{M}/I$ forms a normed algebra.

It also follows from By Prop. 2.1 that $\mathcal{M}/I$ is a Banach algebra if $\mathcal{M}$ is a Banach space.

If $\mathcal{M}$ is unital, then it is easy to check that $1 + I$ is the identity in the quotient algebra $\mathcal{M}/I$. □

Note that in Prop. 3.1 if I is the trivial ideal $\{0\}$, then $\mathcal{M}/I$ is just $\mathcal{M}$ through the obvious identification.

In a unital normed algebra $\mathcal{M}$, an element $a \in \mathcal{M}$ is called **left-invertible** if there is $b \in \mathcal{M}$ such that $ba = 1$. (Such b is called a **left-inverse** of a.) Likewise, a is **right-invertible** if there is $b \in \mathcal{M}$ so that $ab = 1$. (b is called a **right-inverse** of a.)

When a is both left- and right-invertible, we say that a is **invertible**.

Observe that if a has both a left-inverse and a right-inverse, the inverses are equal: let $b, c \in \mathcal{M}$ such that $ba = 1 = ac$. Then

$$b = b \cdot 1 = b(ac) = (ba)c = 1 \cdot c = c.$$

In particular, there is a unique element representing both the left- and the right-inverse.

We call this unique b the **inverse** of a and denote it by a^{-1}.

The set of invertible elements in $\mathcal{M}$ is denoted by $\mathcal{M}^{-1}$.

Note that $\mathcal{M}^{-1}$ contains 1 and is closed under product. *i.e.* $\mathcal{M}^{-1}$ is a subgroup of $\mathcal{M} \setminus \{0\}$ under the product.

Observe that, given a normed algebra $\mathcal{M}$, we have in the nonstandard extension that $^*(\mathcal{M}^{-1}) = (^*\mathcal{M})^{-1}$, hence we express it simply as $^*\mathcal{M}^{-1}$.

Example 3.4. Let X be an infinite dimensional normed linear space. By Example 3.1, $\mathcal{B}_c(X)$ is a nonunital subalgebra of the unital algebra $\mathcal{B}(X)$. For any $T \in \mathcal{B}_c(X)$, if $T \in \mathcal{B}(X)^{-1}$, then, as $1 = T T^{-1}$, the composition of a compact operator with a bounded linear operator, 1 is compact, therefore, by Cor. 2.33, $\dim(X) < \infty$, a contradiction.

i.e. $\mathcal{B}_c(X) \cap \mathcal{B}(X)^{-1} = \emptyset$. $\qquad\qquad\qquad\qquad\qquad\qquad\square$

We remark that the definition of the invertibility of a element is relative to the algebra containing it, as the following shows.

Example 3.5. Consider $S_{\mathbb{C}}$, the unit circle centered at 0, and $\mathcal{M} = C_b(S_{\mathbb{C}})$ as a Banach algebra over $\mathbb{C}$. Let $\mathcal{M}_0 \subset \mathcal{M}$ be the closed subalgebra generated by polynomials over $\mathbb{C}$. Note that $\mathcal{M}_0$ contains the unit of $\mathcal{M}$. Then the identity function z, *i.e.* the function $S_{\mathbb{C}} \ni z \mapsto z \in \mathbb{C}$, is an element of $\mathcal{M}_0$ which is not invertible in $\mathcal{M}_0$ but invertible in $\mathcal{M}$. $\qquad\square$

Lemma 3.1. *Let $\mathcal{M}$ be a unital Banach algebra and $a \in B_{\mathcal{M}}$, i.e. $\|a\| < 1$. Then $(1 + a) \in \mathcal{M}^{-1}$ and $(1 + a)^{-1} = \sum_{n=0}^{\infty}(-a)^n$.*

In general, for $c \in \mathcal{M}^{-1}$ and $a \in \mathcal{M}$, if $\|c^{-1}a\| < 1$ then $(c+a) \in \mathcal{M}^{-1}$.

Proof. Replace a by $-a$, we show that $(1 - a) \in \mathcal{M}^{-1}$ and $(1 - a)^{-1} = \sum_{n=0}^{\infty} a^n$.

First note that $\left\{ \sum_{m=0}^{n} a^m \right\}_{n \in \mathbb{N}}$ is Cauchy in $\mathcal{M}$, and since $\mathcal{M}$ is complete, we have $b := \sum_{n=0}^{\infty} a^n \in \mathcal{M}$.

Note that $ab = ba$, *i.e.* they commute with each other. Then

$$b(1 - a) = (1 - a)b = (1 - a)\sum_{n=0}^{\infty} a^n = 1,$$

in particular $(1 - a) \in \mathcal{M}^{-1}$.

For the second statement, by what was just proved, $(1 + c^{-1}a) \in \mathcal{M}^{-1}$. Therefore, as $c + a = c(1 + c^{-1}a)$, we have

$$(c + a)^{-1} = \left(1 + c^{-1}a\right)^{-1} c^{-1}. \qquad \square$$

Corollary 3.1. *Let $\mathcal{M}$ be a unital normed algebra, then $\mathcal{M}^{-1}$ is open.*

Proof.   Noticing that if $a \in \mathcal{M}$ is invertible, then a remains invertible in the closure $\bar{\mathcal{M}}$, we may assume that $\mathcal{M}$ is a Banach algebra.

Let $a \in \mathcal{M}^{-1}$. Suppose $b \in {}^*\mathcal{M}$ is such that $a \approx b$. Write $b = a + c$ for some $c \approx 0$. So $a^{-1}c \approx 0$ and in particular, $\|a^{-1}c\| < 1$.

Transfer Lem. 3.1, we have $b \in {}^*\mathcal{M}^{-1}$.

That is, $\mu(a) \subset {}^*\mathcal{M}^{-1}$ for every $a \in \mathcal{M}^{-1}$, therefore $\mathcal{M}^{-1}$ is open by Prop. 1.14. $\qquad \square$

We turn to a few remarks about ideals. Note that in a unital normed algebra $\mathcal{M}$, a left ideal $I \subset \mathcal{M}$ that contains a left-invertible element must contain 1, hence $I = \mathcal{M}$. In particular, for an $a \in \mathcal{M}$, the left ideal $\mathcal{M}a = \mathcal{M}$ iff a is left-invertible. Likewise for right ideals. In particular, a proper ideal $I \subset \mathcal{M}$ must satisfy $I \cap \mathcal{M}^{-1} = \emptyset$. It is easy to see that the closure (in $\mathcal{M}$) of an ideal in $\mathcal{M}$ is also an ideal. Since $\mathcal{M}^{-1}$ is open, $I \cap \mathcal{M}^{-1} = \emptyset$ implies $\bar{I} \cap \mathcal{M}^{-1} = \emptyset$, thus the following:

Corollary 3.2. *Let $\mathcal{M}$ be a unital normed algebra and $I \subset \mathcal{M}$ a proper ideal. Then $\bar{I}$, the closure in $\mathcal{M}$, is also a proper ideal.* $\qquad \square$

Given a unital normed algebra $\mathcal{M}$, a proper ideal $I \subset \mathcal{M}$ is called a **maximal ideal** if it satisfies the property that there is no proper ideal $J \subset \mathcal{M}$ such that $I \subset J$. By applying Zorn's Lemma, every proper ideal extends to a maximal ideal. Combining with Cor. 3.2, we state the following:

Corollary 3.3. *In a unital normed algebra, all maximal ideals are closed and every proper ideal extends to a maximal ideal.* $\qquad \square$

Theorem 3.1. *Let $\mathcal{M}$ be a unital normed algebra, then the inverse mapping $\mathcal{M}^{-1} \ni x \mapsto x^{-1} \in \mathcal{M}^{-1}$ is continuous.*

Moreover, the inverse mapping is a homeomorphism on $\mathcal{M}^{-1}$.

Proof.   First of all, by Cor. 3.1, for every $x \in \mathcal{M}^{-1}$, we have $\mu(x) \subset {}^*\mathcal{M}^{-1}$.

Now let $a \in \mathcal{M}^{-1}$ and $a \approx b$, so $b \in {}^*\mathcal{M}^{-1}$ and we need to show that $a^{-1} \approx b^{-1}$.

From $(a - b) \approx 0$, we have $1 - ba^{-1} = (a - b)a^{-1} \approx 0$. Therefore, together with Lem. 3.1,

$$ab^{-1} = (ba^{-1})^{-1} = (1 - (1 - ba^{-1}))^{-1} = \sum_{n=0}^{\infty} (1 - ba^{-1})^n \approx 1. \quad (3.1)$$

On the other hand,

$$b^{-1} - a^{-1} = a^{-1}ab^{-1} - a^{-1} = a^{-1}(ab^{-1})(a - b)a^{-1}, \quad (3.2)$$

hence, by (3.1),

$$\left\| b^{-1} - a^{-1} \right\| \le \left\| a^{-1} \right\| \cdot \left\| ab^{-1} \right\| \cdot \left\| a - b \right\| \cdot \left\| a^{-1} \right\| \approx \left\| a^{-1} \right\|^2 \left\| a - b \right\| \approx 0,$$

as required.

Note that the inverse mapping is bijective on $\mathcal{M}^{-1}$ and its square is just the identity mapping, hence it forms a homeomorphism on $\mathcal{M}^{-1}$. $\qquad \square$

Given a unital normed algebra $\mathcal{M}$, both the product and the inverse mapping on $\mathcal{M}^{-1}$ are continuous, so $\mathcal{M}^{-1}$ forms a topological group.

$\mathcal{M}^{-1}$ is sometimes called the **general linear group** of $\mathcal{M}$, in symbol: $GL(\mathcal{M})$, generalizing the corresponding notion for invertible matrices of fixed dimensions.

The following shows that a nonzero element is invertible in $\mathcal{M}$ iff it is "almost invertible" in $^*\mathcal{M}$.

Proposition 3.2. *Let $\mathcal{M}$ be a unital Banach algebra and $a \in \mathcal{M}$. Then the following are equivalent:*

(i) $a \in \mathcal{M}^{-1}$.

(ii) $\exists b \in {}^*\mathcal{M}^{-1} \left((a \approx b) \wedge \left(\left\| ab^{-1} - 1 \right\| < 1 \right) \right)$.

(iii) $\exists b \in {}^*\mathcal{M}^{-1} \left((a \approx b) \wedge \left(\left\| b^{-1}a - 1 \right\| < 1 \right) \right)$.

Proof. $((i) \Rightarrow (ii))$ and $((i) \Rightarrow (iii))$ are trivial.

$((ii) \Rightarrow (i))$: Suppose $a = b + c$, where $b, c \in {}^*\mathcal{M}$ with $b \in {}^*\mathcal{M}^{-1}$, $\left\| ab^{-1} - 1 \right\| < 1$ and $c \approx 0$.

Then $ab^{-1} = 1 + cb^{-1}$, hence $\left\| cb^{-1} \right\| < 1$. Now by Lem. 3.1, $(1 + cb^{-1}) \in {}^*\mathcal{M}^{-1}$, *i.e.* $ab^{-1} \in {}^*\mathcal{M}^{-1}$, therefore $a \in {}^*\mathcal{M}^{-1}$ which, by transfer, implies $a \in \mathcal{M}^{-1}$.

$((iii) \Rightarrow (i))$ is obtained from a similar proof. $\qquad \square$

Example 3.6. The conditions in (ii) and (iii) in Prop. 3.2 are optimal in the following sense.

Consider $\mathcal{M} = \mathcal{B}(\mathbb{F}^2)$, *i.e.* the unital Banach algebra of 2×2 matrices over $\mathbb{F}$ as bounded linear operators on $\mathbb{F}^2$. Fix $N \in {}^*\mathbb{N} \setminus \mathbb{N}$.

Take $a = \begin{bmatrix} 1 & 0 \\ 0 & 0 \end{bmatrix}$ and $b = \begin{bmatrix} 1 & 0 \\ 0 & N^{-1} \end{bmatrix}$. Then $b^{-1} = \begin{bmatrix} 1 & 0 \\ 0 & N \end{bmatrix}$.

Hence $ab^{-1} - 1 = b^{-1}a - 1 = \begin{bmatrix} 0 & 0 \\ 0 & -1 \end{bmatrix}$, so in this situation we have $a \approx b$, $\|ab^{-1} - 1\| = \|b^{-1}a - 1\| = 1$ but $a \notin \mathcal{M}^{-1}$. $\qquad\qquad \square$

3.1.2   *Spectra*

Let $\mathcal{M}$ be a unital normed algebra. The **spectrum** of an element $a \in \mathcal{M}$ is defined to be

$$\sigma(a) := \{\lambda \in \mathbb{F} \,|\, (a - \lambda) \notin \mathcal{M}^{-1}\}.$$

For emphasis on its dependence on the underlying normed algebra, we sometime write $\sigma_{\mathcal{M}}(a)$.

Given unital normed algebras $\mathcal{M}_0, \mathcal{M}$ such that $\mathcal{M}_0$ is a subalgebra of $\mathcal{M}$ with unit $1 \in \mathcal{M}_0$, it is clear that $\sigma_{\mathcal{M}}(a) \subset \sigma_{\mathcal{M}_0}(a)$ holds for all $a \in \mathcal{M}_0$, since $\mathcal{M}_0^{-1} \subset \mathcal{M}^{-1}$.

We leave it as an exercise to find examples of of such $\mathcal{M}_0, \mathcal{M}$ and $a \in \mathcal{M}_0$ so that $\sigma_{\mathcal{M}}(a) \subsetneq \sigma_{\mathcal{M}_0}(a)$.

Example 3.7.

- Trivially $\sigma(0) = \{0\}$ and for $\forall x \in \mathcal{M} \left(x \in \mathcal{M}^{-1} \Leftrightarrow 0 \notin \sigma(x)\right)$.
- Let $n \in \mathbb{N}$. If $\mathcal{M} = \mathcal{B}(\mathbb{F}^n)$, the Banach algebra of $n \times n$ matrices over $\mathbb{F}$, then the spectrum of a matrix is precisely the set of eigenvalues of the matrix.
- In particular, $\mathcal{B}(\mathbb{R}^2)$ contains a matrix such as $\begin{bmatrix} 0 & 1 \\ -1 & 0 \end{bmatrix}$ whose spectrum is empty. We will show in a moment that that this is never the case for $\mathbb{F} = \mathbb{C}$.
- When $\mathcal{M} = \mathcal{B}(X)$ for some Banach space X, elements in $\mathcal{B}(X)$ are bounded linear operators on X and the $\sigma(T)$ for $T \in \mathcal{B}(X)$, are called the **spectra of bounded linear operators** on X. Given $T \in \mathcal{B}(X)$, we have $\lambda \in \sigma(T)$ iff $(T - \lambda)$ is not invertible. Since $(T - \lambda) \in \mathcal{B}(X)$, it follows from the Inverse Mapping Theorem (Cor. 2.6) that $(T - \lambda) \in \mathcal{B}(X)^{-1}$ iff $(T - \lambda)$ is bijective. That is, the spectrum of T consists of exactly the $\lambda \in \mathbb{F}$ such that $T - \lambda$ is not bijective.

- Given an infinite dimensional normed linear space X, for any compact $T \in \mathcal{B}(X)$ we have by Example 3.4 that $T \notin \mathcal{B}(X)^{-1}$, therefore $0 \in \sigma(T)$. $\square$

Proposition 3.3. *Let $\mathcal{M}$ be a unital Banach algebra and $a \in \mathcal{M}$. Then*

(i) $\sigma(a) \subset \|a\| \, \bar{B}_{\mathbb{F}}$;
(ii) $\sigma(a)$ *is compact.*

Proof. (i): By Lem. 3.1, for every $\lambda \in \mathbb{F}$, if $\lambda \notin \|a\| \, \bar{B}_{\mathbb{F}}$, *i.e.* $|\lambda| > \|a\|$, we have $(a - \lambda) \in \mathcal{M}^{-1}$, hence $\lambda \notin \sigma(a)$.

(ii): Let $\lambda \in \big(\mathbb{F} \backslash \sigma(a)\big)$. Then for any $\alpha \in {}^*\mathbb{F}$, if $\alpha \approx \lambda$, then $(a - \alpha) \approx (a - \lambda)$. Since $(a - \lambda) \in X^{-1}$, it follows from Cor. 3.1 that $(a - \alpha) \in {}^*X^{-1}$, *i.e.* $\alpha \in {}^*\big(\mathbb{F} \backslash \sigma(a)\big)$. Therefore $\sigma(a)$ is closed. Together with (i), we conclude that $\sigma(a)$ is compact. $\square$

Observe that, by Prop. 3.3(i), in the precious example, for a bounded linear operator T on a Banach space and $\alpha \in \mathbb{F}$, if $|\lambda| > \|T\|$, then $(T - \lambda)$ is bijective.

Due to Prop. 3.3, we define the **spectral radius** of an element a in a unital normed algebra with $\sigma(a) \neq \emptyset$ as

$$\rho(a) := \sup\{|\lambda| \mid \lambda \in \sigma(a)\} \quad \Big(= \max\{|\lambda| \mid \lambda \in \sigma(a)\}\Big).$$

So by Prop. 3.3, we have $\rho(a) \leq \|a\|$. We will consider the case when equality does hold. In general, the gap between the two terms could be arbitrary, see Example 3.8 below.

The following is useful for comparing the spectrum of ab with that of ba in a noncommutative unital Banach algebra.

Proposition 3.4. *Let $\mathcal{M}$ be a unital Banach algebra and $a, b \in \mathcal{M}$. Then*

$$\forall \lambda \in (\mathbb{C} \backslash \{0\}) \, \Big(\big(\lambda \in \sigma(ab)\big) \Leftrightarrow \big(\lambda \in \sigma(ba)\big)\Big).$$

In particular, $\rho(ab) = \rho(ba)$.

Proof. Let $0 \neq \lambda \in \mathbb{C}$. If $\lambda \notin \sigma(ab)$, then for some $c \in \mathcal{M}$ we have

$$(ab - \lambda)c = 1 = c(ab - \lambda), \quad \text{hence} \quad abc = 1 + \lambda c = cab,$$

from which we get $babca = ba + \lambda bca = bcaba$, implying

$$(ba - \lambda)\Big(\frac{bca - 1}{\lambda}\Big) = 1 = \Big(\frac{bca - 1}{\lambda}\Big)(ba - \lambda).$$

Therefore $\lambda \notin \sigma(ba)$.

By a symmetry argument, we have $\lambda \notin \sigma(ab)$ iff $\lambda \notin \sigma(ba)$. $\square$

It turns out that any complex Banach space X shares enough common features with $\mathbb{C}$ to admit a generalized version of classical complex analysis to functions taking values in X, *i.e.* X-valued functions. Moreover this generalization of complex analysis gives us access to many useful tools in studying complex Banach algebras. To be more precise and for the ease of references, we briefly mention the following.

Let X be a complex Banach space. Let F be an X-valued function from $\mathbb{C}$, *i.e.* $F : \mathrm{Dom}(F) \to X$ with $\mathrm{Dom}(F) \subset \mathbb{C}$.

- Let $\alpha \in \mathrm{Dom}(F)$. We say that F is **differentiable at** α if
  $$\lim_{\Delta z \to 0} \frac{F(\alpha + \Delta z) - F(\alpha)}{\Delta z} \text{ exists in } X.$$
- F is **analytic at** α if it is differentiable at every point in an open set containing α.
- F is **analytic** if it is analytic at every point in $\mathrm{Dom}(F)$.
- F is **entire** if $\mathrm{Dom}(F) = \mathbb{C}$ and F is analytic (equivalently differentiable) at every $\alpha \in \mathbb{C}$.

Let $\gamma : [0,1] \to \mathbb{C}$ parametrize a curve $\mathcal{C}$ in $\mathbb{C}$.

- $\mathcal{C}$ is **closed** if $\gamma(0) = \gamma(1)$.
- $\mathcal{C}$ is **smooth** if γ is continuously differentiable.
- $\mathcal{C}$ is **piecewise smooth** if γ is continuously differentiable at all but finitely many points in $[0,1]$.
- $\mathcal{C}$ is **simple** if $\forall s, t \in [0,1] \left((\gamma(s) = \gamma(t)) \Rightarrow (s = t) \right)$.

Let $U \subset \mathbb{C}$ be open and $\mathcal{C} \subset U$ be a curve parametrized by a continuous $\gamma : [0,1] \to \mathbb{C}$. Then for a continuous $F : U \to X$, the **line integral** $\int_{\mathcal{C}} F(z)dz$ is defined as the limit of Riemann sums in exactly the same way as in the $\mathbb{C}$-valued case. That is,

$$\int_{\mathcal{C}} F(z)dz = \lim \sum_{k=1}^{n} F(\gamma(t_k))(t_k - t_{k-1}),$$

where the limit is taken over all partitions of $[0,1]$ as $0 = t_0 < t_1 < \cdots < t_n = 1$, $n \in \mathbb{N}$. By continuity, the limit exists and so $\int_{\mathcal{C}} F(z)dz \in X$.

Repeating *verbatim ac litteratim* the classical proofs, but with all $\mathbb{C}$-valued functions replaced by X-valued functions, we have the following results.

Theorem 3.2. (Cauchy's Theorem) *Let X be a complex Banach space and $F : U \to X$ be analytic, where $U \subset \mathbb{C}$ is open connected. Suppose that $\mathcal{C} \subset U$ is a piecewise smooth closed curve.*

Then $\displaystyle\int_{\mathcal{C}} F(z)dz = 0$. □

Theorem 3.3. (Cauchy Integral Formula) *Let X be a complex Banach space and $F : U \to X$ be analytic, where $U \subset \mathbb{C}$ is open. Suppose $\mathcal{C} \subset U$ is a counterclockwise oriented simple closed curve and α is a point inside $\mathcal{C}$. Then*

$$F^{(n)}(\alpha) = \frac{n!}{2\pi i} \int_{\mathcal{C}} \frac{F(z)}{(z-\alpha)^{n+1}} dz, \quad n \in \mathbb{N}.$$

□

As a consequence of the Cauchy's Integral Formula, we also have the following.

Theorem 3.4. (Liouville's Theorem) *Let X be a complex Banach space and $F : \mathbb{C} \to X$ be entire and bounded. Then F is a constant function.* □

Although the above are stated for functions taking values in a complex Banach space, our attention is mostly on those taking values in a unital complex Banach algebra, as the multiplicative inverse is often needed.

Now we apply Liouville's Theorem to obtain a basic result.

Theorem 3.5. *Let $\mathcal{M}$ be a unital complex Banach algebra. Then for every $a \in \mathcal{M}$, $\sigma(a) \neq \emptyset$.*

Proof. Let $a \in \mathcal{M}$ and suppose contrary to the claim that $\sigma(a) = \emptyset$. This means that $(a - \lambda) \in \mathcal{M}^{-1}$ for every $\lambda \in \mathbb{C}$.

So we can define $F : \mathbb{C} \to \mathcal{M}$ by $F(\lambda) = (a - \lambda)^{-1}$.

Now let $\epsilon \in {}^*\mathbb{C}$ such that $0 \neq \epsilon \approx 0$. Apply (3.1) and (3.2) in the proof of Thm. 3.1 to $(a - \lambda)$ and $(a - \lambda - \epsilon)$ in place of a and b, we have

$$\frac{(a - \lambda - \epsilon)^{-1} - (a - \lambda)^{-1}}{\epsilon} = (a - \lambda)^{-1}\left((a - \lambda)(a - \lambda - \epsilon)^{-1}\right)(a - \lambda)^{-1}$$

$$\approx (a - \lambda)^{-2}.$$

Hence, by saturation,

$$\lim_{\Delta z \to 0} \frac{F(\lambda + \Delta z) - F(\lambda)}{\Delta z} = (a - \lambda)^{-2} \in \mathcal{M}.$$

In particular, F is differentiable at every $\lambda \in \mathbb{C}$, *i.e.* F is an entire function.

Observe that, for $\lambda \in \mathbb{C}$, as $|\lambda| \to \infty$,

$$(a - \lambda)^{-1} = \left(\lambda(\lambda^{-1}a - 1) \right)^{-1} = \left(\lambda^{-1}a - 1 \right)^{-1} \lambda^{-1}$$

$$= -\lambda^{-1} \sum_{n=0}^{\infty} \lambda^{-n} a^n \to 0,$$

hence F is a bounded function.

Then by Liouville's Theorem (Thm. 3.4), since F is bounded and entire, it must be a constant function.

But the above calculation shows that $F(\lambda) \to 0$ as $|\lambda| \to \infty$, so F must be the zero function, leading to $a^{-1} = F(0) = 0$, impossible.

So we conclude that $\sigma(a) \neq \emptyset$. $\square$

Recall that a unital associative algebra is called a **division algebra** if every nonzero element is invertible. So $\mathbb{C}$ is a division algebra over itself. In fact it is the only one among unital complex Banach algebras.

Corollary 3.4. (Gelfand-Mazur Theorem) *Let $\mathcal{M}$ be a unital complex Banach algebra. Then $\mathcal{M}$ is a division algebra iff $\mathcal{M} = \mathbb{C}$.*

Proof. For the nontrivial direction, let $\mathcal{M}$ be a unital complex Banach division algebra. Then for any $a \in \mathcal{M}$, by Thm. 3.5, let $\lambda \in \sigma(a)$, then $(a - \lambda) \notin \mathcal{M}^{-1}$, hence $(a - \lambda) = 0$, by assumption, *i.e.* $a \in \mathbb{C}$.

Therefore $\mathcal{M} = \mathbb{C}$. $\square$

Corollary 3.5. *Let $\mathcal{M}$ be a commutative unital complex Banach algebra and $I \subset \mathcal{M}$ a maximal ideal. Then $\mathcal{M}/I \cong \mathbb{C}$, i.e. the quotient algebra is isometrically isomorphic to $\mathbb{C}$ as a Banach algebra.*

Proof. By Cor.3.3, I is a closed proper ideal and so, by Prop.3.1, $\mathcal{M}/I$ is a unital complex Banach algebra which is clearly commutative as well. Let $J \subset (\mathcal{M}/I)$ be any ideal. Let

$$J_0 := \{ x \in \mathcal{M} \,|\, (x + I) \in J \}.$$

Then it is easy to check that $J_0 \subset \mathcal{M}$ forms an ideal. Moreover, since $I = (0 + I) \in J$, we have $\forall x \in I \left((x + I) \in J \right)$, *i.e.* $I \subset J_0$. Unless J is the zero algebra, *i.e.* $J = \{(0 + I)\} = \{I\}$, we have $I \subsetneq J_0$, hence $J_0 = \mathcal{M}$ by I being maximal, resulting $J = (\mathcal{M}/I)$.

In other word, $\mathcal{M}/I$ has no proper ideals, consequently, for every $a \in (\mathcal{M} \setminus I)$, we have $(a + I)(\mathcal{M}/I) = (\mathcal{M}/I)$ implying $(a + I) \in (\mathcal{M}/I)^{-1}$.

Therefore $\mathcal{M}/I$ is a division algebra and so $\mathcal{M}/I \cong \mathbb{C}$ by Cor. 3.4. $\square$

When comparing Thm. 3.5 with the example in Example 3.7, we see that while the spectrum could be empty for an element in a real normed algebra, this is never the case for complex normed algebras.

Consequently, in a unital complex Banach algebra, the spectral radius is defined for all elements.

Due to the availability of tools in spectrum analysis essential for many important results in normed algebras, for the rest of this chapter, we make the following restriction:

$$\boxed{\mathbb{F} = \mathbb{C}.}$$

i.e.

> Unless otherwise stated, all Banach algebras are complex Banach algebras.

For notational clarity when dealing with nonstandard extensions such as $^{*}f$, $^{*}\sigma(^{*}a)$, ..., especially later when we also use $*$ for the involution operation, we declare henceforth the following:

> If the reference is clear from the context, the $*$ sign is allowed to be dropped from the notation for a nonstandard extension.

Given a complex polynomial $p(z) = \sum_{k=0}^{n} \lambda_k z^k$ and an element a in a Banach algebra $\mathcal{M}$, then $\sum_{k=0}^{n} \lambda_k a^k$ is defined and is an element in $\mathcal{M}$ denoted by $p(a)$. In a moment, this will be extended to all analytic functions.

Proposition 3.5. *Let $\mathcal{M}$ be a unital Banach algebra and $a \in \mathcal{M}$. Let $p(z)$ be a polynomial over $\mathbb{C}$. Then $p[\sigma(a)] = \sigma(p(a))$.*

Proof. The results is trivial if p is a constant function, so we assume that $\deg(p) = n \geq 1$.

Let $\lambda \in \mathbb{C}$ and consider the polynomial $q(z) := p(z) - \lambda$. By the Fundamental Theorem of Algebra, $q(z)$ factorizes into $\alpha \prod_{k=1}^{n}(z - \alpha_k)$ for some

$\alpha, \alpha_1, \ldots, \alpha_n \in \mathbb{C}$ with $\alpha \neq 0$. Then we have the following equivalence:

$$\lambda \notin \sigma(p(a)) \Leftrightarrow (p(a) - \lambda) \in \mathcal{M}^{-1} \qquad \Leftrightarrow \prod_{k=1}^{n} (a - \alpha_k) \in \mathcal{M}^{-1}$$

$$\Leftrightarrow (a - \alpha_1), \ldots, (a - \alpha_n) \in \mathcal{M}^{-1} \quad \Leftrightarrow \alpha_1, \ldots, \alpha_n \notin \sigma(a)$$

$$\Leftrightarrow \forall \gamma \in \sigma(a)(q(\gamma) \neq 0) \qquad \Leftrightarrow \forall \gamma \in \sigma(a)(\lambda \neq p(\gamma))$$

$$\Leftrightarrow \lambda \notin p(\sigma(a)).$$

Since this holds for all $\lambda \in \mathbb{C}$, we conclude that $p[\sigma(a)] = \sigma(p(a))$. $\square$

Example 3.8. Let $\mathcal{M}$ be a unital Banach algebra and $a \in \mathcal{M}$ an idempotent element, *i.e.* $a^2 = 0$. Then since $\sigma(a^2) = \{0\}$, so by Prop. 3.5, $\sigma(a) = \{0\}$. *i.e.* $\rho(a) = 0$. In particular, if $\mathcal{M}$ has a nonzero idempotent, then for any $r \in \mathbb{R}^+$, there is $a \in \mathcal{M}$ such that $\rho(a) = 0$ but $\|a\| \geq r$.

As a concrete example, let $\mathcal{M}$ be the Banach algebra of complex 2×2-matrices, then $\begin{bmatrix} 0 & 0 \\ r & 0 \end{bmatrix}$ satisfies the requirement. $\square$

Observe that series in a Banach algebra $\mathcal{M}$ behave just like series in $\mathbb{C}$. For example, the nth root test works in the same way in a Banach algebra $\mathcal{M}$ as in $\mathbb{C}$: let $\{a_n\}_{n \in \mathbb{N}} \subset \mathcal{M}$, then $\sum_{n=0}^{\infty} a_n$ converges in $\mathcal{M}$ if $\limsup_{n \to \infty} \|a_n\|^{1/n} < 1$ and diverges if it is > 1.

Given a function $f : D \to \mathcal{M}$, where $D \subset \mathbb{C}$ and $\mathcal{M}$ is a Banach algebra, as in classical complex analysis, if f is analytic at some $\alpha \in D$, then, near α, f can be represented by some power series $f(z) = \sum_{n=0}^{\infty} (z - \alpha)^n a_n$ for some $\{a_n\}_{n \in \mathbb{N}} \subset \mathcal{M}$. Moreover, by the nth root test, the power series has radius of convergence given by $\left(\limsup_{n \to \infty} \|a_n\|^{1/n} \right)^{-1}$.

Next we prove a useful formula for calculating the spectral radius.

Theorem 3.6. *Let $\mathcal{M}$ be a unital Banach algebra and $a \in \mathcal{M}$.*
Then $\lim_{n \to \infty} \|a^n\|^{1/n}$ exists and $\rho(a) = \lim_{n \to \infty} \|a^n\|^{1/n}$.

Proof.   First note that, by Prop. 3.5, we have for all $n \in \mathbb{N}$ that

$$\sigma(a^n) = \sigma(a)^n \quad (i.e. \ \ \{\lambda^n \mid \lambda \in \sigma(a)\}.)$$

Hence, for any $\lambda \in \sigma(a)$ and $n \in \mathbb{N}$, we have $|\lambda|^n = |\lambda^n| \leq \|a^n\|$, by Prop. 3.3(i). In particular, $|\lambda| \leq \|a^n\|^{1/n}$.

Therefore

$$\rho(a) \leq \liminf_{n \to \infty} \|a^n\|^{1/n}. \tag{3.3}$$

Now let $\lambda \in \mathbb{C}$ such that $|\lambda| > \rho(a)$. As $\lambda \notin \sigma(a)$, we get $(a - \lambda) \in \mathcal{M}^{-1}$. By Lem. 3.1 and $\lambda \neq 0$,

$$(a - \lambda)^{-1} = \lambda^{-1} (\lambda^{-1} a - 1)^{-1} = -\lambda^{-1} \sum_{n=0}^{\infty} (\lambda^{-1} a)^n.$$

As the series converges, it follows from the nth root test that $\limsup_{n \to \infty} |\lambda^{-1}| \, \|a_n\|^{1/n} \leq 1$. So we have shown that

$$\forall \lambda \in \mathbb{C} \left(\big(|\lambda| > \rho(a) \big) \Rightarrow \big(\limsup_{n \to \infty} \|a_n\|^{1/n} \leq |\lambda| \big) \right),$$

hence

$$\limsup_{n \to \infty} \|a^n\|^{1/n} \leq \rho(a). \tag{3.4}$$

Now (3.3) and (3.4) together show that the limit $\lim_{n \to \infty} \|a^n\|^{1/n}$ exists and equals $\rho(a)$. $\qquad \square$

As remarked on p.190, in general, the spectrum of an element depends on the invertibility relative to the subalgebra containing the element. As the number $\lim_{n \to \infty} \|a^n\|^{1/n}$ in Thm. 3.6 is independent of the subalgebra chosen, we have the following result for the spectral radius.

Corollary 3.6. *Let $\mathcal{M}$ be a unital Banach algebra and let $\mathcal{M}_0 \subset \mathcal{M}$ be a closed subalgebra such that $1 \in \mathcal{M}_0$.*
Then $\rho_{\mathcal{M}_0}(a) = \rho_{\mathcal{M}}(a)$ holds for all $a \in \mathcal{M}_0$.
That is the spectral radius does not depend on the invertibility relative to the subalgebra in which the element is contained. $\qquad \square$

In a unital normed algebra over $\mathbb{F}$, for any element a and a polynomial p over $\mathbb{F}$, $p(a)$ is defined in an obvious way as an element of the algebra. Hence we can define the polynomial of an element from any unital normed algebra. Now we want to show that in the case of unital complex Banach algebras, this can be extended to analytic functions by the following procedure.

Let $\mathcal{M}$ be a unital Banach algebra and $a \in \mathcal{M}$. Let $f : U \to \mathbb{C}$ be a analytic function with open domain $U \subset \mathbb{C}$ such that $\sigma(a) \subset U$. Then we define $f(a) \in \mathcal{M}$ as follows.

By Prop. 3.3 and Thm. 3.5, $\sigma(a)$ is a nonempty compact subset of $\mathbb{C}$. In general, $\sigma(a)$ may not be connected and even if it is, it need not be **simply connected**, *i.e.* it could have a hole inside. But in the case $\sigma(a)$ is simply connected, there is a simple closed smooth curve $\mathcal{C} \subset U$ in counterclockwise direction such that $\sigma(a)$ is strictly inside $\mathcal{C}$. Then we simply define

$$f(a) := \frac{1}{2\pi i} \int_{\mathcal{C}} f(z)(z - a)^{-1} dz \in \mathcal{M}.$$

Note that as $z \in \mathcal{C}$, $z \notin \sigma(a)$, hence $(z - a) \in \mathcal{M}^{-1}$. Moreover, the function $z \mapsto (z - a)^{-1} \in \mathcal{M}$ is a continuous function defined on an open set which includes $\mathcal{C}$.

Observe that in the above if $\mathcal{C}'$ is another simple closed smooth curve satisfying the same requirements, then $\int_{\mathcal{C}-\mathcal{C}'} f(z)(z - a)^{-1}dz = 0$, as a consequence of Cauchy's Theorem (Thm. 3.2), hence $f(a)$ is well-defined.

For the general case where $\sigma(a)$ is not necessarily simply connected, we use the fact that $\sigma(a)$ is compact to get a finite family of simple closed smooth curves $\mathcal{C}_1, \ldots, \mathcal{C}_n \subset U$ in counterclockwise direction such that

- For $z \in \sigma(a)$, the winding numbers of the $\mathcal{C}_i$'s around z sum up to 1;
- for $z \in \mathbb{C} \setminus U$, the winding numbers of the $\mathcal{C}_i$'s around z sum up to 0.

(Recall that the **winding number** of a closed piecewise smooth curve $\mathcal{C}$ around a point $z \in \mathbb{C} \setminus \mathcal{C}$ is the number of times it winds around z as measured in the counterclockwise direction. Moreover, the number equals to $\frac{1}{2\pi i} \int_{\mathcal{C}} (w - z)^{-1} dw$.)

Then we define

$$f(a) := \frac{1}{2\pi i} \sum_{i=1}^{n} \int_{\mathcal{C}_i} f(z)(z - a)^{-1}dz \in \mathcal{M}.$$

As in the case where $\sigma(a)$ is simply connected, the above $f(a)$ is well-defined and is independent of the choice of the $\mathcal{C}_i$'s as long as they satisfy the required properties.

We remark that for $n \in \mathbb{N}$, if $f(z) = z^n$, then $f(a) = a^n$. Therefore the above definition extends the polynomial functions on $\mathcal{M}$.

Furthermore, if f is a power series $\sum_{n=0}^{\infty} \lambda_n z^n$ for some $\lambda_n \in \mathbb{C}$ and $\rho(a)$ is less than the radius of convergence of the series, then $\sum_{n=0}^{\infty} \lambda_n \|a^n\|$ converges and $f(a) = \sum_{n=0}^{\infty} \lambda_n a^n$.

As a special case, in a unital Banach algebra $\mathcal{M}$, the **exponential function** is defined for any $a \in \mathcal{M}$

$$e^a := \sum_{n=0}^{\infty} \frac{a^n}{n!}.$$

As another example, suppose $\sigma(a) \subset B_{\mathbb{C}}$, i.e. $\rho(a) < 1$, then

$$\sqrt{1 + a} = \sum_{n=0}^{\infty} \binom{\frac{1}{2}}{n} a^n = \sum_{n=0}^{\infty} \frac{(-1)^n (2n)!}{(1 - 2n)2^{2n}(n!)^2} a^n,$$

by using the branch of the square root function $\sqrt{1 + z}$ that contains the positive real roots.

The following is left as an exercise to check.

Theorem 3.7.   (Riesz Functional Calculus) *Let $\mathcal{M}$ be a unital Banach algebra. Then for any $a \in \mathcal{M}$ and analytic $f, g : U \to \mathbb{C}$, where $U \subset \mathbb{C}$ is open and $\sigma(a) \subset U$, the following holds:*

(i) $\forall \lambda \in \mathbb{C} \left((f + \lambda g)(a) = f(a) + \lambda g(a) \right)$.

(ii) $(fg)(a) = f(a)g(a) = (gf)(a)$.

(iii) *If $f = 1$, is the constant unit function, then $f(a) = 1$.*

(iv) *If $f = \mathrm{id}$, is the identity function, then $f(a) = a$.*

(v) *Suppose $f_n : U \to \mathbb{C}$, $n \in \mathbb{N}$, are analytic and converge uniformly to f on U, then $f(a) = \lim_{n \to \infty} f_n(a)$.* $\square$

The following is a generalization of Prop. 3.5.

Corollary 3.7.   (The Spectral Mapping Theorem) *Let $\mathcal{M}$ be a unital Banach algebra, $a \in \mathcal{M}$ and $f : U \to \mathbb{C}$ be analytic, where U is open with $\sigma(a) \subset U \subset \mathbb{C}$.*
  Then $f[\sigma(a)] = \sigma(f(a))$.

Proof.   ($\subset$): Let $\lambda \in \sigma(a)$.
  Suppose $f(\lambda) \notin \sigma(f(a))$, then $\left(f(a) - f(\lambda) \right) \in \mathcal{M}^{-1}$.
  Note that $\left(f(z) - f(\lambda) \right) = (z - \lambda)g(z)$ for some analytic $g : U \to \mathbb{C}$ since $\left(f(z) - f(\lambda) \right)(z - \lambda)^{-1}$ is analytic in U outside a neighborhood around λ and $f(z) - f(\lambda)$ expands into a power series in $(z - \lambda)$ when z is near λ. By Thm. 3.7 and the commutativity between $(a - \lambda)$ and $g(a)$, we have

$$\left(f(a) - f(\lambda) \right) = (a - \lambda)g(a) = g(a)(a - \lambda).$$

Hence

$$\left(f(a) - f(\lambda) \right)^{-1} g(a)(a - \lambda) = 1 = (a - \lambda)g(a)\left(f(a) - f(\lambda) \right)^{-1}$$

and $(a - \lambda) \in \mathcal{M}^{-1}$ as it possesses both a left- and a right-inverse. (See p.187.)
  Therefore $\lambda \notin \sigma(a)$, a contradiction.

  ($\supset$): Let $\lambda \in \sigma(f(a))$. Suppose that $\lambda \notin f[\sigma(a)]$. Then there is an open $V \subset \mathbb{C}$ such that $\sigma(a) \subset V$ and $\forall z \in V \left(f(z) \neq \lambda \right)$.
  Therefore the function $\left(f(z) - \lambda \right)^{-1}$ is analytic on V and it follows from Thm. 3.7 that $\left(f(a) - \lambda \right)^{-1}$ is defined and is the inverse of $(f(a) - \lambda)$ in $\mathcal{M}$, *i.e.* $\lambda \notin \sigma(f(a))$, a contradiction. $\square$

200   *Nonstandard Methods in Functional Analysis*

Example 3.9.

- From Example 3.7, for an infinite dimensional unital Banach space X, if $T \in \mathcal{B}(X)$ is compact, it is necessary that $0 \in \sigma(T)$. Hence, for any $T \in \mathcal{B}(X)$, e^T is not compact, since $0 \notin e^{\sigma(T)} = \sigma(e^T)$ by the Spectral Mapping Theorem (Cor. 3.7).
- In an unital Banach space $\mathcal{M}$, for $a \in \mathcal{M}^{-1}$, we have $0 \notin \sigma(a)$. Hence there is a neighborhood including $\sigma(a)$ on which the reciprocal function is analytic. Therefore it holds for any $\lambda \in \mathbb{C}$ that

$$(a - \lambda) \in \mathcal{M}^{-1} \text{ iff } \lambda \notin \sigma(a) \text{ iff } \lambda^{-1} \notin \sigma(a^{-1}) \text{ iff } (a^{-1} - \lambda^{-1}) \in \mathcal{M}^{-1}.$$

$\square$

The following useful formula is obtained by a simple application of the Spectral Mapping Theorem similar to the above example.

Corollary 3.8. *Let $\mathcal{M}$ be a unital Banach algebra, $a \in \mathcal{M}$ and $\lambda \in \mathbb{C} \setminus \sigma(a)$.*
Then $\mathrm{dist}(\lambda, \sigma(a)) = \rho((a - \lambda)^{-1})^{-1}$.

Proof. Since $\lambda \notin \sigma(a)$, there is a neighborhood of $\sigma(a)$ on which the function $f(z) = (z - \lambda)^{-1}$ is analytic.
Then By the Spectral Mapping Theorem (Cor. 3.7),

$$\sigma((a - \lambda)^{-1}) = \left(\sigma(a) - \lambda\right)^{-1}, \quad i.e. \ \{(\alpha - \lambda)^{-1} \mid \alpha \in \sigma(a)\}.$$

Hence

$$\rho((a - \lambda)^{-1}) = \sup\{|\alpha - \lambda|^{-1} \mid \alpha \in \sigma(a)\}$$

$$= \frac{1}{\inf\{|\alpha - \lambda| \mid \alpha \in \sigma(a)\}} = \frac{1}{\mathrm{dist}(\lambda, \sigma(a))}.$$

$\square$

Corollary 3.9. *Let $\mathcal{M}$ be a unital Banach algebra and let $\mathcal{M}_0 \subset \mathcal{M}$ be a closed subalgebra such that $1 \in \mathcal{M}_0$. Let $\lambda \in \mathbb{C} \setminus \sigma_{\mathcal{M}_0}(a)$.*
Then $\mathrm{dist}(\lambda, \sigma_{\mathcal{M}_0}(a)) = \mathrm{dist}(\lambda, \sigma_{\mathcal{M}}(a))$.

Proof. Since $\lambda \in \mathbb{C} \setminus \sigma_{\mathcal{M}_0}(a)$, we have $\lambda \in \mathbb{C} \setminus \sigma_{\mathcal{M}}(a)$. Then by Cor. 3.6, $\rho_{\mathcal{M}_0}((a - \lambda)^{-1})^{-1} = \rho_{\mathcal{M}}((a - \lambda)^{-1})^{-1}$, hence the conclusion follows from Cor. 3.8. $\square$

As a straightforward consequence of Thm. 3.7, in a unital Banach algebra $\mathcal{M}$, for any $a \in \mathcal{M}$, $e^a \in \mathcal{M}^{-1}$ and

$$(e^a)^{-1} = \sum_{n=0}^{\infty} \frac{(-a)^n}{n!}.$$

Moreover, for $a, b \in \mathcal{M}$ with $ab = ba$, we let $e^a \approx \sum_{n=0}^{N} \dfrac{a^n}{n!}$ and $e^b \approx \sum_{n=0}^{N} \dfrac{b^n}{n!}$, where $N \in {}^*\mathbb{N} \setminus \mathbb{N}$, then

$$e^a e^b \approx \sum_{n,m=0}^{N} \frac{a^n b^m}{n! m!} = \sum_{k=0}^{2N} \sum_{\substack{n+m=k \\ 0 \leq n,m \leq N}} \binom{k}{n} \frac{a^n b^m}{k!} \approx \sum_{n=0}^{N} \frac{(a+b)^n}{n!} \approx e^{a+b}.$$

In particular, e^a and e^b commutes as well.

In the above, an easy fact was used: In ${}^*\mathcal{M}$ it holds that $x_1 y_1 \approx x_2 y_2$ whenever $x_1 \approx x_2$ and $y_1 \approx y_2$. More will be dealt with in the next subsection.

Now the following is defined:

$$\exp(\mathcal{M}) := \left\{ e^{a_1} \cdots e^{a_n} \mid a_1, \ldots, a_n \in \mathcal{M},\ n \in \mathbb{N} \right\}.$$

We let $e^{\mathcal{M}}$ denote the set $\{e^a \mid a \in \mathcal{M}\}$. So for a commutative unital Banach algebra $\mathcal{M}$, we get $e^{\mathcal{M}} = \exp(\mathcal{M})$.

Clearly $\exp(\mathcal{M})$ is closed under finite products. It is also closed under inverse: for an element $e^{a_1} \cdots e^{a_n}$ from $\exp(\mathcal{M})$, its inverse is simply $e^{-a_n} \cdots e^{-a_1}$. We have seen that $\exp(\mathcal{M}) \subset \mathcal{M}^{-1}$, thus $\exp(\mathcal{M})$ is a subgroup of $\mathcal{M}^{-1}$ under the multiplication.

Moreover, $\exp(\mathcal{M})$ is a **clopen** subset of $\mathcal{M}^{-1}$, *i.e.* both open and closed. The little fact below will be used to prove it.

Lemma 3.2. *Let $\mathcal{M}$ be an internal unital Banach algebra. Let $a, b \in \mathcal{M}$ such that $a \in \mathcal{M}^{-1}$ with $a^{-1} \in \mathrm{Fin}(\mathcal{M})$ and $a \approx b$.*

Then $\sigma\big(a^{-1}b\big) \subset \mu(1)$, the monad of $1 \in {}^*\mathbb{C}$.

Proof. From $b \approx a$ and $a^{-1} \in \mathrm{Fin}(\mathcal{M})$ we have

$$\left\| a^{-1}b - 1 \right\| = \left\| a^{-1}(b - a) \right\| \leq \left\| a^{-1} \right\| \left\| b - a \right\| \approx 0,$$

i.e. $(a^{-1}b - 1) \approx 0$. Hence, by Prop.3.3, $\sigma\big(a^{-1}b - 1\big) \subset \mu(0)$. Let $f(z) = z - 1$, then by the Spectral Mapping Theorem (Cor. 3.7),

$$\sigma\big(a^{-1}b\big) - 1 = f\big[\sigma(a^{-1}b)\big] = \sigma\big(f(a^{-1}b)\big) = \sigma\big(a^{-1}b - 1\big) \subset \mu(0).$$

That is, $\sigma\big(a^{-1}b\big) \subset \mu(1)$.   □

First we show that $\exp(\mathcal{M})$ is open.

Let $a \in \exp(\mathcal{M})$ and $b \in {}^*\mathcal{M}$ such that $b \approx a$. Applying Lem. 3.2 to ${}^*\mathcal{M}$, we have $\sigma\big(a^{-1}b\big) \subset \mu(1)$. In particular, the principal logarithm function $\mathrm{Log}(z)$ is analytic in an open neighborhood of $\sigma\big(a^{-1}b\big)$.

Then we have $b = a\big(a^{-1}b\big) = a\, e^{\mathrm{Log}(a^{-1}b)} \in \exp({}^*\mathcal{M})$.

Therefore $\exp(\mathcal{M})$ is open.

Next we show that $\exp(\mathcal{M})$ is a closed subset of $\mathcal{M}^{-1}$.

So for any $a \in \mathcal{M}^{-1}$ and $b \in {}^*\mathcal{M}$, if $a \approx b$ and $b \in \exp({}^*\mathcal{M})$, we must show that $a \in \exp(\mathcal{M})$.

Clearly, as $b \in \exp({}^*\mathcal{M})$, we have $b \in {}^*\mathcal{M}^{-1}$.

Also, write $c = a^{-1}(a - b)$, then $b = a(1 - c)$ and from $b \approx a$ we have $0 \approx c \in {}^*\mathcal{M}$. Then, by Lem. 3.1, $(1 - c) \in {}^*\mathcal{M}^{-1}$ with $\big\|(1 - c)^{-1}\big\| = \big\|\sum_{n=0}^{\infty} c^n\big\| \approx 1$.

Hence $b^{-1} = (1 - c)^{-1}a^{-1} \approx a^{-1}$. Consequently, $b^{-1} \in \mathrm{Fin}({}^*\mathcal{M})$.

Then by Lem. 3.2, we have $\sigma\big(b^{-1}a\big) \subset \mu(1)$. As above, we can apply the principal logarithm and obtain

$$a = b\big(b^{-1}a\big) = b\, e^{\mathrm{Log}(b^{-1}a)} \in \exp({}^*\mathcal{M}),$$

and so $a \in \exp(\mathcal{M})$ by transfer.

Therefore $\exp(\mathcal{M})$ is a closed subset of $\mathcal{M}^{-1}$.

(Note that the above did not show that $\exp(\mathcal{M})$ is a closed subset of $\mathcal{M}$, which is false.)

$\exp(\mathcal{M})$ is a path-connected set, since for any element $e^{a_1} \cdots e^{a_n}$ from $\exp(\mathcal{M})$, the function

$$[0, 1] \ni t \mapsto e^{ta_1} \cdots e^{ta_n}$$

is a continuous path in $\mathcal{M}^{-1}$ from $1 = e^0$ to $e^{a_1} \cdots e^{a_n}$.

A subset X of a topological space is connected if X is not the union of more than one disjoint nonempty open subsets of X. Path-connected sets are connected. A maximal connected subset of X is called a **connected component** of X. So a connected clopen subset of X is a connected component.

The following summarize what was just proved:

Theorem 3.8. *Let $\mathcal{M}$ be an unital Banach algebra.*

Then $\exp(\mathcal{M})$ is a clopen connected component of $\mathcal{M}^{-1}$. □

3.1.3   Nonstandard hulls

We define in this subsection the nonstandard hull of an internal normed algebra and study its properties including those about invertibility and the spectrum.

Given an internal normed linear algebra $\mathcal{M}$ over $\mathbb{F}$, the nonstandard hull construction from p.81 given a Banach space $\widehat{\mathcal{M}}$.

For $a_1, a_2, b_1, b_2 \in \text{Fin}(\mathcal{M})$ such that $a_1 \approx a_2$ and $b_1 \approx b_2$, we have $\|a_1 b_1 - a_2 b_1\| \leq \|a_1 - a_2\| \|b_1\| \approx 0$ and $\|a_2 b_1 - a_2 b_2\| \leq \|a_2\| \|b_1 - b_2\| \approx 0$, hence $a_1 b_1 \approx a_2 b_2$, i.e. $\widehat{a_1 b_1} = \widehat{a_2 b_2}$.

Therefore the following product is well-defined for $\widehat{a}, \widehat{b} \in \widehat{\mathcal{M}}$:

$$\widehat{a}\,\widehat{b} := \widehat{ab}.$$

Clearly this product gives $\widehat{\mathcal{M}}$ a ring structure.

For $\widehat{a}, \widehat{b} \in \widehat{\mathcal{M}}$ and $\lambda \in \mathbb{F}$, we have

- $(\widehat{\lambda a})\widehat{b} = (\widehat{\lambda a})\,\widehat{b} = \widehat{\lambda ab} = \lambda\left(\widehat{ab}\right) = \lambda(\widehat{a}\,\widehat{b}) = \widehat{\lambda ab} = \widehat{a\lambda b} = \widehat{a}(\widehat{\lambda b}) = \widehat{a}(\lambda \widehat{b})$,
- $\|\widehat{a}\widehat{b}\| = \|\widehat{ab}\| \approx \|ab\| \leq \|a\| \|b\| \approx \|\widehat{a}\| \|\widehat{b}\|$,

therefore $\widehat{\mathcal{M}}$ forms a Banach algebra over $\mathbb{F}$.

Moreover, in case $\mathcal{M}$ is unital with unit 1, then

$$\widehat{1}\widehat{a} = \widehat{1 \cdot a} = \widehat{a} = \widehat{a \cdot 1} = \widehat{a}\widehat{1}$$

holds for all $\widehat{a} \in \widehat{\mathcal{M}}$, hence $\widehat{\mathcal{M}}$ is also unital. For convenience, the unit $\widehat{1}$ is denoted simply by 1 too.

It is also clear that if $\mathcal{M}$ is commutative, so is $\widehat{\mathcal{M}}$.

For a normed algebra $\mathcal{M}$, it is easy to see that the canonical embedding of $\mathcal{M}$ into $^*\mathcal{M}$ as Banach spaces gives the identification of $\mathcal{M}$ as a subalgebra of $^*\mathcal{M}$. This will always be assumed and $\mathcal{M} \subset \widehat{\mathcal{M}}$ is written with this meaning.

Example 3.10. Let X be an internal normed linear space, then by Prop. 2.16, $\widehat{\mathcal{B}(X)}$ embeds into $\mathcal{B}(\widehat{X})$ as a Banach space. It is straightforward to check that it also embeds as a subalgebra. So we write $\widehat{\mathcal{B}(X)} \subset \mathcal{B}(\widehat{X})$ as a subalgebra with this identification assumed. In general, it is a proper subalgebra.   □

A word about a notation. For an internal unital normed algebra $\mathcal{M}$ and $a \in \mathcal{M}^{-1}$, $a^{-1} \in \text{Fin}(\mathcal{M})$ is equivalent to $a = b^{-1}$ for some $b \in \text{Fin}(\mathcal{M})$, hence we also write $a \in (\text{Fin}(\mathcal{M}))^{-1}$ in this case.

Hence $(\text{Fin}(\mathcal{M}))^{-1}$ is the set of elements in $\mathcal{M}$ having finite inverses.

Proposition 3.6. Let $\mathcal{M}$ be an internal unital Banach algebra and $a \in \mathcal{M}$.

(i) If $a \in (\text{Fin}(\mathcal{M}))^{-1}$, then $\mu(a) \subset \mathcal{M}^{-1}$ and $(\mu(a))^{-1} \subset \mu(a^{-1})$.

(ii) If $a \in \left(\text{Fin}(\mathcal{M}) \cap (\text{Fin}(\mathcal{M}))^{-1}\right)$, then $(\mu(a))^{-1} = \mu(a^{-1})$.

Proof. (i): Let $b \in \mu(a)$, *i.e.* $b \in \mathcal{M}$ and $b \approx a$. Write $c := b - a \approx 0$.

Since $a^{-1} \in \text{Fin}(\mathcal{M})$, we have $a^{-1}c \approx 0$. By Lem. 3.1, $(1+a^{-1}c) \in \mathcal{M}^{-1}$. Hence

$$b = a + c = a(1 + a^{-1}c) \in \mathcal{M}^{-1}.$$

Since $b \in \mu(a)$ is arbitrary, we get $\mu(a) \subset \mathcal{M}^{-1}$.

To show that $(\mu(a))^{-1} \subset \mu(a^{-1})$, let $b \in \mu(a)$ and prove that $b^{-1} \approx a^{-1}$. So, still with $c = b - a$, notice the following:

$$a^{-1} - b^{-1} = a^{-1}(1 - ab^{-1}) = a^{-1}\left(1 - (ba^{-1})^{-1}\right)$$

$$= a^{-1}\left(1 - (1 + ca^{-1})^{-1}\right) = a^{-1}\left(1 - \sum_{n=0}^{\infty}(ca^{-1})^n\right)$$

$$= a^{-1}\sum_{n=1}^{\infty}(ca^{-1})^n \approx 0,$$

where we apply the second last equality Lem. 3.1 to $ca^{-1} \approx 0$.

Therefore $a^{-1} \approx b^{-1}$.

(ii): From (i), we already have $(\mu(a))^{-1} \subset \mu(a^{-1})$.

For the other inclusion, by noticing that since $a \in (\text{Fin}(\mathcal{M}) \cap \mathcal{M}^{-1})$, we have $a^{-1} \in (\text{Fin}(\mathcal{M}))^{-1}$, and so (i) gives

$$(\mu(a^{-1}))^{-1} \subset \mu((a^{-1})^{-1}) = \mu(a).$$

Therefore

$$\mu(a^{-1}) = \left((\mu(a^{-1}))^{-1}\right)^{-1} \subset (\mu(a))^{-1}. \qquad \square$$

The following characterizes the invertible elements in a nonstandard hull.

Proposition 3.7. *Let $\mathcal{M}$ be an internal unital Banach algebra. Then for any $a \in \text{Fin}(\mathcal{M})$,*

$$\widehat{a} \in (\widehat{\mathcal{M}})^{-1} \quad \textit{iff} \quad a \in (\text{Fin}(\mathcal{M}))^{-1}.$$

Proof. ($\Leftarrow$): If $a \in (\text{Fin}(\mathcal{M}))^{-1}$, then since $a^{-1} \in \text{Fin}(\mathcal{M})$, we have $\widehat{a^{-1}} \in \widehat{\mathcal{M}}$. But

$$\widehat{a}\,(\widehat{a^{-1}}) = \widehat{a(a^{-1})} = 1 = \widehat{(a^{-1})a} = (\widehat{a^{-1}})\,\widehat{a},$$

so $\widehat{a} \in (\widehat{\mathcal{M}})^{-1}$ with $\widehat{a}^{-1} = \widehat{a^{-1}}$.

($\Rightarrow$): Suppose $\widehat{a} \in \left(\widehat{\mathcal{M}}\right)^{-1}$, then $ab \approx 1 \approx ba$ holds for some $b \in \text{Fin}(\mathcal{M})$. By Lem. 3.1, as $(ab - 1) \approx 0$, $ab = \left(1 + (ab - 1)\right) \in \mathcal{M}^{-1}$, we let

$$c := \sum_{n=0}^{\infty} (ab - 1)^n \text{ to be the inverse of } ab.$$

Note that $c \approx 1$ and

$$a(bc) = (ab)c = 1,$$

i.e. a is right-invertible.

Likewise, as $(ba - 1) \approx 0$, $ba = \left(1 + (ba - 1)\right) \in \mathcal{M}^{-1}$, with the inverse of ba denoted by $c' := \sum_{n=0}^{\infty}(ba - 1)^n$, we have

$$(c'b)a = c'(ba) = 1,$$

i.e. a is left-invertible.

Now by the remarks on p.187, $a \in \mathcal{M}^{-1}$ and $a^{-1} = bc$, the right-inverse above. Since $c \approx 1$, we have $bc \approx b \in \text{Fin}(\mathcal{M})$, so $a^{-1} \in \text{Fin}(\mathcal{M})$. $\square$

Corollary 3.10. *Let $\mathcal{M}$ be an internal unital Banach algebra.*
Then for any $a \in \left(\text{Fin}(\mathcal{M}) \cap \left(\text{Fin}(\mathcal{M})\right)^{-1}\right)$ and $b \in \mathcal{M}$ with $b \approx a$, we
have $b \in \left(\text{Fin}(\mathcal{M}) \cap \left(\text{Fin}(\mathcal{M})\right)^{-1}\right)$.

Proof. We have $\widehat{b} = \widehat{a} \in \widehat{\mathcal{M}}$, so the conclusion follows by noticing that $b \in \text{Fin}(\mathcal{M})$, since $b \approx a$, and by applying Prop. 3.7 to $\widehat{b}$. $\square$

By combining Prop. 3.7 and Cor. 3.10, we can write:

Corollary 3.11. *Let $\mathcal{M}$ be an internal unital Banach algebra.*
Then $\left(\widehat{\mathcal{M}}\right)^{-1} = \left(\text{Fin}(\mathcal{M}) \cap \left(\text{Fin}(\mathcal{M})\right)^{-1}\right)^{\wedge}$. $\square$

As an example, take $\mathcal{M} = \mathbb{C}^2$ under the supremum norm and $a = (1, \epsilon)$, where ϵ is a nonzero infinitesimal. Then $\|\widehat{a}\| = 1$ and $\widehat{a}$ is not invertible, but a is invertible.

Contrary to the dependence of the spectrum on the underlying Banach algebra as remarked on p.190, we mention the following straightforward result.

Proposition 3.8. *Let $\mathcal{M}$ be a unital Banach algebra.*
Then $\forall x \in \mathcal{M} \left(x \in \mathcal{M}^{-1} \Leftrightarrow x \in \left(\widehat{^*\mathcal{M}}\right)^{-1}\right)$.
Consequently, $\sigma_{\mathcal{M}}(a) = \sigma_{\widehat{^*\mathcal{M}}}(a)$ holds for every $a \in \mathcal{M}$.

Proof. As $\mathcal{M} \subset {}^*\widehat{\mathcal{M}}$, we have $\mathcal{M}^{-1} \subset \left({}^*\widehat{\mathcal{M}}\right)^{-1}$, so one implication is clear.

For the other implication, let $a \in \mathcal{M} \cap \left({}^*\widehat{\mathcal{M}}\right)^{-1}$, then $a \in \left(\mathrm{Fin}({}^*\mathcal{M})\right)^{-1}$, by Prop. 3.7. Hence for some $r \in \mathbb{R}^+$, we have

$$\exists y \in {}^*\mathcal{M}\left((\|y\| \leq r) \wedge (ay = 1 = ya)\right).$$

Now by transfer, we get

$$\exists y \in \mathcal{M}\left((\|y\| \leq r) \wedge (ay = 1 = ya)\right),$$

i.e. $a \in \mathcal{M}^{-1}$. $\qquad\qquad\square$

Recall that an element $a \neq 0$ in a ring is a **left zero-divisor** if there is $b \neq 0$ such that $ab = 0$. Similarly a is a **right zero-divisor** if there is $b \neq 0$ such that $ba = 0$. We say that a is a **zero divisor** if $a \neq 0$ and is either a left zero divisor or a right zero divisor. Clearly if a in a unital ring is a zero divisor it cannot be invertible, for otherwise $ab = 0$ or $ba = 0$ holds only when $b = 0$.

The following gives a sufficient condition for the invertibility in a nonstandard hull.

Proposition 3.9. *Let $\mathcal{M}$ be an internal unital Banach algebra. Let $\widehat{a} \in \widehat{M}$ be a nonzero divisor, where $a \in \mathrm{Fin}(\mathcal{M})$.*

Suppose $\lim\limits_{n\to\infty} \left\|(\widehat{a}-1)^n\right\| = 0$. Then $\widehat{a} \in \left(\widehat{\mathcal{M}}\right)^{-1}$.

Proof. By saturation, there is $N \in {}^*\mathbb{N} \setminus \mathbb{N}$ such that $(a-1)^{N+1} \approx 0$. Let $c := \sum_{n=0}^{N}(1-a)^n$. Then we have

$$ca = ac = \left(1 - (1-a)\right)\sum_{n=0}^{N}(1-a)^n = 1 - (1-a)^{N+1} \approx 1.$$

If $\|c\|$ is infinite and we could set $b = \|c\|^{-1}c$, getting $\|b\| = 1$ such that $ab = ba \approx 0$, implying $\widehat{ab} = \widehat{ab} = 0$, contradicting to $\widehat{a}$ being a nonzero divisor.

Hence $c \in \mathrm{Fin}(\mathcal{M})$ and $ac = ca \approx 1$, *i.e.* $\widehat{ac} = \widehat{ac} = 1$.

Therefore $\widehat{a} \in \left(\widehat{\mathcal{M}}\right)^{-1}$. $\qquad\qquad\square$

3.1.4 *Notes and exercises*

Thm. 3.6 was proved by A. Beurling and I.M. Gelfand hence the formula is called the *Beuling-Gelfand Formula*. The result is also attributed to S. Mazur.

By Thm. 3.6, we see that in a unital Banach algebra $\mathcal{M}$, an element is such that $\sigma(a) = \{0\}$ iff $\lim_{n \to \infty} \|a^n\|^{1/n} = 0$. Such elements are called *quasinilpotent*. So nilpotent elements (*i.e.* those having a vanishing power) are quasinilpotent.

EXERCISES

(1) Does Cor. 3.4 hold for the real case? That is, if $\mathcal{M}$ be a unital real Banach division algebra, is it necessary that $\mathcal{M} = \mathbb{R}$? If not, can such $\mathcal{M}$ be infinite dimensional?

(2) In a unital Banach algebra $\mathcal{M}$, is the function $\mathcal{M} \ni x \mapsto x^{-1} \in \mathcal{M}^{-1}$ uniformly continuous?

(3) Find unital Banach algebras $\mathcal{M}_0, \mathcal{M}$ such that $\mathcal{M}_0$ is a subalgebra of $\mathcal{M}$ with unit $1 \in \mathcal{M}_0$, so that $\sigma_{\mathcal{M}}(a) \subsetneq \sigma_{\mathcal{M}_0}(a)$ holds for some $a \in \mathcal{M}_0$.

(4) Prove Thm. 3.7.

(5) Use the Riesz functional calculus to prove the Cayley-Hamilton Theorem: Let M be an $n \times n$-matrix over $\mathbb{C}, n = 1, 2, \ldots$. Then $p(M) = 0$, the $n \times n$ zero matrix, where $p(z)$ is the characteristic polynomial defined by $p(z) = \det(zI - M)$ with I denoting the $n \times n$ identity matrix.

(6) Find a unital Banach space $\mathcal{M}$ in which there are elements a, b so that $e^{a+b} \neq e^a e^b$.

3.2 C^*-Algebras

One sees more stars in C^*-algebras than in van Gogh's famous painting.

Given a (not necessarily unital) Banach algebra $\mathcal{M}$, a mapping

$$\mathcal{M} \ni x \mapsto x^* \in \mathcal{M}$$

is called an *involution* on $\mathcal{M}$ if the following properties are satisfied:

- $\forall x \in \mathcal{M}\left((x^*)^* = x\right)$;
- $\forall \alpha \in \mathbb{F} \; \forall x, y \in \mathcal{M}\left((\alpha x + y)^* = \bar{\alpha}x^* + y^*\right)$;
- $\forall x, y \in \mathcal{M}\left((xy)^* = y^* x^*\right)$.

That is, an involution is an involutive antilinear antimultiplicative mapping.

Fortunately, the $*$ symbol used for an involution is clearly distinguishable from the $*$ used for the nonstandard extension.

A ***Banach $*$-algebra*** is a Banach algebra with a fixed involution.

If a Banach $*$-algebra $\mathcal{M}$ further satisfies

- $\forall x \in \mathcal{M}\left((\|x^*x\| = \|x\|^2)\right)$,

we call $\mathcal{M}$ a C^*-***algebra***.

Note in particular that all C^*-algebras are assumed to be complete.

This section is about C^*-algebras and their representations as concrete operator algebras through the Gelfand transform and the GNS construction.

3.2.1   *Examples and basic properties*

An involution $*$ on a Banach algebra $\mathcal{M}$ is a bijection on $\mathcal{M}$ with $0^* = 0$.

If $\mathcal{M}$ is unital, the above implies that

$$\forall x \in \mathcal{M}\left(1^*x^* = x^* 1^* = x^*\right),$$

i.e. $1^* = 1$. Consequently, for a unital Banach $*$-algebra $\mathcal{M}$ and $a \in \mathcal{M}^{-1}$,

$$(a^{-1})^*a^* = (aa^{-1})^* = 1^* = 1 = 1^* = (a^{-1}a)^* = a^*(a^{-1})^*,$$

i.e. $\left(a^*\right)^{-1} = (a^{-1})^*$. Since $(a^*)^* = a$ holds for all $a \in \mathcal{M}$, we see that the involution is bijective on $\mathcal{M}^{-1}$.

For a in a unital Banach $*$-algebra $\mathcal{M}$, we have $(a^n)^* = (a^*)^n$ for all $n \in \mathbb{Z}$.

For an element a in a C^*-algebra $\mathcal{M}$, we have $\|a\|^2 = \|a^*a\| \le \|a^*\|\,\|a\|$, hence $\|a\| \le \|a^*\|$. Since $(a^*)^* = a$, we also have $\|a^*\| \le \|a\|$. Therefore

$\forall x \in \mathcal{M} \left(\|x^*\| = \|x\| \right)$, *i.e.* the involution on a C^*-algebra is an isometry. In particular, the involution is a homeomorphism on $\mathcal{M}$. From the previous paragraph, we see also that the involution is a homeomorphism on $\mathcal{M}^{-1}$.

Also for an element a in a C^*-algebra $\mathcal{M}$, we have

$$a = 0 \quad \text{iff} \quad \|a\|^2 = 0 \quad \text{iff} \quad \|(a^*)\,a\| = 0 \quad \text{iff} \quad (a^*)\,a = 0.$$

Even if we do not require the existence of the unit element of norm 1, no two distinct norms can be admitted in a C^*-algebra. Suppose $\mathcal{M}$ forms a C^*-algebra under $\|\cdot\|_1$ and under $\|\cdot\|_2$, with the property (by Prop. 2.4) that for some $k \in \mathbb{N}^+$, $\|x\|_1 \le k \|x\|_2$ for all $x \in \mathcal{M}$. Then it holds for all $x \in \mathcal{M}$ that $\|x\|_1^2 = \|x^*x\|_1 \le k \|x^*x\|_2 = k \|x\|_2^2, \ldots$, hence

$$\forall n \in \mathbb{N} \left(\|x\|_1 \le k^{1/2^n} \|x\|_2 \right), \quad i.e. \quad \forall x \in \mathcal{M} \left(\|x\|_1 \le \|x\|_2 \right).$$

So, by symmetry, $\|\cdot\|_1$ and $\|\cdot\|_2$ are the same.

For a in a unital Banach *-algebra $\mathcal{M}$ and $\lambda \in \mathbb{C}$, we have $(a - \lambda) \in \mathcal{M}^{-1}$ iff $(a^* - \bar{\lambda}) \in \mathcal{M}^{-1}$, hence $\sigma(a^*) = \overline{\sigma(a)}$, *i.e.* $\{\bar{\lambda} \mid \lambda \in \sigma(a)\}$.

Furthermore, given a unital C^*-algebra $\mathcal{M}$ and $a \in \mathcal{M}$, suppose for some $\lambda_n \in \mathbb{C}, n \in \mathbb{N}$, that $\sum_{n=0}^{\infty} \lambda_n a^n$ converges, then by the continuity of the involution,

$$\left(\sum_{n=0}^{\infty} \lambda_n a^n \right)^* = \sum_{n=0}^{\infty} \bar{\lambda}_n (a^*)^n.$$

In particular, $\left(e^a \right)^* = e^{a^*}$ holds for every a in a unital C^*-algebra $\mathcal{M}$.

Example 3.11.

- The most basic C^*-algebra is simply $\mathbb{C}$ itself, where the involution is just the conjugate operation.
- Let X be a complex Hilbert space, then the $\mathcal{B}(X)$, the Banach algebra of bounded linear operators on X, forms a unital C^*-algebra where, for each $T \in \mathcal{B}(X)$, T^* is the adjoint of T given in Prop. 2.46. $\mathcal{B}(X)$ is noncommutative when $\dim(X) > 1$.
  Also $\mathcal{B}_c(X)$ forms a C^*-subalgebra of $\mathcal{B}(X)$, but it is not unital unless X is finite dimensional.
- When $X = \mathbb{C}^n, n \in \mathbb{N}$, the matrix algebra $\mathcal{B}(\mathbb{C}^n)$ of $n \times n$ matrices over $\mathbb{C}$ forms a C^*-algebra, where the involution is just the conjugate transpose of the matrix.
- Let Ω be a topological space. Then the commutative unital complex Banach algebra $C_b(\Omega)$ forms a C^*-algebra, where for each $f \in C(\Omega)$, the involution is defined by $f^* : x \mapsto \overline{f(x)}, \ x \in \Omega$. We simply write $f^* = \bar{f}$. It is easy to check that the C^* requirements are satisfied.

- If Ω is a locally compact space, $C_0(\Omega)$ similarly forms a commutative C^*-algebra. If Ω is not compact, $C_0(\Omega)$ would not be unital. As it will become clear in the next subsection, we will be mostly interested in the case when Ω is a compact topological space. Recall that $C_0(\Omega) = C_b(\Omega) = C(\Omega)$ when Ω is a compact.
- For a compact space Ω, we define $B(\Omega)$ to be the space of bounded Borel functions $\Omega \to \mathbb{C}$. Then under pointwise addition, multiplication, complex conjugate and the supremum norm, $B(\Omega)$ forms a unital C^*-algebra. Note that $C(\Omega)$ is a unital C^*-subalgebra of $B(\Omega)$. □

C^*-algebras above are called ***concrete C^*-algebras***.

We will see later that any commutative unital C^*-algebra can be represented as $C_b(\Omega)$ for some compact topological space Ω and any unital C^*-algebra can be represented as a C^*-subalgebra of $\mathcal{B}(X)$ for some complex Hilbert space X.

An ***ideal*** of a C^*-algebra is a (not necessarily unital) C^*-subalgebra which is also an ideal as in a Banach algebra as before. One can also define left- and right-ideal similarly as before. The quotient algebra construction in Prop. 3.1 extends to the C^*-algebra case. The following is easy to check.

Proposition 3.10. *Let $\mathcal{M}$ be a C^*-algebra and $I \subset \mathcal{M}$ a closed proper ideal. Then the quotient algebra $\mathcal{M}/I$ forms a C^*-algebra under the involution given by*

$$(x + I)^* := (x^* + I), \quad x \in \mathcal{M}.$$

Moreover, if $\mathcal{M}$ is unital, so is $\mathcal{M}/I$. □

In a C^*-algebra $\mathcal{M}$, it is easy to check that the closure of an ideal of $\mathcal{M}$ is an ideal of $\mathcal{M}$ in the C^*-algebra sense. Also in the C^*-algebra setting, maximal ideals are closed proper ideals and each proper ideal extends to a maximal ideal as in Cor. 3.3.

Given a C^*-algebra $\mathcal{M}$ and $X \subset \mathcal{M}$, the ***C^*-algebra generated by*** X, written as $\mathcal{M}_X$, is the intersection of all C^*-subalgebras of $\mathcal{M}$ that include X. If $\mathcal{M}$ is unital, it is understood that $1 \in \mathcal{M}_X$. When $X = \{a\}$ for some $a \in \mathcal{M}$, we call it the C^*-algebra generated by a and write $\mathcal{M}_a$. So $\mathcal{M}_a$ coincides with the closure of the linear span of elements of the form $(a^*)^{n_1} a^{m_1} \cdots (a^*)^{n_k} a^{m_k}$, $k \in \mathbb{N}$, $n_1, m_1, \ldots, n_k, m_k \in \mathbb{N}$.

Here is a list some common types of elements in a C^*-algebra $\mathcal{M}$.

- $a \in \mathcal{M}$ is ***self-adjoint*** if $a = a^*$.
- The set of self-adjoint elements of $\mathcal{M}$ is denoted by $\operatorname{Re}(\mathcal{M})$.

- $a \in \mathcal{M}$ is **positive** if $a = b^2$ for some $b \in \mathrm{Re}(\mathcal{M})$.
- The set of positive elements of $\mathcal{M}$ is denoted by $\mathcal{M}^+$.
- $a \in \mathcal{M}$ is a **projection** if $a = a^* = a^2$.
- The set of projections in $\mathcal{M}$ is denoted by $\mathrm{Proj}(\mathcal{M})$.
- $a \in \mathcal{M}$ is a **partial isometry** if $a^*a \in \mathrm{Proj}(\mathcal{M})$.
- $a \in \mathcal{M}$ is **normal** if $a^*a = aa^*$.

Observe that if $\mathcal{M}_0$ is a C^*-subalgebra of $\mathcal{M}$, then $\mathrm{Re}(\mathcal{M}_0) = \mathcal{M}_0 \cap \mathrm{Re}(\mathcal{M})$, $\mathrm{Proj}(\mathcal{M}_0) = \mathcal{M}_0 \cap \mathrm{Proj}(\mathcal{M})$ and elements in $\mathcal{M}_0$ which are normal remain normal in $\mathcal{M}$ *etc.* By Thm. 3.10 and Thm. 3.16(iv) below, at least in the case of unital C^*-algebras, we also have $\mathcal{M}_0^+ = \mathcal{M}_0 \cap \mathcal{M}^+$.

In other words, all these notions are independent of a particular choice of the C^*-algebra that contains the element.

The terminology partial isometry is further justified by the property given in Ex. 4 on p.241.

In a unital C^*-algebra $\mathcal{M}$, the following are also defined:

- $a \in \mathcal{M}$ is an **isometry** if $a^*a = 1$.
- $a \in \mathcal{M}$ is an **coisometry** if $aa^* = 1$.
- $a \in \mathcal{M}$ is **unitary** if $a^*a = aa^* = 1$.
- The set of unitary elements of $\mathcal{M}$ is denoted by $\mathcal{U}(\mathcal{M})$.

In particular, for any $a \in \mathcal{M}$, a^*a is always self-adjoint, projections are positive, positive elements are self-adjoint, self-adjoint elements are normal and unitary elements are also normal.

Of course, if $\mathcal{M}$ is commutative, all elements are normal. In a noncommutative C^*-algebra $\mathcal{M}$, the significance of $a \in \mathcal{M}$ being normal is that $\mathcal{M}_a$ is commutative. Note also that if $a \in \mathcal{M}$ is normal and $b \in \mathcal{M}$ commutes with a, then b commutes with any $c \in \mathcal{M}_a$.

If $a \in \mathcal{M}$ is normal and $\alpha \in \mathbb{C}$ then it is clear that $(a + \alpha)$ is normal.

If $a \in \mathcal{M}^{-1}$ is normal, then a^{-1} is normal since

$$a^{-1}(a^{-1})^* = a^{-1}(a^*)^{-1} = (a^*a)^{-1} = (aa^*)^{-1} = (a^*)^{-1}a^{-1} = (a^{-1})^*a^{-1}.$$

Example 3.12.

- In $\mathcal{B}(\mathbb{C}^2)$, the algebra of complex 2×2-matrices, $\begin{bmatrix} 0 & 0 \\ 1 & 0 \end{bmatrix}$ is an example of a element which is not normal. Similar result holds for $\mathcal{B}(\mathbb{C}^n)$, $n \in \mathbb{N}$.
- In the infinite dimensional case, let $S \in \mathcal{B}(\ell_2)$ be the **unilateral shift**, *i.e.* for every $\xi = \{\xi_n\}_{n \in \mathbb{N}} \in \ell_2$,

$$S(\xi)_0 := 0, \quad S(\xi)_n := \xi_{n-1} \quad \text{if} \quad n \in \mathbb{N}^+.$$

Then for every $\xi \in \ell_2$, we have $S^*(\xi)_n := \xi_{n+1}$ for all $n \in \mathbb{N}$, hence $S^* S$ is the identity on ℓ_2 while SS^* is the orthogonal projection onto the closed subspace $\{\xi \in \ell_2 \,|\, \xi_0 = 0\}$.

In particular, S is not normal. □

Clearly real linear combinations of self-adjoint elements are self-adjoint and since the involution is continuous, $\mathrm{Re}(\mathcal{M})$ is a closed real linear subspace of $\mathcal{M}$.

In a unital C^*-algebra $\mathcal{M}$, for any $a \in \mathrm{Re}(\mathcal{M})$, $\left(e^{ia}\right)^* = e^{-ia^*} = e^{-ia}$, hence e^{ia} is unitary.

On p.103 projections have already been defined for the special case for operators in $\mathcal{B}(X)$, where X is a Banach space. See also Prop. 2.38 Prop. 2.39 for what has been discussed about projections in $\mathcal{B}(X)$, where X is a Hilbert space.

For $T \in \mathcal{B}(X)$, where X is a Hilbert space, $T^*T = 1$ means for all $x \in X$ that $\langle T(x), T(x) \rangle = \langle x, T^*T(x) \rangle = \langle x, x \rangle$, by Prop. 2.46. Hence the usage of the term isometry for the above definition is justified.

An element $a \in \mathcal{M}$ is an isometry iff a^* is a coisometry and *vice versa*. It is unitary iff it is both an isometry and a coisometry.

Also, if $a \in \mathcal{M}$ is unitary, then a is an isometry, $a^{-1} = a^*$ and $\|a\| = 1$.

In fact it is easily seen that $\mathcal{U}(\mathcal{M}) \subset \mathcal{M}^{-1}$ and forms a closed subgroup of $\mathcal{M}^{-1}$. Moreover, $\mathcal{U}(\mathcal{M})$ is closed under the involution. $\mathcal{U}(\mathcal{M})$ is also called the **unitary group** of $\mathcal{M}$.

Note that the following results lend justification to the use of the symbols $\mathcal{U}(\mathcal{M})$, $\mathrm{Re}(\mathcal{M})$ etc.

Theorem 3.9. *Let $\mathcal{M}$ be a unital C^*-algebra.*

(i) *If $a \in \mathcal{U}(\mathcal{M})$, then $\sigma(a) \subset S_{\mathbb{C}}$.*

(ii) *If $a \in \mathrm{Re}(\mathcal{M})$, then $\sigma(a) \subset \mathbb{R}$.*

(iii) *If $a \in \mathcal{M}^+$, then $\sigma(a) \subset [0, \infty)$.*

(iv) *If $a \in \mathrm{Proj}(\mathcal{M})$, then $\sigma(a) \subset \{0, 1\}$.*

(v) *If $a \in \mathrm{Proj}(\mathcal{M})$, then $(1 - a) \in \mathrm{Proj}(\mathcal{M})$.*

(vi) *Every $a \in \mathcal{M}$ decomposes as $a = a_1 + ia_2$, where $a_1, a_2 \in \mathrm{Re}(\mathcal{M})$ are unique. Moreover, $a_1 = \dfrac{a + a^*}{2}$ and $a_2 = \dfrac{a - a^*}{2i}$.*

(vii) *If $a \in \mathcal{M}$ is normal, then $\rho(a) = \|a\|$.*
In particular, $\forall x \in \mathcal{M} \left(\|x\| = \sqrt{\|x^* x\|} = \sqrt{\rho(x^* x)} \right).$
Moreover, for normal a, we have $a = 0$ iff $\sigma(a) = \{0\}$.

Consequently, it holds in a commutative unital C^*-algebra $\mathcal{M}$ that

$$\forall x \in \mathcal{M}\left(\rho(x) = \|x\|\right) \quad and \quad \forall x \in \mathcal{M}\left((x=0) \Leftrightarrow (\sigma(x) = \{0\})\right).$$

(viii) *Let $a \in \mathcal{M}$. Then a is a partial isometry iff $a = aa^*a$.*
Consequently, a is a partial isometry iff a^* is a partial isometry.

Proof. (i): Let $a \in \mathcal{U}(\mathcal{M})$. Then $\|a\| = 1$, so $\sigma(a) \subset \bar{B}_{\mathbb{C}}$. Since $a^* \in \mathcal{U}(\mathcal{M})$, we also have $\sigma(a^*) \subset \bar{B}_{\mathbb{C}}$. Moreover, as $a^{-1} = a^*$,

$$\left(\sigma(a)\right)^{-1} = \sigma\left(a^{-1}\right) = \sigma(a^*) \subset \bar{B}_{\mathbb{C}}.$$

From $\sigma(a), \left(\sigma(a)\right)^{-1} \subset \bar{B}_{\mathbb{C}}$, we obtain $\sigma(a) \subset S_{\mathbb{C}}$.

(ii): Let $a \in \text{Re}(\mathcal{M})$. As remarked earlier, $e^{ia} \in \mathcal{U}(\mathcal{M})$, hence we have $\sigma(e^{ia}) \subset S_{\mathbb{C}}$ by (i). But, by the Spectral Mapping Theorem (Cor. 3.7), we have $\sigma(e^{ia}) = \{e^{i\lambda} \mid \lambda \in \sigma(a)\}$, implying that $\sigma(a) \subset \mathbb{R}$.

(iii): Let $a \in \mathcal{M}^+$, so $a = b^2$ for some $b \in \text{Re}(\mathcal{M})$. Then by (ii) and the Spectral Mapping Theorem (Cor. 3.7) we have $\sigma(a) = \sigma(b)^2 \subset [0,\infty)$.

(iv): If $a \in \mathcal{M}$ is a projection, then, by (ii) or (iii), $\sigma(a) \subset \mathbb{R}$. Let $r \in \mathbb{R} \setminus \{0, 1\}$. Then by using $a^2 = a$, clearly $(a-r)(a-1+r) = r(1-r) \neq 0$, so $(a - r) \in \mathcal{M}^{-1}$, i.e. $r \notin \sigma(a)$. Hence $\sigma(a) \subset \{0, 1\}$.

(v): If $a \in \text{Proj}(\mathcal{M})$, then $(1 - a)^* = (1 - a^*) = (1 - a)$ and $(1 - a)^2 = 1 - a - a + a^2 = 1 - a - a + a = (1 - a)$, so $(1 - a) \in \text{Proj}(\mathcal{M})$.

(vi): For $a \in \mathcal{M}$, if we define $a_1 := \dfrac{a + a^*}{2}$ and $a_1 := \dfrac{a - a^*}{2i}$, then clearly $a_1, a_2 \in \text{Re}(\mathcal{M})$ and $a = a_1 + ia_2$.

On the other hand if for some $a_1, a_2 \in \text{Re}(\mathcal{M})$ that $a = a_1 + ia_2$, then $a + a^* = a_1 + ia_2 + a_1 - ia_2 = 2a_1$ and similarly $a - a^* = 2ia$, so a_1 and a_2 must have the above forms.

(vii): First note that if $c \in \text{Re}(\mathcal{M})$, then $\|c^2\| = \|c^*c\| = \|c\|^2$, so it follows by induction that $\left\|c^{2^n}\right\| = \|c\|^{2^n}$ holds for all $n \in \mathbb{N}$.

Let $N \in {}^*\mathbb{N} \setminus \mathbb{N}$, then by Thm. 3.6,

$$\rho(c) \approx \left\|c^{2^N}\right\|^{1/2^N} = \|c\|^{2^N/2^N} = \|c\|, \quad \text{hence } \rho(c) = \|c\|.$$

Now let $a \in \mathcal{M}$ be normal. Note that by a being normal, $(a^n)^*a^n = (a^*a)^n$ holds for all $n \in \mathbb{N}$. Let $N \in {}^*\mathbb{N} \setminus \mathbb{N}$, then by Thm. 3.6 again,

$$\rho(a) \approx \left\|a^{2^N}\right\|^{2^{-N}} = \left\|a^{2^N}\right\|^{2(2^{-N-1})} = \left\|\left(a^{2^N}\right)^* \left(a^{2^N}\right)\right\|^{2^{-N-1}}$$

$$= \left\|(a^*a)^{2^N}\right\|^{2^{-N-1}} = \|a^*a\|^{2^N 2^{-N-1}} = \sqrt{\rho(a^*a)} = \sqrt{\|a\|^2} = \|a\|,$$

where equalities on the last line follows from $(a^*a) \in \mathrm{Re}(\mathcal{M})$ and the above result for self-adjoint elements.

(viii): Let $a \in \mathcal{M}$. If a is a partial isometry, then $a^*a \in \mathrm{Proj}(\mathcal{M})$, so $(a^*a) = (a^*a)^2 = (a^*a)^3$, hence

$$(a - aa^*a)^*(a - aa^*a) = (a^*a) - (a^*a)^2 - (a^*a)^2 + (a^*a)^3 = 0$$

so $\|a - aa^*a\|^2 = \|(a - aa^*a)^*(a - aa^*a)\| = 0$, *i.e.* $a = aa^*a$.

Conversely, if $a = aa^*a$, then $(a^*a)^2 = a^*(aa^*a) = a^*a$ and $(a^*a)^* = a^*a$, hence $a^*a \in \mathrm{Proj}(\mathcal{M})$.

Note that $a = aa^*a$ iff $a^* = a^*aa^*$, so a is a partial isometry iff a^* is a partial isometry. $\qquad\square$

Partial converse of the statements in the above theorem will be dealt with in the next subsection.

In a unital C^*-algebra $\mathcal{M}$, by Thm. 3.9(vi), for $a \in \mathcal{M}$, we let $\mathrm{Re}(a)$ denote the **real part** of a, *i.e.* $\dfrac{a + a^*}{2}$ and $\mathrm{Im}(a)$ denote the **imaginary part** of a, *i.e.* $\dfrac{a - a^*}{2i}$. This is of course an extension of the use of the same notation for $\mathbb{C}$, which we regard as a subalgebra of $\mathcal{M}$.

Corollary 3.12. *Let $\mathcal{M}$ be an internal unital C^*-algebra and $p \in \mathrm{Proj}(\mathcal{M})$. Then*

(i) *$p \approx 0$ iff $p = 0$.*
(ii) *$p \approx 1$ iff $p = 1$.*

Proof.   Consider only the nontrivial direction.

(i): Since $p^2 = p^* = p$, we have $\|p\|^2 = \|p^*p\| = \|p^2\| = \|p\|$.
If $p \approx 0$, then $\|p\| \neq 1$, therefore $\|p\| = 0$, *i.e.* $p = 0$.

(ii): By Thm. 3.9(v), $(1 - p) \in \mathrm{Proj}(\mathcal{M})$.
If $p \approx 1$, then $(1 - p) \approx 0$, so $(1 - p) = 0$ by (i), *i.e.* $p = 1$. $\qquad\square$

Theorem 3.10.   *Let $\mathcal{M}$ be a unital C^*-algebra and $\mathcal{M}_0 \subset \mathcal{M}$ be a C^*-subalgebra such that $1 \in \mathcal{M}_0$. (i.e. a unital C^*-subalgebra.) Then*

(i) *$\mathcal{M}_0^{-1} = \mathcal{M}^{-1} \cap \mathcal{M}_0$.*
(ii) *$\forall x \in \mathcal{M}_0 \left(\sigma_{\mathcal{M}}(x) = \sigma_{\mathcal{M}_0}(x) \right)$.*

Proof.   Notice (ii) implies for any $a \in \mathcal{M}_0$ that $a \in \mathcal{M}_0^{-1}$ iff $0 \notin \sigma_{\mathcal{M}_0}(a)$ iff $0 \notin \sigma_{\mathcal{M}}(a)$ iff $a \in \mathcal{M}^{-1}$, hence (i) follows from (ii).

For the proof of (ii), as remarked earlier, $\sigma_{\mathcal{M}}(a) \subset \sigma_{\mathcal{M}_0}(a)$ holds trivially for all $a \in \mathcal{M}_0$, we only need to prove the other inclusion.

First consider any $c \in \mathrm{Re}(\mathcal{M}_0)$. Let $\lambda \in \mathbb{C} \setminus \sigma_{\mathcal{M}}(c)$. So $(c - \lambda) \in \mathcal{M}^{-1}$. Choose any $\alpha \in {}^*\mathbb{C} \setminus {}^*\mathbb{R}$ such that $\alpha \approx \lambda$. By Thm. 3.9(ii), $\sigma_{\mathcal{M}_0}(c) \subset \mathbb{R}$, so $\alpha \notin {}^*\sigma_{\mathcal{M}_0}(c)$, $i.e.$ $(c - \alpha) \in {}^*\mathcal{M}_0^{-1}$.

By Thm. 3.1, $\mathcal{M} \ni (c - \lambda)^{-1} \approx (c - \alpha)^{-1} \in {}^*\mathcal{M}_0$. Since $\mathcal{M}_0$ is a closed subspace of $\mathcal{M}$, we have $(c - \lambda)^{-1} \in \mathcal{M}_0$, $i.e.$ $\lambda \notin \sigma_{\mathcal{M}_0}(c)$.

Hence we have proved that $\sigma_{\mathcal{M}}(c) \supset \sigma_{\mathcal{M}_0}(c)$ for any $c \in \mathrm{Re}(\mathcal{M}_0)$.

Next let $c \in \mathcal{M}_0$ be arbitrary and let $\lambda \in \mathbb{C} \setminus \sigma_{\mathcal{M}}(c)$.

So $(c - \lambda) \in \mathcal{M}^{-1}$. Let $a := (c - \lambda) \in \mathcal{M}_0$ and $b := (c - \lambda)^{-1} \in \mathcal{M}$. From $ab = 1 = ba$ we also have $a^*b^* = 1 = b^*a^*$, hence

$$(a^*a)(bb^*) = a^*(ab)b^* = a^*b^* = 1 = ba = b(b^*a^*)a = (bb^*)(a^*a).$$

$i.e.$ $(a^*a) \in \mathcal{M}^{-1}$. Since $(a^*a) \in \mathrm{Re}(\mathcal{M}_0)$, we have by what has just been proved that

$$(a^*a) \in \mathcal{M}^{-1} \text{ iff } 0 \notin \sigma_{\mathcal{M}}\big((a^*a)\big) \text{ iff } 0 \notin \sigma_{\mathcal{M}_0}\big((a^*a)\big) \text{ iff } (a^*a) \in \mathcal{M}_0^{-1},$$

hence $(a^*a) \in \mathcal{M}_0^{-1}$ and so $(bb^*) \in \mathcal{M}_0$.

Then $b = b(b^*a^*) = (bb^*)a^* \in \mathcal{M}_0$, $i.e.$ $(c - \lambda)^{-1} \in \mathcal{M}_0$, $i.e.$ $\lambda \notin \sigma_{\mathcal{M}_0}(c)$. Therefore $\sigma_{\mathcal{M}}(c) \supset \sigma_{\mathcal{M}_0}(c)$ holds for any $c \in \mathcal{M}_0$. $\square$

The results in Thm. 3.10 depends crucially on the C^*-properties and should be compared with remarks on p.190 and the results in Prop. 3.8.

To illustrate what has been discussed so far, consider the following list of some elementary properties about any concrete C^*-algebra $C(\Omega)$, where Ω is a compact space. In fact it will be shown in the next subsection that all properties about commutative unital C^*-algebras are already captured by the $C(\Omega)$'s.

Example 3.13. Let Ω be a compact topological space. As remarked already, $C(\Omega)$ is a commutative unital C^*-algebra, with involution given by $f^* = \bar{f}$ for $f \in C(\Omega)$ and $C(\Omega)$ contains 1, the constant unit function.

(i) Let $f \in C(\Omega)$. then $f \in C(\Omega)^{-1}$ iff f is invertible as a function iff $\forall \omega \in \Omega\,(f(\omega) \neq 0)$. If $f \in C(\Omega)^{-1}$, then f^{-1} is the function $\mathbb{C} \ni z \mapsto 1/f(z)$.

(ii) Let $f \in C(\Omega)$, then for any $\lambda \in \mathbb{C}$, $\lambda \in \sigma(f)$ iff $(f - \lambda) \notin C(\Omega)^{-1}$ iff $\exists \omega \in \Omega\,\big((f(\omega) - \lambda) = 0\big)$ iff $\lambda \in f[\Omega]$, $i.e.$ $\sigma(f) = f[\Omega]$.
    Consequently, $\rho(f) = \|f\|$.

(iii) $f \in \mathrm{Re}\big(C(\Omega)\big)$ iff $f^* = f$ iff $\bar{f} = f$ iff $f[\Omega] \subset \mathbb{R}$ iff $\sigma(a) \subset \mathbb{R}$.

(iv) Likewise, $f \in C(\Omega)^+$ iff $f = g^2$ for some $g \in \mathrm{Re}\big(C(\Omega)\big)$ iff $f = |f|$ iff $f[\Omega] \subset [0, \infty)$ iff $\sigma(a) \subset [0, \infty)$.

(v) In particular, $f \in C(\Omega)^+$ iff $\exists g \in C(\Omega) \, (f = g^*g)$.

(vi) Let $f \in C(\Omega)^+$ and $r \in \mathbb{R}+$, since $f[\Omega] \subset [0, \infty)$, there is a unique $g \in C(\Omega)^+$, $f = g^r$. In fact g is the positive r^{th} root of f. In particular, for $f \in C(\Omega)^+$, $\sqrt{f} \in C(\Omega)^+$, the positive root and $|f| = \sqrt{f^*f} \in C(\Omega)^+$.

(vii) $C(\Omega)^+$ is closed, contains $0, 1$ and for any $f, g \in C(\Omega)^+$ and $r \in [0, \infty)$, we have $(f + rg)[\Omega] \subset [0, \infty)$ and $(fg)[\Omega] \subset [0, \infty)$, so $(f + rg), (fg) \in C(\Omega)^+$. In particular, $\forall r \in [0, \infty) \, (rf \in C(\Omega)^+)$ and $\frac{1}{2}(f+g) \in C(\Omega)^+$. This property is what we call a ***cone***, so $C(\Omega)^+$ forms a closed cone in $C(\Omega)$. Furthermore, $C(\Omega)^+ \cap (-C(\Omega)^+) = \{0\}$, where $-C(\Omega)^+$ is the negative cone $\{-f \mid f \in C(\Omega)^+\}$.

(viii) For $f \in \text{Re}(C(\Omega))$, let $f_1 = \max\{f, 0\}$ and $f_2 = -\min\{f, 0\}$, then $f_1, f_2 \in C(\Omega)^+$, $f_1 f_2 = 0$ and $f = f_1 - f_2$. Such decomposition is unique.

(ix) $f \in C(\Omega)$ is unitary iff it is an isometry iff $f^*f = 1$ iff $|f| = 1$ iff $f[\Omega] \subset S_{\mathbb{C}}$.

(x) $f \in C(\Omega)$ is a projection iff $f = \bar{f} = f^2$ iff $f \in \text{Re}(C(\Omega))$ and $f = f^2$ iff $f[\Omega] = \{0, 1\}$ iff $\sigma(f) = \{0, 1\}$.

(xi) For every $f \in C(\Omega)^{-1}$ (hence $0 \notin f[\Omega]$), we can write $f = |f| \, e^{i\text{Arg}(f)}$, where Arg is the principal argument function. Let $u = e^{i\text{Arg}(f)}$, then we see that $u \in C(\Omega)$ and is unitary, and $f = |f| \, u$ can be viewed as the ***polar decomposition*** of f.

(xii) Observe that for $f \in C(\Omega)$ the real part $\text{Re}(f)$ in the sense of a C^*-algebra agrees with the same notion in the sense of a complex-valued function. Likewise for $\text{Im}(f)$. Moreover,
$$\text{Re}(C(\Omega)) = \text{Re}[C(\Omega)] = \{\text{Re}(f) \mid f \in C(\Omega)\} = \{f \in C(\Omega) \mid f[\Omega] \subset \mathbb{R}\}$$
and the same equations for $\text{Im}[C(\Omega)]$ hold similarly. Note that $\text{Re}[C(\Omega)] = \text{Im}[C(\Omega)]$.

(xiii) For any $\theta \in C(\mathbb{C})$ and $f \in C(\Omega)$, obviously $(\theta \circ f)$ is continuous, hence $\theta(f) \in C(\Omega)$. When θ is analytic on $\sigma(f)$ *i.e.* on $f[\Omega]$, this agrees with the $\theta(f)$ obtained from the Riesz functional calculus. For example, let $\sigma(f) \subset U$ and $C_1, \ldots, C_n \subset U$ be as given on p.198 and $\theta : U \to \mathbb{C}$ be analytic, then the Riesz functional calculus gives the function
$$\Omega \ni \omega \mapsto \frac{1}{2\pi i} \sum_{i=1}^{n} \int_{C_i} \frac{\theta(z)}{z - f(\omega)} dz,$$
which is the same as the function $\Omega \ni \omega \mapsto \theta(f(\omega))$ by the Cauchy Integral Formula.

Consequently, the ***Continuous Functional Calculus*** just described is an extension of the Riesz functional calculus.   □

We end this subsection with the following result about the $C(\Omega)$ which will be an important tool in the next subsection.

Theorem 3.11.   (Stone-Weierstrass Theorem) *Let Ω be a compact topological space and $\mathcal{M} \subset C(\Omega)$ be a C^*-subalgebra with $1 \in \mathcal{M}$. Furthermore, suppose that $\mathcal{M}$ separates points in Ω, i.e.*

$$\forall \omega_1, \omega_2 \in \Omega \left((\omega_1 \neq \omega_2) \Rightarrow \left(\exists f \in \mathcal{M}(f(\omega_1) \neq f(\omega_2)) \right) \right).$$

Then $\mathcal{M} = C(\Omega)$.

Proof.   First we need the following:

CLAIM 1: For any $\omega_1 \neq \omega_2$ from Ω and α_1, α_2 from $\mathbb{C}$, there is $g \in \mathcal{M}$ such that $g(\omega_1) = \alpha_1$ and $g(\omega_2) = \alpha_2$.

By $\mathcal{M}$ separating points in Ω, let $h \in \mathcal{M}$ such that $h(\omega_1) \neq h(\omega_2)$. If we define $g : \Omega \to \mathbb{C}$ to be the function

$$\Omega \ni \omega \mapsto \alpha_2 + \frac{\alpha_1 - \alpha_2}{h(\omega_1) - h(\omega_2)} \big(h(\omega) - h(\omega_2) \big),$$

then $g(\omega_1) = \alpha_1$ and $g(\omega_2) = \alpha_2$.

Moreover, since $1 \in \mathcal{M}$ we have $g \in \mathcal{M}$ and the Claim is proved.

CLAIM 2: (i) Let $g \in \mathcal{M} \cap C(\Omega)^+$, then $\sqrt{g} \in \mathcal{M}$.
  (ii) Suppose $g_1, g_2 \in \mathcal{M} \cap \mathrm{Re}\big(C(\Omega)\big)$. Then

$$\min \big\{ g_1, g_2 \big\} \in \mathcal{M} \quad \text{and} \quad \max \big\{ g_1, g_2 \big\} \in \mathcal{M}.$$

For any $g \in \mathcal{M} \cap C(\Omega)^+$, we have $\sigma(g) \subset [0, \infty)$ by Thm. 3.9(iii). Now for each $n \in \mathbb{N}^+$, let θ be the analytic function corresponding to a branch of the function $z \mapsto \sqrt{z + n^{-1}}$ that includes the positive roots for $z \in [0, \infty)$. So the domain of θ contains $\sigma(g)$ in its interior, therefore $\theta(g) \in \mathcal{M}$ by the Riesz functional calculus. By Example 3.13(xiii), $\theta(g) = \sqrt{g + n^{-1}}$, the composition of the positive square root with $(g + n^{-1})$. Moreover,

$$\left| \sqrt{g + n^{-1}} - \sqrt{g} \right| = \left| \frac{n^{-1}}{\sqrt{g + n^{-1}} + \sqrt{g}} \right| \leq \frac{1}{\sqrt{n}},$$

hence $\sqrt{g + n^{-1}} \to \sqrt{g}$ in $C(\Omega)$ as $n \to \infty$.

Since $\mathcal{M}$ is closed, $\sqrt{g} \in \mathcal{M}$ and (i) is proved.

For (ii), let $g_1, g_2 \in \mathcal{M} \cap \mathrm{Re}\big(C(\Omega)\big)$ and note that

$$\min \big\{ g_1, g_2 \big\} = \frac{1}{2}\Big(g_1 + g_2 - |g_1 - g_2| \Big) = \frac{1}{2}\Big(g_1 + g_2 - \sqrt{(\bar{g}_1 - \bar{g}_2)(g_1 - g_2)} \Big).$$

Hence, by $(\bar{g}_1 - \bar{g}_2)(g_1 - g_2) \in \mathcal{M} \cap C(\Omega)^+$ (Example 3.13(v)) and (i), we have $\min\{g_1, g_2\} \in \mathcal{M}$.

Consequently, $\max\{g_1, g_2\} = -\min\{-g_1, -g_2\} \in \mathcal{M}$.

So the Claim is proved.

Now, to prove the theorem, let $f \in C(\Omega)$. We would like to find $g \in {}^*\mathcal{M}$ such that $g \approx {}^*f$, then it follows from $\mathcal{M}$ being closed that $f \in \mathcal{M}$.

As both $C(\Omega)$ and $\mathcal{M}$ are C^*-algebras, we have $\mathrm{Re}(f), \mathrm{Im}(f) \in C(\Omega)$, $\mathrm{Re}(C(\Omega)) \subset C(\Omega)$ and $\mathrm{Re}(\mathcal{M}) \subset \mathcal{M}$, we may assume that $f[\Omega] \subset \mathbb{R}$, *i.e.* $f \in \mathrm{Re}(C(\Omega))$.

Observe that for any finite list of open sets $V_1, \ldots, V_k \subset \Omega$, $k \in \mathbb{N}$, there is an internal family $\{U_\omega \mid \omega \in {}^*\Omega\}$ of *open subsets of $^*\Omega$ such that for each $\omega \in {}^*\Omega$, $\omega \in U_\omega$ and $U_\omega \subset {}^*V_i$ whenever $\omega \in {}^*V_i$, where $i = 1, \ldots, k$.

Hence it follows from saturation that there is an internal family $\{U_\omega \mid \omega \in {}^*\Omega\}$ of *open subsets of $^*\Omega$ such that for each $\omega \in {}^*\Omega$, $\omega \in U_\omega$ and for each open subset $V \subset \Omega$, if $\omega \in {}^*V$ then $U_\omega \subset {}^*V$, *i.e.* $U_\omega \subset \mu(\omega)$.

Since $^*\Omega$ is *compact and $\{U_\omega \mid \omega \in {}^*\Omega\}$ covers $^*\Omega$, there is a hyperfinite subfamily of $\{U_\omega \mid \omega \in {}^*\Omega\}$ that still covers $^*\Omega$. That is, for some $N \in {}^*\mathbb{N}$, there is $\{\omega_n \mid n < N\} \subset {}^*\Omega$ and a family $\{U_n \mid n < N\}$ of *open subsets of $^*\Omega$ such that

- $\forall n < N \left(\omega_n \in U_n \subset \mu(\omega_n) \right)$;
- $^*\Omega \subset \bigcup_{n < N} U_n$.

Note $^\circ\omega_n \in \Omega$ for any $n < N$, by the compactness of Ω.

Now we are ready to construct the required $g \in {}^*\mathcal{M}$. Fix $\epsilon \in \mathbb{R}^+$.

CLAIM 3: For $n, m < N$, there is $g_{nm} \in {}^*\mathcal{M} \cap {}^*\mathrm{Re}(C(\Omega))$ such that

- $\forall \omega \in U_n \left(|g_{nm}(\omega) - {}^*f(\omega)| \leq \epsilon \right)$;
- $\forall \omega \in U_m \left(g_{nm}(\omega) \leq {}^*f(\omega) + \epsilon \right)$.

If $\omega_n \approx \lambda_m$, we simply let g_{nm} be the constant function $^*f(\omega_n)$, which is in $^*\mathrm{Re}[\mathcal{M}]$ since $1 \in \mathcal{M}$, then the both conditions follows from the continuity of f.

If $\omega_n \not\approx \lambda_m$, we apply Claim 1 to get $h \in \mathcal{M}$ such that $h(^\circ\omega_n) = f(^\circ\omega_n)$ and $h(^\circ\lambda_m) = f(^\circ\lambda_m)$. Since $f \in \mathrm{Re}(C(\Omega))$, replace h by $\mathrm{Re}(h)$, we can assume that $h \in \mathrm{Re}(C(\Omega))$. Now take $g_{nm} := {}^*h$, we see the required conditions are satisfied by the continuity of h and f. So the Claim is proved.

By transferring Claim 2 and use it hyperfinitely many times, we define for each $n < N$ the function $g_n := \min\{g_{nm} \mid m < N\} \in {}^*\mathcal{M} \cap {}^*\mathrm{Re}(C(\Omega))$.

Then the following are satisfied:

- $\forall \omega \in U_n \left(\left| g_n(\omega) - {}^*f(\omega) \right| \leq \epsilon \right);$
- $\forall \omega \in {}^*\Omega \left(g_n(\omega) \leq {}^*f(\omega) + \epsilon \right).$

By Claim 2 again, we define $g_\epsilon := \max\{ g_n \mid n < N \} \in {}^*\mathcal{M}$. Then

- $\forall \omega \in {}^*\Omega \left(\left| g_\epsilon(\omega) - {}^*f(\omega) \right| \leq \epsilon \right).$

Since $\epsilon \in \mathbb{R}^+$ is arbitrary, we apply saturation to obtain $g \in {}^*\mathcal{M}$ such that $g \approx {}^*f$ and the theorem is proved. $\qquad\square$

Corollary 3.13. (Tietze Extension Theorem) *Let $\tilde{\Omega}$ be a compact topological space and $\Omega \subset \tilde{\Omega}$ be a closed subspace.*
Then every continuous $f : \Omega \to \mathbb{C}$ extends to a continuous $F : \tilde{\Omega} \to \mathbb{C}$.

Proof. Note that Ω is compact. Let $\pi : C(\tilde{\Omega}) \to C(\Omega)$ be given by $F \mapsto F \restriction_\Omega$ and let $\mathcal{M} = \pi[\tilde{\Omega}]$. So we need to show that $\mathcal{M} = C(\Omega)$.

By Prop. 1.24, Ω is normal, so, by Urysohn's Lemma (Thm. 1.17), $\mathcal{M}$ separates points in Ω. Clearly, $\mathcal{M}$ is a unital normed subalgebra of $C(\Omega)$ and is closed under the involution.

We leave it as an exercise to show that $\mathcal{M}$ is closed, hence $\mathcal{M}$ is a unital C^*-subalgebra of $C(\Omega)$.

Therefore the required conclusion follows from the Stone-Weierstrass Theorem (Thm. 3.11). $\qquad\square$

3.2.2 The Gelfand transform

We will show that the Gelfand transform represents a commutative unital C^*-algebra as the algebra of continuous functions: $C(\Omega)$ for some compact topological space Ω.

Given a unital Banach algebra $\mathcal{M}$, a nonzero linear functional $\theta : \mathcal{M} \to \mathbb{C}$ which is multiplicative, i.e. $\forall x, y \in \mathcal{M} \left(\theta(xy) = \theta(x)\theta(y) \right)$, is called a **character** of $\mathcal{M}$. We let $\hom(\mathcal{M})$ denote the set of characters of $\mathcal{M}$. (The choice of the symbol suggests that characters are also regarded as homomorphisms of the ring structure.)

Immediately, we see that $\forall \theta \in \hom(\mathcal{M}) \left(\theta(1) = 1 \right)$.

Note in particular that $\theta(xy) = \theta(yx)$ for any $x, y \in \mathcal{M}$.

Moreover, if $\theta \in \hom(\mathcal{M})$ and $a \in \mathcal{M} \setminus \{0\}$, then $\left(a - \theta(a) \right) \notin \mathcal{M}^{-1}$. For otherwise, there is $b \in \mathcal{M}^{-1}$ such that $\left(a - \theta(a) \right)b = 1$, then we have an absurdity:

$$1 = \theta\Big((a - \theta(a))b\Big) = \theta(ab - \theta(a)b) = \theta(ab) - \theta(a)\theta(b) = 0.$$

In particular, $\theta(a) \in \sigma(a)$, hence $|\theta(a)| \leq \rho(a) \leq \|a\|$ by Prop. 3.3. Since this holds for any $a \in \mathcal{M}$, together with $\theta(1) = 1$, we see that $\|\theta\| = 1$. In particular, θ is bounded.

Therefore $\mathrm{hom}(\mathcal{M}) \subset S_{\mathcal{M}'}$.

We equip $\mathrm{hom}(\mathcal{M})$ with the weak*-topology inherited from $\mathcal{M}'$.

Let $\theta \in \mathcal{M}'$ and $\phi \in {}^*\mathrm{hom}(\mathcal{M})$ such that $\theta \approx \phi$ in the weak*-topology. Then for any $a, b \in \mathcal{M}$, using the weak*-open sets given by a, a, ab, we have $\theta(ab) \approx \phi(ab), \theta(a) \approx \phi(a)$ and $\theta(b) \approx \phi(b)$, hence $\theta(ab) \approx \phi(ab) = \phi(a)\phi(b) \approx \theta(a)\theta(b)$, giving $\theta(ab) = \theta(a)\theta(b)$. Therefore $\mathrm{hom}(\mathcal{M})$ is weak*-closed and consequently, by the Alaoglu's Theorem (Thm. 2.16), we have:

Theorem 3.12. *Let $\mathcal{M}$ be a unital C^*-algebra, then $\mathrm{hom}(\mathcal{M})$ is weak*-compact.* $\square$

Now we concentrate on the commutative case for the time being.

Given a commutative unital C^*-algebra $\mathcal{M}$, we define

$$\wedge : \mathcal{M} \to C(\Omega), \quad \text{where } \Omega = \mathrm{hom}(\mathcal{M}),$$

by the evaluation mapping

$$\hat{a}(\theta) = \theta(a), \quad \text{for each } a \in \mathcal{M} \text{ and } \theta \in \mathrm{hom}(\mathcal{M}).$$

$\hat{a}$ is called the **Gelfand transform** of a.

As it should be distinguishable from the context, there is no confusion of the meaning of $\hat{a}$ here with that of an element in the nonstandard hull.

If we apply the same definition of the Gelfand transform to the group algebra of a discrete Abelian topological group (Example 3.2), we get the Fourier transform, hence the usage of the same wedge symbol.

Note that the weak*-topology is the weakest topology under which all the $\hat{a}$, $a \in \mathcal{M}$, are continuous. In particular, the range of the Gelfand transform is indeed in $C(\Omega)$.

Observe that the Gelfand transform corresponds to the canonical embedding into the bidual $\mathcal{M}''$ given by Prop. 2.23, but with the domain of the linear functionals $\mathcal{M}''$ restricted to $\mathrm{hom}(\mathcal{M})$.

We let $\mathfrak{M}(\mathcal{M})$ denote the family of all maximal ideals of a Banach algebra $\mathcal{M}$.

Lemma 3.3. *Let $\mathcal{M}$ be a commutative unital C^*-algebra. Then the following mapping is bijective:*

$$\mathrm{Ker} : \mathrm{hom}(\mathcal{M}) \to \mathfrak{M}(\mathcal{M}), \quad \text{given by} \quad \theta \mapsto \mathrm{Ker}(\theta).$$

Proof. Let $\theta \in \text{hom}(\mathcal{M})$. Then clearly, $\text{Ker}(\theta)$ forms an ideal of $\mathcal{M}$. As θ is linear and its range is $\mathbb{C}$, having dimension 1, $\text{Ker}(\theta)$ is maximal. Hence the range of Ker is indeed $\mathfrak{M}(\mathcal{M})$.

Next let $I \subset \mathcal{M}$ be a maximal ideal, let $\theta_0 : \mathcal{M} \to \mathcal{M}/I$ be the homomorphism $\theta_0(x) := (x + I)$, $x \in \mathcal{M}$. By Cor. 3.5, let $\pi : \mathcal{M}/I \to \mathbb{C}$ be the isometric isomorphism. Then for $\theta := \pi \circ \theta_0$, we see that $\theta \in \text{hom}(\mathcal{M})$ and $\text{Ker}(\theta) = I$. Therefore Ker is surjective.

Now suppose $\theta_1, \theta_2 \in \text{hom}(\mathcal{M})$ and $\text{Ker}(\theta_1) = \text{Ker}(\theta_2)$. Then for any $a \in \mathcal{M}$, we have $\big(a - \theta_1(a)\big) \in \text{Ker}(\theta_1) = \text{Ker}(\theta_2)$, so $\theta_2\big(a - \theta_1(a)\big) = 0$, hence $\theta_2(a) = \theta_1(a)$, therefore $\theta_1 = \theta_2$. So Ker is injective. $\qquad\square$

Observe where commutativity is used in the above proof.

Theorem 3.13. *Let $\mathcal{M}$ be a commutative unital C^*-algebra and $a \in \mathcal{M}$. Then $\sigma(a) = \{\theta(a) \,|\, \theta \in \text{hom}(\mathcal{M})\}$.*

Proof. The inclusion $\supset$ follows from a remark on p.220.

For the other inclusion, let $\lambda \in \sigma(a)$. So $(a - \lambda) \notin \mathcal{M}^{-1}$, hence $(a - \lambda)\mathcal{M}$ is a proper ideal of $\mathcal{M}$, extendable to some maximal ideal I. Then by Lem. 3.3, let $\theta \in \text{hom}(\mathcal{M})$ such that $I = \text{Ker}(\theta)$ and we get $\theta\big((a - \lambda)\big) = 0$, i.e. $\theta(a) = \lambda$. $\qquad\square$

Given a unital C^*-algebra $\mathcal{M}$, $\theta \in \text{hom}(\mathcal{M})$ and $a \in \text{Re}(\mathcal{M})$, by considering the commutative unital C^*-algebra $\mathcal{M}_a$ generated by a, noticing $\theta \restriction_{\mathcal{M}_a} \in \text{hom}(\mathcal{M}_a)$ and applying Thm. 3.9 and Thm. 3.13, we conclude that $\theta(a) = \theta \restriction_{\mathcal{M}_a}(a) \in \mathbb{R}$. Likewise, $\theta(a) \in [0, \infty)$ whenever $a \in \mathcal{M}^+$, etc.

Corollary 3.14. *Let $\mathcal{M}$ be a unital C^*-algebra. Let $\theta \in \text{hom}(\mathcal{M})$ and $a \in \mathcal{M}$. Then $\theta(a^*) = \overline{\theta(a)}$.*

Proof. By Thm. 3.9, write $a = a_1 + ia_2$, where $a_1, a_2 \in \text{Re}(\mathcal{M})$.

The above remark implies that $\theta(a_1), \theta(a_2) \in \mathbb{R}$, hence
$$\theta(a^*) = \theta(a_1 - ia_2) = \theta(a_1) - i\theta(a_2) = \overline{\theta(a_1) + i\theta(a_2)} = \overline{\theta(a_1 + ia_2)} = \overline{\theta(a)}.$$
$$\square$$

Corollary 3.15. *Let $u \in \mathcal{U}(\mathcal{M})$, where $\mathcal{M}$ is a commutative unital C^*-algebra. Then it holds for any $a \in \mathcal{M}$ that $\rho(a) = \rho(au)$.*

Proof. By Thm. 3.9(i) and Thm. 3.13, for any $\theta \in \text{hom}(\mathcal{M})$, we have $\theta(u) \in S_{\mathbb{C}}$, hence $|\theta(au)| = |\theta(a)|\,|\theta(u)| = |\theta(a)|$.

Therefore by Thm. 3.13 again,
$$\rho(au) = \sup\big\{\,|\theta(au)| \,\big|\, \theta \in \text{hom}(\mathcal{M})\big\}$$
$$= \sup\big\{\,|\theta(a)| \,\big|\, \theta \in \text{hom}(\mathcal{M})\big\} = \rho(a). \qquad\square$$

Corollary 3.16. *Let $\mathcal{M}$ be a unital C^*-algebra and $a, b \in \mathrm{Re}(\mathcal{M})$.*

(i) *If $ab = ba$ then $\sigma(a + b) \subset (\sigma(a) + \sigma(b))$.*

(ii) *If $ab = 0 = ba$ then $\sigma(a + b) \subset (\sigma(a) \cup \sigma(b)) \subset (\sigma(a + b) \cup \{0\})$.*

Proof.   (i): Under the given assumptions, $\mathcal{M}_{\{a,b\}}$ is a commutative unital C^*-subalgebra of $\mathcal{M}$.

By Thm. 3.10(ii), we have $\sigma_{\mathcal{M}_{\{a,b\}}}(a) = \sigma(a)$, $\sigma_{\mathcal{M}_{\{a,b\}}}(b) = \sigma(b)$ and $\sigma_{\mathcal{M}_{\{a,b\}}}(a + b) = \sigma(a + b)$, so we assume without loss of generality that $\mathcal{M} = \mathcal{M}_{\{a,b\}}$.

Then it follows from Thm. 3.13 that

$$\sigma(a + b) = \{\theta(a + b) \mid \theta \in \mathrm{hom}(\mathcal{M})\}$$
$$= \{\theta(a) + \theta(b) \mid \theta \in \mathrm{hom}(\mathcal{M})\} \subset (\sigma(a) + \sigma(b)).$$

(ii): We continue with the setting in (i). Suppose $ab = 0 = ba$.

Then for every $\theta \in \mathrm{hom}(\mathcal{M})$, as $\theta(a)\theta(b) = \theta(ab) = 0$, we cannot have $\theta(a) \neq 0 \neq \theta(b)$.

Hence, for every $\theta \in \mathrm{hom}(\mathcal{M})$, we have

$$\text{either} \quad \theta(a + b) = \theta(a) \quad \text{or} \quad \theta(a + b) = \theta(b).$$

In any case $\theta(a + b) \in (\sigma(a) \cup \sigma(b))$. So we conclude by Thm. 3.13 that $\sigma(a + b) \subset (\sigma(a) \cup \sigma(b))$.

For the other inclusion, for any $\theta \in \mathrm{hom}(\mathcal{M})$, if $\theta(a) \neq 0$, then $\theta(a) = \theta(a + b)$, i.e. $\theta(a) \in \sigma(a + b)$. Likewise if $\theta(b) \neq 0$, then $\theta(b) \in \sigma(a + b)$. Therefore, by Thm. 3.13, we have $(\sigma(a) \cup \sigma(b)) \subset (\sigma(a + b) \cup \{0\})$. $\square$

A mapping $\theta : \mathcal{M} \to \tilde{\mathcal{M}}$ between two unital C^*-algebras $\mathcal{M}$ and $\tilde{\mathcal{M}}$ is a ***homomorphism*** if it is a homomorphism in the algebra and C^* sense, *i.e.* $\forall x, y \in \mathcal{M} \quad \forall \alpha \in \mathbb{C}$

- $(\theta(x + \alpha y) = \theta(x) + \alpha\theta(y)) \wedge (\theta(xy) = \theta(x)\theta(y)) \wedge (\theta(1) = 1)$
- $\theta(x^*) = (\theta(x))^*$.

If in addition, θ is injective, it is a ***isomorphism***. In this case we also say that θ **embeds** $\mathcal{M}$ into $\tilde{\mathcal{M}}$ and θ is a C^*-**embedding**.

Note that we do not require a *isomorphism to be surjective. When there is need, we would specifically mention whether the *isomorphism is onto or just into the target algebra. Note also that an isometric *homomorphism between unital C^*-algebras must be a *isomorphism. It will be shown in Thm. 3.15 for a *homomorphism, isometry is the same as *isomorphism.

By Cor. 3.14, a character of a unital C^*-algebra is a *homomorphism into $\mathbb{C}$.

Now we prove that every commutative unital C^*-algebra is representable as an algebra of continuous functions on some compact set. Hence we have a classification of all commutative unital C^*-algebras.

Theorem 3.14. *Let $\mathcal{M}$ be a commutative unital C^*-algebra.*

*Then the Gelfand transform $\wedge : \mathcal{M} \to C(\Omega)$, where $\Omega = \hom(\mathcal{M})$, is an isometric *isomorphism onto $C(\Omega)$.*

*(In fact isometry follows from being a *isomorphism, by Thm. 3.15.)*

Proof. Let $a, b \in \mathcal{M}$, $\alpha \in \mathbb{C}$, and $\theta \in \Omega$. Then

$$(\widehat{a + \alpha b})(\theta) = \theta(a + \alpha b) = \theta(a) + \alpha\theta(b) = \hat{a}(\theta) + \alpha\hat{b}(\theta),$$

$$(\widehat{ab})(\theta) = \theta(ab) = \theta(a)\theta(b) = (\hat{a}(\theta))(\hat{b}(\theta)), \quad \widehat{a^*}(\theta) = \theta(a^*) = \overline{\theta(a)} = \overline{\hat{a}(\theta)},$$

where the last equality follows from Cor. 3.14.

So the Gelfand transform is a *homomorphism.

Moreover, for $a \in \mathcal{M}$,

$$\|\hat{a}\|^2 = \max_{\theta \in \Omega} |\theta(a)|^2 = \max_{\theta \in \Omega} \left(\overline{\theta(a)}\, \theta(a)\right) = \max_{\theta \in \Omega} \left(\theta(a^*)\, \theta(a)\right)$$

$$= \max_{\theta \in \Omega} \theta(a^*a) = \rho(a^*a) = \|a^*a\| = \|a\|^2,$$

where the last line follows from Thm. 3.13 and Thm. 3.9(vii). So we get $\|\hat{a}\| = \|a\|$.

Therefore the Gelfand transform is an isometry.

Consequently, for $a, b \in \mathcal{M}$,

$$\hat{a} = \hat{b} \quad \text{iff} \quad 0 = \|\hat{a} - \hat{b}\| = \|\widehat{a - b}\| \quad \text{iff} \quad 0 = \|a - b\| \quad \text{iff} \quad a = b,$$

i.e. the Gelfand transform is injective, *i.e.* a *isomorphism.

So the Gelfand transform embeds $\mathcal{M}$ onto a unital C^*-subalgebra of $C(\Omega)$. Moreover, it separates points: for $\theta_1 \neq \theta_2$ in $\Omega = \hom(\mathcal{M})$, there is $a \in \mathcal{M}$ such that $\theta_1(a) \neq \theta_2(a)$, hence $\hat{a}(\theta_1) \neq \hat{a}(\theta_2)$.

Now the surjectivity follows from the Stone-Weierstrass Theorem (Thm. 3.11) and we conclude that the Gelfand transform is a *isomorphism of $\mathcal{M}$ onto $C(\Omega)$. $\square$

For a not necessarily commutative unital C^*-algebra, Thm. 3.14 is still applicable to $\mathcal{M}_a$, the C^*-subalgebra generated by $a \in \mathcal{M}$, when a is normal, as $1 \in \mathcal{M}_a$ and is a commutative C^*-algebra of $\mathcal{M}$. By the Gelfand

transform in Thm. 3.14, from Thm. 3.13 and Example 3.13(ii), by recalling
Thm. 3.10(ii) that $\sigma_{\mathcal{M}}$ is the same as σ_{M_a}, we have:

Corollary 3.17. *Let $\mathcal{M}$ be a unital C^*-algebra. Then for every normal element $a \in \mathcal{M}$ and $\Omega := \mathrm{hom}(\mathcal{M}_a)$,*

$$\sigma_{\mathcal{M}}(a) = \hat{a}[\Omega] = \sigma_{C(\Omega)}(\hat{a}).$$

$\square$

Consider a commutative unital C^*-algebra $\mathcal{M}$, $\Omega = \mathrm{hom}(\mathcal{M})$ and $a \in \mathcal{M}$. Then for every $\theta \in C(\sigma(a))$, by the Gelfand transform in Thm. 3.14 and Cor. 3.17 there is an unique $b \in \mathcal{M}$ such that $\hat{b} = \theta(\hat{a}) \in C(\Omega)$.

We define this unique b as $\theta(a)$. This is referred to as the ***Continuous Functional Calculus*** on $\mathcal{M}$. It is more convenient than the Riesz functional calculus, even for a simple example such as the positive root, $\sqrt{|a|}$ is defined for every $a \in \mathcal{M}$, but is not available from analytic functions, as 0 is a singularity for any branch of the square root function.

However, as in Example 3.13(xiii),the continuous functional calculus extends consistently the Riesz functional calculus: Suppose θ is analytic on $\sigma(a)$. As the argument for the general case is similar, we assume that $\sigma(a)$ is simply connected and let a simple closed smooth counterclockwise curve $\mathcal{C}$ be in the interior of the domain of f with $\sigma(a)$ inside and be parametrized by $\gamma : [0,1] \to \mathbb{C}$. By viewing $\theta(a)$ as given by the Riesz functional calculus, we have at each $\omega \in \Omega$ the value of the Gelfand transform of $\theta(a)$ given by

$$\widehat{\theta(a)}(\omega) = \omega\big(\theta(a)\big) = \omega\left(\frac{1}{2\pi i} \int_{\mathcal{C}} \theta(z)(z-a)^{-1}dz \right)$$

$$= \frac{1}{2\pi i}\, \omega\left(\lim \sum_{k=1}^{n} \theta(\gamma(t_k))(\gamma(t_k) - a)^{-1}(t_k - t_{k-1}) \right)$$

$$= \frac{1}{2\pi i}\left(\lim \sum_{k=1}^{n} \theta(\gamma(t_k))(\gamma(t_k) - \omega(a))^{-1}(t_k - t_{k-1}) \right)$$

$$= \frac{1}{2\pi i} \int_{\mathcal{C}} \theta(z)(z - \omega(a))^{-1}dz = \theta\big(\omega(a)\big),$$

where the limit is taken over all partitions of $[0,1]$ as $0 = t_0 < t_1 < \cdots < t_n = 1$, $n \in \mathbb{N}$. Equalities on the third and fourth lines are justified by the convergence of the Riemann sums and the continuity of θ.

As $\theta\big(\omega(a)\big) = \theta\big(\hat{a}(\omega)\big) = (\theta \circ \hat{a})(\omega) = \widehat{\theta(a)}(\omega)$, the value at ω of the Gelfand transform of $\theta(a)$ viewed by the continuous functional calculus is the same as that viewed by the Riesz functional calculus, hence both views of $\theta(a)$ are the same by the *isomorphism of the Gelfand transform.

By Cor. 3.17, we also have

$$\sigma_{\mathcal{M}}\big(\theta(a)\big) = \widehat{\theta(a)}[\Omega] = \theta(\hat{a})[\Omega] = \theta[\hat{a}[\Omega]] = \theta[\sigma_{C(\Omega)}(\hat{a})] = \theta[\sigma_{\mathcal{M}}(a)].$$

In the case of noncommutative unital C^*-algebra $\mathcal{M}$, the continuous functional calculus can still be defined for normal elements $a \in \mathcal{M}$, for the C^*-subalgebra $\mathcal{M}_a$ generated by a is commutative and contains 1.

Note that then the Gelfand transform is defined on $\mathcal{M}_a$, *i.e.*

$$\wedge : \mathcal{M}_a \to C\big(\mathrm{hom}(\mathcal{M}_a)\big).$$

We also have $\sigma_{\mathcal{M}_a}\big(\theta(a)\big) = \theta[\sigma_{\mathcal{M}_a}(a)]$, but by Thm. 3.10(ii), $\sigma_{\mathcal{M}}$ is the same as $\sigma_{\mathcal{M}_a}$, so $\sigma_{\mathcal{M}}\big(\theta(a)\big) = \theta[\sigma_{\mathcal{M}}]$ holds.

We thus summarize all these as follows.

Corollary 3.18. (The Continuous Functional Calculus) *Let $\mathcal{M}$ be a unital C^*-algebra. Then for every normal element $a \in \mathcal{M}$ and $f \in C(\sigma(a))$, an element $f(a) \in \mathcal{M}$ is uniquely definable from*

$$\widehat{f(a)} = f(\hat{a}), \quad \text{where } \wedge \text{ is the Gelfand transform on } \mathcal{M}_a.$$

(Hence $\|f(a)\|_{\mathcal{M}} = \|f\|_{C(\sigma(a))}$.)

This is in agreement with the Riesz functional calculus when f is analytic on $\sigma(a)$. Moreover we have the **Spectral mapping Theorem** *for continuous functions:*

$$\sigma\big(f(a)\big) = f[\sigma(a)].$$

If $\mathcal{M}$ is commutative, all the above hold for every $a \in \mathcal{M}$. $\square$

The following shows that a *isomorphism is automatically an isometry, *i.e.* the topological property is already encoded in the algebraic properties of the C^*-algebra.

Theorem 3.15. *Let $\mathcal{M}$ and $\tilde{\mathcal{M}}$ be unital C^*-algebras. Let $\pi : \mathcal{M} \to \tilde{\mathcal{M}}$.*

(i) *If π is a *homomorphism, then $\pi \in \mathcal{B}(\mathcal{M}, \tilde{\mathcal{M}})$ with $\|\pi\| \leq 1$, i.e. a contraction.*

(ii) *If π is a *isomorphism, then π is an isometry.*

Proof. (i): Firstly, π is clearly linear.

To show that $\|\pi\| \leq 1$, we first note that by being a *homomorphism, $\pi[\mathcal{M}^{-1}] \subset \tilde{\mathcal{M}}^{-1}$. Hence $\forall x \in \mathcal{M}\,\big(\sigma_{\tilde{\mathcal{M}}}(\pi(x)) \subset \sigma_{\mathcal{M}}(x)\big)$.

If $c \in \mathcal{M}$ is normal, then $\pi(c)$ is also normal. So, by Thm. 3.9(vii), we have both $\|c\| = \rho(c)$ and $\|\pi(c)\| = \rho(\pi(c))$. Moreover, by the above,

$$\rho(\pi(c)) = \max\{|\lambda| \mid \lambda \in \sigma_{\tilde{\mathcal{M}}}(\pi(c))\} \leq \max\{|\lambda| \mid \lambda \in \sigma_{\mathcal{M}}(c)\} = \rho(c).$$

Therefore, it holds for any $a \in \mathcal{M}$ that

$$\|\pi(a)\|^2 = \|\pi(a)^* \pi(a)\| = \|\pi(a^* a)\| \leq \|a^* a\| = \|a\|^2, \quad i.e. \ \|\pi(a)\| \leq \|a\|,$$

showing $\|\pi\| \leq 1$ and in particular $\pi \in \mathcal{B}(\mathcal{M}, \tilde{\mathcal{M}})$.

(ii): We need to show that $\forall x \in \mathcal{M} \left(\|\pi(x)\| = \|x\| \right)$.

It suffices to show that $\forall x \in \mathrm{Re}(\mathcal{M}) \left(\|\pi(x)\| = \|x\| \right)$. Because this implies for any $a \in \mathcal{M}$ that (since $(a^* a) \in \mathrm{Re}(\mathcal{M})$)

$$\|\pi(a)\|^2 = \|(\pi(a))^* \pi(a)\| = \|\pi(a^* a)\| = \|a^* a\| = \|a\|^2.$$

So fix $a \in \mathrm{Re}(\mathcal{M})$ and show that $\|\pi(a)\| = \|a\|$.

CLAIM: $\hom(\mathcal{M}_a) = \{(\theta \circ \pi \!\restriction_{\mathcal{M}_a}) \mid \theta \in \hom(\tilde{\mathcal{M}}_{\pi(a)})\}$.

PROOF OF THE CLAIM: First note that, by (i), as a contraction, π is continuous in norm. Therefore, as a *isomorphism, we have $\pi[\mathcal{M}_a] \subset \tilde{\mathcal{M}}_{\pi(a)}$. For convenience, denote $\Omega := \hom(\mathcal{M}_a)$ and let $\Omega_0 := \{(\theta \circ \pi \!\restriction_{\mathcal{M}_a}) \mid \theta \in \hom(\tilde{\mathcal{M}}_{\pi(a)})\}$.

First note that Ω_0 is weak*-compact in Ω : Let $\phi \in {}^*\Omega_0$, i.e. for some $\theta \in {}^*\hom(\tilde{\mathcal{M}}_{\pi(a)})$ we have $\phi = (\theta \circ \pi \!\restriction_{\mathcal{M}_a})$. Then since $\hom(\tilde{\mathcal{M}}_{\pi(a)})$ is weak*-compact, for some $\psi \in \hom(\tilde{\mathcal{M}}_{\pi(a)})$ we have $\theta \approx \psi$ in the weak*-topology. Then, in the weak*-topology, $\phi \approx (\psi \circ \pi \!\restriction_{\mathcal{M}_a}) \in \Omega_0$.

If $\theta \in \hom(\tilde{\mathcal{M}}_{\pi(a)})$, then clearly, $(\theta \circ \pi \!\restriction_{\mathcal{M}_a}) \in \hom(\mathcal{M}_a)$, so the inclusion $\Omega_0 \subset \Omega$ is trivial.

Now suppose there is some $\theta \in \Omega \setminus \Omega_0$. Since Ω is compact in the weak*-topology, it is a Tychonoff space (Cor. 1.8), so there is $F \in C(\Omega)$ such that $F(\theta) = 1$ and $F[\Omega_0] = \{0\}$.

By Thm. 3.14, there is $c \in \mathcal{M}_a$ such that $F = \hat{c}$, using the Gelfand transform on $\mathcal{M}_a$. Then

$$\theta(c) = \hat{c}(\theta) = F(\theta) = 1 \quad \text{and} \quad \phi(c) = \hat{c}(\phi) = F(\phi) = 0 \quad \text{for all } \phi \in \Omega_0.$$

The former implies that $c \neq 0$. By Thm. 3.13, the latter implies that $\sigma_{\tilde{\mathcal{M}}_{\pi(a)}}(\pi(c)) = \{0\}$, hence $\pi(c) = 0$ by Thm. 3.9(vii) and $\pi(c)$ being normal. ($\mathcal{M}_a$ is commutative and π is a *isomorphism, so $\pi[\mathcal{M}_a]$ is commutative. In particular, $\pi(c)$ is normal.)

Therefore we have a contradiction to π being injective from $c \neq 0$ but $\pi(c) = 0$. The Claim is proved.

Now as both a and $\pi(a)$ are self-adjoint (and hence normal), we get from Thm. 3.13 and the Claim that

$$\|\pi(a)\|_{\tilde{\mathcal{M}}} = \|\pi(a)\|_{\tilde{\mathcal{M}}_{\pi(a)}} = \rho_{\tilde{\mathcal{M}}_{\pi(a)}}(\pi(a)) = \max_{\theta \in \hom(\tilde{\mathcal{M}}_{\pi(a)})} \theta(\pi(a))$$

$$= \max_{\theta \in \hom(\mathcal{M}_a)} \theta(a) = \rho_{\mathcal{M}_a}(a) = \|a\|_{\mathcal{M}_a} = \|a\|_{\mathcal{M}}.$$

Therefore (ii) is proved. $\qquad\qquad\qquad\qquad\qquad\qquad\qquad\qquad\qquad \square$

By Thm. 3.15(ii), for a unital algebra with an involution operation, there is at most one norm turning it into a unital C^*-algebra.

In Thm. 3.15(i), if we let $I = \mathrm{Ker}(\pi)$, then it is easy to see that I is a closed ideal of $\mathcal{M}$. By Prop. 3.10, $\mathcal{M}/I$ is a unital C^*-algebra. Let $\pi_I : \mathcal{M}/I \to \tilde{\mathcal{M}}$ be given by $\pi_I(x + I) = \pi(x)$, where $x \in \mathcal{M}$. Then π_I is a *isomorphism, hence $\pi_I[\mathcal{M}]$ is a unital C^*-algebra, by Thm. 3.15(ii). Since $\pi[\mathcal{M}] = \pi_I[\mathcal{M}]$, we have the following:

Corollary 3.19. *Let* $\pi : \mathcal{M} \to \tilde{\mathcal{M}}$ *be a *homomorphism between unital C^*-algebras $\mathcal{M}$ and $\tilde{\mathcal{M}}$. Then $\pi[\mathcal{M}]$ is a unital C^*-subalgebra of $\tilde{\mathcal{M}}$ and, in particular, $\pi[\mathcal{M}]$ is closed.* $\qquad\square$

With the availability of the continuous functional calculus, we are ready to prove some useful properties motivated by Example 3.13.

The following contains the converse of some assertions in Thm. 3.9.

Theorem 3.16. *Let $\mathcal{M}$ be a unital C^*-algebra and $a \in \mathcal{M}$ be normal. Then the following hold.*

(i) *Let $f \in C(\sigma(a))$. Then $f(a)$ is normal.*
(ii) $a \in \mathcal{U}(\mathcal{M})$ *iff* $\sigma(a) \subset S_{\mathbb{C}}$.
(iii) $a \in \mathrm{Re}(\mathcal{M})$ *iff* $\sigma(a) \subset \mathbb{R}$.
(iv) $a \in \mathcal{M}^+$ *iff* $\sigma(a) \subset [0, \infty)$.
(v) $a \in \mathcal{M}^+$ *iff* $a \in \mathrm{Re}(\mathcal{M})$ *and* $\big\| \|a\| - a \big\| \le \|a\|$.
(vi) $\pm a \in \mathcal{M}^+$ *iff* $a = 0$.
(vii) $a \in \mathrm{Proj}(\mathcal{M})$ *iff* $\sigma(a) \subset \{0, 1\}$.

Proof. (i): Since $f(a) \in \mathcal{M}_a$, we have $\big(f(a)\big)^* \in \mathcal{M}_a$. As $\mathcal{M}_a$ is commutative, $f(a)$ commutes with $\big(f(a)\big)^*$.

(ii): One direction was given in Thm. 3.9(i). For the other one, suppose $\sigma(a) \subset S_{\mathbb{C}}$, then $\hat{a}[\Omega] = \sigma(\hat{a}) \subset S_{\mathbb{C}}$, by Cor. 3.17, where $\Omega := \mathrm{hom}(\mathcal{M}_a)$. Then for any $\omega \in \Omega$, $(\bar{\hat{a}} \circ a)(\omega) = 1$, Hence $(\bar{\hat{a}} \circ a) = 1$, *i.e.* $\widehat{a^*a} = \hat{a}^*\hat{a} = 1$, therefore, by the Gelfand transform, $a^*a = 1$.

By a being normal, also $aa^* = 1$, so $a \in \mathcal{U}(\mathcal{M})$.

(iii): Again, by Thm. 3.9(ii), we consider the other direction and let $\sigma(a) \subset \mathbb{R}$. Let $f \in C(\mathbb{C})$ be the function $f(z) = z - \bar{z}$. So

$$f\big[\sigma(a)\big] = \{\lambda - \bar{\lambda} \,|\, \lambda \in \sigma(a)\} = \{0\}.$$

By Cor. 3.18, $f\big[\sigma(a)\big] = \sigma\big(f(a)\big)$, so $\sigma\big(f(a)\big) = \{0\}$. By (i), $f(a)$ is normal, so it follows from Thm. 3.9(vii) that $f(a) = 0$, implying $a - a^* = 0$, *i.e.* $a \in \mathrm{Re}(\mathcal{M})$.

(iv): By Thm. 3.9(iii), we consider the other direction and let $\sigma(a) \subset [0, \infty)$. Let $f(z) = \sqrt{z}$ be a branch of the square root function that includes the positive roots. By Cor. 3.18, $\sigma(f(a)) \subset [0, \infty)$, so $f(a) \in \mathrm{Re}(\mathcal{M})$ according to (i) and (iii). Furthermore, $(f(a))^2 = f^2(a) = a$, therefore $a \in \mathcal{M}^+$.

(v): Let $\Omega := \mathrm{hom}(\mathcal{M}_a)$.

Suppose $a \in \mathcal{M}^+$. Then clearly $a \in \mathrm{Re}(\mathcal{M})$. By (iv), $\sigma(a) \subset [0, \infty)$, hence $\hat{a}[\Omega] \subset [0, \infty)$, by Cor. 3.17. So $\forall \omega \in \Omega \left(0 \leq \hat{a}(\omega) \leq \|\hat{a}\| \right)$, implying $\left\| \|\hat{a}\| - \hat{a} \right\| \leq \|\hat{a}\|$, therefore $\left\| \|a\| - a \right\| \leq \|a\|$, by Cor. 3.17 and Cor. 3.18.

For the converse, we have $\sigma(\hat{a}) \subset \mathbb{R}$ and similarly $\left\| \|\hat{a}\| - \hat{a} \right\| \leq \|\hat{a}\|$. It follows then $\hat{a}(\omega) \geq 0$ for all $\omega \in \Omega$, hence $\sigma(\hat{a}) \subset [0, \infty)$ and thus $\sigma(a) \subset [0, \infty)$, therefore $a \in \mathcal{M}^+$ by (iv).

(vi): For the nontrivial direction, if $\pm a \in \mathcal{M}^+$, then by (iv), we have

$$\sigma(a) \subset [0, \infty) \quad \text{and} \quad -[\sigma(a)] = \sigma(-a) \subset [0, \infty),$$

hence $\sigma(a) = \{0\}$. Therefore, by Thm. 3.9(vii), $a = 0$.

(vii): One direction is Thm. 3.9(iv). So suppose $\sigma(a) \subset \{0, 1\}$. Then by (iv), $a^* = a$. Let $f \in C(\mathbb{C})$ be the function $f(z) = z - z^2$. By Cor. 3.18, $\{0\} = f[\sigma(a)] = \sigma(f(a)) = \sigma(a - a^2)$, so $a - a^2 = 0$ by Thm. 3.9(vii), *i.e.* $a = a^* = a^2$ and so $a \in \mathrm{Proj}(\mathcal{M})$. $\square$

Corollary 3.20. *Let $\mathcal{M}$ be a unital C^*-algebra and $a \in \mathcal{M}^+$.*
   Then $(\|a\| - a) \in \mathcal{M}^+$.

Proof. By Thm. 3.16(iv) and Thm. 3.9(vii), $\sigma(a) \subset [0, \|a\|]$. Then by Cor. 3.18, $\sigma(\|a\| - a) = (\|a\| - \sigma(a)) \subset [0, \|a\|]$.
   Hence $(\|a\| - a) \in \mathcal{M}^+$ by Thm. 3.16(iv) again. $\square$

Because of Thm. 3.16, we can decompose a self-adjoint element as the sum of a positive and a negative element. Note the resemblance between this and the Hahn-Jordan Decomposition of a signed measure given on p. 35.

Corollary 3.21. *Let $\mathcal{M}$ be a unital C^*-algebra and $a \in \mathrm{Re}(\mathcal{M})$.*
   Then there are unique $a_1, a_2 \in \mathcal{M}^+$ such that

$$a = a_1 - a_2 \quad \text{and} \quad a_1 a_2 = 0 = a_2 a_1.$$

Moreover, $\|a_1\|, \|a_2\| \leq \|a\|$.

Proof.   Note that a is normal. Let $f_1, f_2 \in C(\mathbb{C})$ be given by

$$f_1(z) := \max\{\operatorname{Re}(z), 0\} \quad \text{and} \quad f_2(z) := -\min\{\operatorname{Re}(z), 0\}.$$

Define $a_1 := f_1(a)$ and $a_2 := f_2(a)$.

Then $f_1(a)$ and $f_2(a)$ are both normal, since they are elements of $\mathcal{M}_a$ which is a commutative unital C^*-algebra. Moreover,

$$\sigma(a_1) = f_1[\sigma(a)] \subset [0, \infty) \quad \text{and} \quad \sigma(a_2) = f_2[\sigma(a)] \subset [0, \infty),$$

so it follows from Thm. 3.16(iv) that $a_1, a_2 \in \mathcal{M}_a^+$, so $a_1, a_2 \in \mathcal{M}^+$.

Also, $f_1 + f_2$ is the identity function, therefore

$$\widehat{a_1 + a_2} = \left(f_1(a) + f_2(a) \right)^\wedge = f_1(\hat{a}) + f_2(\hat{a}) = \hat{a}$$

and it follows from the Gelfand transform (for $\mathcal{M}_a$) that $a_1 + a_2 = a$.

Furthermore, $a_1 a_2 = 0$, since

$$\widehat{a_1 a_2} = \hat{a}_1 \hat{a}_2 = \widehat{f_1(a)} \widehat{f_2(a)} = f_1(\hat{a}) f_2(\hat{a}) = (f_1 \circ f_2)(\hat{a}) = 0.$$

Note that a_1 commutes with a_2, since both are elements of $\mathcal{M}_a$ and a is normal.

As for the uniqueness, suppose $a = b_1 - b_2$ for some $b_1, b_2 \in \mathcal{M}^+$ with $b_1 b_2 = 0 = b_2 b_1$. Since $b_1 a = b_1^2 = a b_1$ and $b_2 a = -b_2 = a b_2$, all elements in $\mathcal{M}_a \cup \{b_1, b_2\}$ commute with each other, by the remark on p.211.

Let $\tilde{\mathcal{M}}$ be the C^*-subalgebra of $\mathcal{M}$ generated by $\mathcal{M}_a \cup \{b_1, b_2\}$, then $1 \in \tilde{\mathcal{M}}$ and $\tilde{\mathcal{M}}$ is commutative. From the Gelfand transform for $\tilde{\mathcal{M}}$, we have $\hat{a}_1, \hat{a}_2, \hat{b}_1, \hat{b}_2 : \operatorname{hom}(\tilde{\mathcal{M}}) \to [0, \infty)$. From $a = a_1 - a_2 = b_1 - b_2$ and $a_1 a_2 = 0 = b_1 b_2$, we have $\hat{a}_1 - \hat{a}_2 = \hat{b}_1 - \hat{b}_2$ and $\hat{a}_1 \hat{a}_2 = 0 = \hat{b}_1 \hat{b}_2$. The latter implies $\nexists \omega \in \operatorname{hom}(\tilde{\mathcal{M}}) \left((\hat{a}_1(\omega) > 0) \wedge (\hat{a}_2(\omega) > 0) \right)$ and similarly for $\hat{b}_1, \hat{b}_2$. Hence we conclude that $\hat{a}_1 = \hat{b}_1$ and $\hat{a}_2 = \hat{b}_2$, therefore $a_1 = b_1$ and $a_2 = b_2$.

Finally, by Thm. 3.9(vii), $\|a_1\| = \rho(a_1)$. By Cor. 3.18, since $\sigma(a_1) = f_1[\sigma(a)] \subset [0, \|a\|]$, we get $\|a_1\| \le \|a\|$. Likewise, $\|a_2\| \le \|a\|$. $\square$

Positive elements plays an important rôle in unital C^*-algebras by reflecting their key algebraic structures. So we continue with the followings.

Theorem 3.17. *Let $\mathcal{M}$ be a unital C^*-algebra. Then:*

(i) $\mathcal{M}^+$ *forms a closed cone and* $\mathcal{M}^+ \cap (-\mathcal{M}^+) = \{0\}$.

(ii) $(\mathcal{M}^+)^r = \mathcal{M}^+$ *holds for all* $r \in (0, \infty)$.

(iii) $\forall x \in \mathcal{M} \left((x \in \mathcal{M}^+) \Leftrightarrow (\exists y \in \mathcal{M} (x = y^* y)) \right)$.

Proof. (i): Clearly $r\mathcal{M}^+ \subset \mathcal{M}^+$ holds for all $r \in [0, \infty)$. So it suffices to show that $\mathcal{M}^+ \cap \bar{B}_\mathcal{M}$ is closed convex. But by Thm. 3.16(v),

$$\mathcal{M}^+ \cap \bar{B}_\mathcal{M} = \text{Re}(\mathcal{M}) \cap \{ x \in \mathcal{M} \mid (\|x\| \le 1) \wedge (\|1 - x\| \le 1) \},$$

with both sets on the right side being closed convex, hence $\mathcal{M}^+$ is closed convex.

The other statement about $\mathcal{M}^+$ follows form Thm. 3.16(vi).

(ii): Let $a \in \mathcal{M}^+$ and $r \in (0, \infty)$. By the Gelfand transform for $\mathcal{M}_a$, we have by Cor. 3.18 that $\sigma(a^r) = (\sigma(a))^r \subset [0, \infty)$, hence $a^r \in \mathcal{M}^+$ by Thm. 3.16(iv). Therefore $(\mathcal{M}^+)^r \subset \mathcal{M}^+$.

Consequently, $(\mathcal{M}^+)^{1/r} \subset \mathcal{M}^+$, so we also get $\mathcal{M}^+ \subset (\mathcal{M}^+)^r$.

(iii): The direction ($\Rightarrow$) is trivial.

For the nontrivial direction, first we need the following:

CLAIM: Let $c \in \mathcal{M}$. Then $(c^* c + c c^*) \in \mathcal{M}^+$.

By Thm. 3.9(vi), decompose $c = c_1 + i c_2$, where $c_1, c_2 \in \text{Re}(\mathcal{M})$. Then

$$c^* c + c c^* = 2(c_1^2 + c_2^2) \in \mathcal{M}^+$$

by $c_1^2, c_2^2 \in \mathcal{M}^+$ and (i) and so the Claim is proved.

Now let $a = c^* c$ for some $c \in \mathcal{M}$. We need to show that $a \in \mathcal{M}^+$.

Clearly $a \in \text{Re}(\mathcal{M})$. Therefore, by Cor. 3.21, we have the decomposition $a = a_1 - a_2$ for some $a_1, a_2 \in \mathcal{M}^+$ with $a_1 a_2 = 0 = a_2 a_1$. Then

$$-(c a_2)^* (c a_2) = -a_2 c^* c a_2 = -a_2 a a_2 = a_2^3 \in \mathcal{M}^+,$$

by (ii). Consequently, by the Claim and (i), we get

$$(c a_2)(c a_2)^* = \Big(\big((c a_2)^* (c a_2) + (c a_2)(c a_2)^*\big) - (c a_2)^* (c a_2) \Big) \in \mathcal{M}^+.$$

So, by Thm. 3.16(iv), $\sigma((c a_2)(c a_2)^*) \subset [0, \infty)$. Hence, by Prop. 3.4, we also have $\sigma((c a_2)^* (c a_2)) \subset [0, \infty)$, *i.e.* $((c a_2)^* (c a_2)) \in \mathcal{M}^+$, by Thm. 3.16(iv) again.

Note that $(c a_2)^* (c a_2) = a_2 c^* c a_2 = a_2 a a_2 = -a_2^3$, so we have $\pm a_2^3 \in \mathcal{M}^+$ and hence $a_2^3 = 0$ by Thm. 3.16(vi). Since $a_2 \in \mathcal{M}^+$, it is necessary that $a_2 = 0$, by Thm. 3.9(vii).

Therefore we conclude that $a = a_1 \in \mathcal{M}^+$. $\square$

We define a natural partial ordering on $\mathcal{M}$ by letting $a \le b$, where $a, b \in \mathcal{M}$, if $(b - a) \in \mathcal{M}^+$ and $a < b$ if $(b - a) \in \mathcal{M}^+ \setminus \{0\}$.

Then, because of Thm. 3.17(i), $\le$ is a partial ordering: it is clearly reflexive, as $0 \in \mathcal{M}^+$; it is antisymmetric, for if $a \le b \le a$, then $(b - a) \in$

$M^+ \cap (-\mathcal{M}^+)$, hence $a = b$; and it is transitive, for if $a \leq b \leq c$ then $(c - a) = (c - b) + (b - a) \in \mathcal{M}^+$, i.e. $a \leq c$.

Note that $0 \leq x$ is the same as $x \in \mathcal{M}^+$.

Proposition 3.11. *Let $\mathcal{M}$ be a unital C^*-algebra.*

(i) $\forall x \in \mathcal{M}^+ \left(x \leq \|x\| \right)$.
(ii) $\forall x \in \mathcal{M}^+ \, \forall y \in \mathcal{M} \left((y^* x y) \in \mathcal{M}^+ \right)$.
(iii) $\forall x, y \in \mathcal{M} \left((xy)^*(xy) \leq \|x\|^2 \, y^* y \right)$.

Proof. (i): Let $a \in \mathcal{M}^+$. Then $\sigma(a) \subset [0, \|a\|)$, by Prop. 3.3 and Thm. 3.16(iv). Now apply Cor. 3.18 to the function $f(z) = \|a\| - z$, we have $\sigma\left(\|a\| - a \right) = f[\sigma(a)] \subset [0, \|a\|)$.

So it follows from Thm. 3.16(iv) that $0 \leq \left(\|a\| - a \right)$, i.e. $a \leq \|a\|$.

(ii): Let $a \in \mathcal{M}^+$ and $b \in \mathcal{M}$. By Thm. 3.17(ii), let $c \in \mathcal{M}^+$ such that $c^2 = c^* c = a$, then

$$b^* a b = b^* c^* c b = (cb)^*(cb) \in \mathcal{M}^+,$$

by Thm. 3.17(iii).

(iii): Let $a, b \in \mathcal{M}$. By Thm. 3.17(iii), $b^* b \in \mathcal{M}^+$, so it follows from (i) that $b^* b \leq \|b^* b\| = \|b\|^2$.

Then by (ii), we get $0 \leq a^* \left(\|b\|^2 - b^* b \right) a = \|b\|^2 a^* a - a^* b^* b a$, hence $(ba)^*(ba) \leq \|b\|^2 a^* a$. □

The following gives some information about the partial ordering $\leq$ within a monad.

Proposition 3.12. *Let $\mathcal{M}$ be an internal unital C^*-algebra and $a, b \in \mathcal{M}$ are such that $a \approx b$.*

Then there is $c \in \mathcal{M}$ such that $c \approx a \approx b$ and $a, b \leq c$.

Proof. Let $d = a - b \approx 0$. Then by Cor. 3.21 there are $d_1, d_2 \in \mathcal{M}^+$ such that $d = d_1 - d_2$ and $d_1 \approx 0 \approx d_2$.

Now define $c = a + d_1 + d_2$. Then $c \approx a \approx b$.

Moreover, as required, $(c - a) = (d_1 + d_2) \in \mathcal{M}^+$ and $(c - b) = 2d_1 \in \mathcal{M}^+$, by Thm. 3.17(i). □

As a consequence of Thm. 3.17(iii) and the continuous functional calculus (Cor. 3.18), in a unital C^*-algebra $\mathcal{M}$, we define the **modulus** of an element $a \in \mathcal{M}$ as

$$|a| := \sqrt{aa^*} \in \mathcal{M}^+.$$

232   *Nonstandard Methods in Functional Analysis*

Proposition 3.13. *Let $\mathcal{M}$ be a unital C^*-algebra.*

(i) *If $a \in \text{Re}(\mathcal{M})$, then $|a| = a_1 + a_2$, where $a_1, a_2 \in \mathcal{M}^+$ with $a_1 a_2 = 0 = a_2 a_1$ is the decomposition given by Cor. 3.21.*

(ii) *(Polar Decomposition) If $a \in \mathcal{M}^{-1}$, then there is $u \in \mathcal{U}(\mathcal{M})$ such that $a = |a| \, u$. Moreover this decomposition is unique in the sense that if there is $b \in \mathcal{M}^{-1}$ and $v \in \mathcal{U}(\mathcal{M})$ such that $a = bv$, then $b = |a|$ and $v = u$.*

Proof.   We give part of the proof of (ii), the rest is left as an exercise. Since $a \in \mathcal{M}^{-1}$, $\sigma(|a|) \subset (0, \infty)$, hence $|a| \in \mathcal{M}^{-1}$. Let $u := |a|^{-1} a$. Note that $\left(|a|^{-1} \right)^* = \left(|a|^* \right)^{-1} = |a|^{-1}$. So

$$u^* u = a^* |a|^{-2} a = a^* (aa^*)^{-1} a = 1 \quad \text{and} \quad uu^* = |a|^{-1} |a|^2 |a|^{-1} = 1,$$

therefore u is unitary and $a = |a| \, u$. $\square$

One should compare the above Polar Decomposition for invertible elements with Example 3.13(xi).

Observe that, as remarked earlier, if $a \in \text{Re}(\mathcal{M})$, then $e^{ia} \in \mathcal{U}(\mathcal{M})$, hence Prop. 3.13 implies that $\left| e^{ia} \right| = 1$. On the other hand, not every unitary element has the exponential form and Prop. 3.13 is as close as one can get for the decomposition.

We end this subsection with an application of the representation given by the Gelfand transform.

Example 3.14. (Stone-Čech compactification of a Tychonoff space.)
Let X be a Tychonoff space. Apply Thm. 3.14 to $\mathcal{M} = C_b(X)$, we see that $C_b(X)$ is *isomorphic to $C(\Omega)$, where $\Omega = \text{hom}(C_b(X))$. (Recall a *isomorphism is automatically a isometry.)

For each $x \in X$, let θ_x denote the functional $C_b(X) \ni f \mapsto f(x)$ given by the evaluation. It is easy to see that θ_x is nonzero, linear and multiplicative. Moreover $|f(x)| \leq \|f\|$, for all $f \in C_b(X)$, so θ_x is a character, *i.e.* $\theta_x \in \Omega$. Clearly, the mapping $X \ni x \mapsto \theta_x$ is an bijection onto its image. Neighborhoods of θ_x are generated by subsets of the form $\{\phi \in \Omega \mid |\phi(f) - f(x)| < \epsilon\}$ for some $f \in C_b(X)$ and $\epsilon \in \mathbb{R}^+$, whose restriction on $\{\theta_y \mid y \in X\}$ has the form $\{\theta_y \mid (y \in X) \wedge \left(|f(y) - f(x)| < \epsilon \right)\}$, hence X is homeomorphic to $\{\theta_y \mid y \in X\}$ and so we identify $X \subset \Omega$ as a subspace.

Moreover, X is dense in Ω. For if not, let $\phi \in \Omega \setminus X$. Then by Prop. 1.24 and Urysohn's Lemma (Thm. 1.17), there is $F \in C(\Omega)$ such that $F(\phi) = 1$ and $F[X] = \{0\}$. By Thm. 3.14, F has the form $\hat{f}$ for some $f \in \mathcal{M} = C_b(X)$, so $F[X] = \{0\}$ implies for all $x \in X$ that $0 = \hat{f}(x) = \theta_x(f) = f(x)$, hence $f = 0$, the zero function, which implies $F = 0$, a contradiction.

Note that if $f : X \to [0,1]$ is continuous, then $f \in C_b(X) = \mathcal{M}$, hence $\hat{f} : \Omega \to \mathbb{C}$ is continuous. For each $x \in X$, $f(x) = \theta_x(f) = \hat{f}(\theta_x) = \hat{f}(x)$, through the identification of x with θ_x, hence $\hat{f}$ extends f. Since X is dense in Ω and $\hat{f}$ is continuous, f and $\hat{f}$ have the same compact range $[0,1]$, $i.e.$ $\hat{f} : \Omega \to [0,1]$. So if we let $\beta f := \hat{f}$ then the first two requirements in the definition in §1.6.6 are satisfied.

Therefore, by Prop. 1.29, the Stone-Čech compactification of the Tychonoff space X is just $\hom(C_b(X))$.

Moreover, by Prop. 1.30, for any Tychonoff space X and compact topological space Y, any continuous $f : X \to Y$ extends uniquely to a continuous function $\beta f : \hom(C_b(X)) \to Y$.

If $X = \mathbb{N}$, we have $C_b(\mathbb{N}) = \ell_\infty$, so ℓ_∞ is *isomorphic to $C(\beta\mathbb{N})$ and $\beta\mathbb{N}$ can be identified with $\hom(\ell_\infty)$.

Furthermore, by Example 2.1, $ba(\mathbb{N}) = C(\beta\mathbb{N})'$.   □

3.2.3   The GNS construction

For a general unital C^*-algebra, we will show that it can be represented as a subalgebra of some $\mathcal{B}(H)$, the algebra of operators on a complex Hilbert space H, via the Gelfand-Naimark-Segal construction.

Unless specified otherwise, $\mathcal{M}$ always denote a not necessarily commutative C^*-algebra throughout this subsection.

A linear functional $\phi : \mathcal{M} \to \mathbb{C}$ is called a *positive functional* if $\forall x \in \mathcal{M}^+ \left(\phi(x) \in [0, \infty) \right)$.

Note that if ϕ_1, ϕ_2 are positive functionals and $r \in \mathbb{R}^+$, then $(\phi_1 + r\phi_2)$ is a positive functional.

Almost identical to the proof of Thm. 3.9(vi) and the proof of Cor. 3.21, one can show the following.

Proposition 3.14. *Every $\phi \in \mathcal{M}'$ is a linear combination of positive functionals. (In fact only four positive functionals are enough.)*   □

Proposition 3.15. *Let $\phi : \mathcal{M} \to \mathbb{C}$ be a positive functional. Then*

 (i) $\forall x \in \mathrm{Re}(\mathcal{M}) \left(\phi(x) \in \mathbb{R} \right)$;
 (ii) $\forall x \in \mathcal{M} \left(\phi(x^*) = \overline{\phi(x)} \right)$.

Proof.   (i): If $a \in \mathrm{Re}(\mathcal{M})$, decompose it as $a_1 - a_2$, where $a_1, a_2 \in \mathcal{M}^+$

as given by Cor. 3.21. Then $\phi(a) = \phi(a_1) - \phi(a_2) \in \mathbb{R}$, since $\phi(a_1), \phi(a_2) \in [0, \infty)$ by ϕ being positive.

(ii): If $a \in \mathcal{M}$, decompose it as $a_1 + ia_2$, where $a_1, a_2 \in \mathrm{Re}(\mathcal{M})$ as given by Thm. 3.9(vi). Then by (i),

$$\phi(a) = \phi(a_1) + i\phi(a_2) = \overline{\phi(a_1)} - i\overline{\phi(a_2)} = \overline{\phi(a_1 - ia_2)}.$$

By $a_1^* = a_1$ and $a_2^* = a_2$, we have $a^* = a_1 - ia_2$, therefore $\phi(a^*) = \overline{\phi(a)}$. $\square$

Note that if ϕ is a positive functional and $a, b \in \mathrm{Re}(\mathcal{M})$ such that $a \leq b$ then $\phi(a) \leq \phi(b)$.

Lemma 3.4. *Let* $\phi : \mathcal{M} \to \mathbb{C}$ *be a positive functional.*

Define $\langle \cdot, \cdot \rangle_\phi : \mathcal{M}^2 \to \mathbb{C}$ *by* $\langle x, y \rangle_\phi := \phi(y^* x)$, $x, y \in \mathcal{M}$. *Then* $\langle \cdot, \cdot \rangle_\phi$ *forms a pre-inner product on* $\mathcal{M}$.

Proof. Let $a \in \mathcal{M}$, then $\langle a, a \rangle_\phi := \phi(a^* a) \in [0, \infty)$, since $a^* a \in \mathcal{M}^+$ by Thm. 3.17(iii).

For $a_1, a_2, b \in \mathcal{M}$ and $\alpha \in \mathbb{C}$, clearly we have

$$\langle a_1 + \alpha a_2, b \rangle_\phi = \langle a_1, b \rangle_\phi + \alpha \langle a_2, b \rangle_\phi.$$

For $a, b \in \mathcal{M}$, by Prop. 3.15(ii), we get

$$\langle a, b \rangle_\phi = \phi(b^* a) = \phi((a^* b)^*) = \overline{\phi(a^* b)} = \overline{\langle b, a \rangle_\phi}.$$

Therefore $\langle \cdot, \cdot \rangle_\phi$ forms a pre-inner product. $\square$

We continue to use the notation $\langle \cdot, \cdot \rangle_\phi$ throughout.

Theorem 3.18. $\phi : \mathcal{M} \to \mathbb{C}$ *is a positive functional iff* $\phi \in \mathcal{M}'$ *and* $\|\phi\| = \phi(1)$.

Proof. ($\Rightarrow$) : Let $a \in S_\mathcal{M}$. So $\|a^* a\| = \|a\|^2 = 1$ and $\phi(a^* a) \geq 0$ since $a^* a \geq 0$. In particular, $0 \leq \phi(a^* a) \leq \|a\|$.

By the Cauchy-Schwarz Inequality (Thm. 2.25(i)),

$$|\phi(a)| = |\phi(1 \cdot a)| = |\langle a, 1 \rangle_\phi| \leq \sqrt{\langle a, a \rangle_\phi}\sqrt{\langle 1, 1 \rangle_\phi} = \sqrt{\phi(a^* a)}\sqrt{\phi(1)}.$$

Hence $|\phi(a)|^2 \leq \|\phi\| \phi(1)$ for any $a \in S_\mathcal{M}$. Hence $\|\phi\| \leq \phi(1)$. As $\|1\| = 1$, we get $\|\phi\| = \phi(1)$.

Note in particular that $\phi \in \mathcal{M}'$.

($\Leftarrow$) : Consider $\phi \neq 0$, replace ϕ by $\|\phi\|^{-1} \phi$ if necessary, we assume that $\|\phi\| = 1$.

Let $a \in \mathcal{M}^+$. So $\sigma(a) \subset [0, \|a\|]$, by Thm. 3.16(iv) and Thm. 3.9(vii). We will show that $\phi(a) \in [0, \|a\|]$. If this is not the case, $\phi(a)$ is separated from a ball that includes $[0, \|a\|]$. *i.e.* there is $\alpha \in \mathbb{C}$ and $r \in \mathbb{R}^+$ such that

$$\phi(a) \notin B_{\mathbb{C}}(\alpha, r) \supset [0, \|a\|].$$

Since $(a - \alpha)^* = (a - \bar{\alpha})$, $(a - \alpha)$ is normal and Thm. 3.9(vii) applies. Moreover, by the continuous functional calculus (Cor. 3.18), $\sigma(a - \alpha) = \sigma(a) - \alpha$, hence $\rho(a - \alpha) = \text{dist}(\alpha, \sigma(a))$. Therefore

$$r < |\phi(a) - \alpha| = |\phi(a) - \alpha\phi(1)| = |\phi(a - \alpha)|$$
$$\leq \|a - \alpha\| = \rho(a - \alpha) = \text{dist}(\alpha, \sigma(a)) \leq r,$$

impossible. So $\phi(a) \in [0, \|a\|]$. In particular, $\phi(a) \geq 0$. □

A positive functional $\phi : \mathcal{M} \to \mathbb{C}$ is called a **state** if $\phi(1) = 1$. By Thm. 3.18, states are positive functionals belonging to the unit sphere $S_{\mathcal{M}'}$. Note that by Prop. 3.14, linear functionals on $\mathcal{M}$ are linear combination of states.

If ϕ_1, ϕ_2 are states and $r \in [0, 1]$, Then $(r\phi_1 + (1 - r)\phi_2)$ is a positive functional. Moreover, $(r\phi_1 + (1 - r)\phi_2)(1) = 1$, hence $(r\phi_1 + (1 - r)\phi_2)$ is a state. Let $\mathbb{S}(\mathcal{M})$ denote the set of states on $\mathcal{M}$, so $\mathbb{S}(\mathcal{M})$ is convex subset of $S_{\mathcal{M}'}$. It is not hard to see that $\mathbb{S}(\mathcal{M})$ is weak*-closed subset of $S_{\mathcal{M}'}$, hence it is weak*-compact, by the Alaoglu's Theorem (Thm. 2.16).

By the Krein-Milman Theorem (Thm. 2.32), $\mathbb{S}(\mathcal{M}) = \overline{\text{conv}}(E)$ (the weak*-closure), where $E \subset \mathbb{S}(\mathcal{M})$ is the set of extreme points. Such extreme points are called **pure states**. So each state is infinitely close to a convex combination of pure states, for this reason, non-pure states are called **mixed states**.

Note that by the Hahn-Banach Theorem, the identity mapping on $\mathbb{C}$ extends to a state on $\mathcal{M}$. In particular, states exist for every unital C^*-algebra. In fact, by Thm. 3.18, $\text{hom}(\mathcal{M}) \subset \mathbb{S}(\mathcal{M})$.

Example 3.15.

- Let $\mathcal{M} = \mathcal{B}(\mathbb{C}^n)$ be the C^*-algebra of $n \times n$ matrices over $\mathbb{C}$, $n \in \mathbb{N}$. Let $P \in \mathcal{M}$ be a matrix with positive real entries such that the trace $\text{Tr}(P) = 1$. (*i.e.* the main diagonal entries sum up to 1.) Then the functional given by $\mathcal{M} \ni M \mapsto \text{Tr}(MP)$ is a state.

- Let $\mathcal{M} = C_{\text{b}}(\Omega)$, where Ω is a locally compact space. Let μ be a Borel probability measure on Ω. Then the functional given by $\mathcal{M} \ni f \mapsto \int_{\Omega} f d\mu$ is a state. It is a pure state if μ is a Dirac delta measure, *i.e.* for some $a \in \Omega$, $\mu(\{a\}) = 1$. □

It turns out that positive elements are normed by pure states.

Proposition 3.16. *Let $\mathcal{M}$ be a unital C^*-algebra and $a \in \mathcal{M}^+$.*
Then there is a state ϕ such that $\phi(a) = \|a\|$.
Moreover, such ϕ can be chosen from pure states.

Proof.   By Thm. 3.9(vii), $\rho(a) = \|a\|$, and by Thm. 3.16(iv), $\sigma(a) \subset$
$[0, \infty)$, so $\|a\| \in \sigma(a)$. Since $\sigma_{\mathcal{M}_a}(a) = \sigma(a)$ (Thm. 3.10(ii)), $\|a\| \in \sigma_{\mathcal{M}_a}(a)$.
Now Thm. 3.13 implies that $\|a\| = \theta(a)$ for some $\theta \in \mathrm{hom}(\mathcal{M}_a)$.
By the Hahn-Banach Theorem, θ extends to some $\phi \in \mathcal{M}'$ with $\|\phi\| = \|\theta\|_{\mathcal{M}_a}$. So
$$\|\phi\| = \|\theta\|_{\mathcal{M}_a} = 1 \quad \text{and} \quad \phi(1) = \theta(1) = 1,$$
hence $\phi \in \mathbb{S}(\mathcal{M})$, by Thm. 3.18.

Therefore $C := \{\phi \in \mathbb{S}(\mathcal{M}) \mid \phi(a) = \|a\|\} \neq \emptyset$.

It is clear that C is a closed convex subset of $\mathbb{S}(\mathcal{M})$, hence it is weak*-compact by the Alaoglu's Theorem (Thm. 2.16). Then it follows from the Krein-Milman Theorem (Thm. 2.32), C contains extreme points. Let $\phi \in C$ be an extreme point.

Then ϕ is also an extreme point in $\mathbb{S}(\mathcal{M})$: For if $\phi = (\phi_1 + \phi_2)/2$, where $\phi_1, \phi_2 \in \mathbb{S}(\mathcal{M})$, then
$$\|a\| = \phi(a) = \frac{1}{2}\Big(\phi_1(a) + \phi_2(a)\Big) \leq \frac{1}{2}\Big(\|a\| + \|a\|\Big) = \|a\|$$
and as ϕ_1, ϕ_2 are positive, we must have $\phi_1(a) = \|a\| = \phi_2(a)$, hence $\phi_1, \phi_2 \in C$ implying $\phi_1 = \phi_2 = \phi$, therefore ϕ is an extreme point in $\mathbb{S}(\mathcal{M})$. By definition, such ϕ is a pure state. $\square$

In the following, **GNS** stands for "Gelfand-Naimark-Segal" and we now describe the construction of representing an arbitrary C^*-algebra $\mathcal{M}$ due to these discoverers.

By a **GNS representation** of $\mathcal{M}$ we refers to a pair (π, H), where H is a complex Hilbert space and $\pi : \mathcal{M} \to \mathcal{B}(H)$ is a *homomorphism.

Note in particular the requirement that $\pi(1) = 1$. Also, by Cor. 3.19, $\pi[\mathcal{M}]$ is a unital C^*-subalgebra of $\mathcal{B}(H)$.

When π in a GNS representation (π, H) is an *isomorphism, (π, H) is also called **faithful**. (Equivalently, the *homomorphism π is injective, a *monomorphism*.)

Given a GNS representation (π, H) and $\xi \in H$, $\pi[\mathcal{M}][\xi]$, *i.e.* the set $\{\pi(a)(\xi) \mid a \in \mathcal{M}\}$, is a subspace of H. If there is $\xi \in H$ such that $\pi[\mathcal{M}][\xi]$ is dense in H, then (π, H) is called a **cyclic GNS representation** and ξ is called a **cyclic** element for (π, H).

Example 3.16. Let (π, H) be a cyclic GNS representation of $\mathcal{M}$ with a cyclic element $\xi \in S_H$. Then the functional $\mathcal{M} \ni x \mapsto \langle \pi(x)(\xi), \xi \rangle_H \in \mathbb{C}$ is a positive functional. $\qquad\qquad\qquad\qquad\qquad\qquad\qquad\qquad$ $\square$

The converse of the above holds:

Theorem 3.19. (The GNS construction.) *Let $\mathcal{M}$ be a unital C^*-algebra and $\phi \in \mathcal{M}'$ be a positive functional.*

Then there is a cyclic GNS representation (π, H) of $\mathcal{M}$ with a cyclic element $\xi \in S_H$ such that $\forall x \in \mathcal{M}$ $\left(\phi(x) = \langle \pi(x)(\xi), \xi \rangle_H \right)$.

Proof. Let $\phi \in \mathcal{M}'$ be a positive functional and define the pre-inner product $\langle \cdot, \cdot \rangle_\phi$ on $\mathcal{M}$ as given by Lem. 3.4. Let $H := \widehat{\mathcal{M}}$, the nonstandard hull w.r.t. $\langle \cdot, \cdot \rangle_\phi$, as in Prop. 2.36 and the remarks preceding it.

So H is a Hilbert space.

Now define $\pi : \mathcal{M} \to \mathcal{B}(H)$ as follows: for each $a \in \mathcal{M}$, we let $\pi(a)$ be the linear operator in $\mathcal{B}(H)$ given by

$$\pi(a)(\eta) := \widehat{ax}, \quad \text{where } \eta = \widehat{x} \in H.$$

(Not to be confused with the Gelfand transform: $\widehat{x}$ here denotes an element in the nonstandard hull $\widehat{\mathcal{M}}$, for some $x \in \text{Fin}(^*\mathcal{M})$.)

(Recall also the convention stated on p.195 regarding to the suppression of the $*$ signs.)

First of all, π is well-defined: let $a \in \mathcal{M}$ and let $c_1, c_2, \epsilon \in \text{Fin}(^*\mathcal{M})$ such that $c_1 = c_2 + \epsilon$ with $\epsilon \approx 0$. *i.e.* $\widehat{c_1} = \widehat{c_2}$ and $0 \approx \|\epsilon\|_\phi^2 = \langle \epsilon, \epsilon \rangle_\phi = \phi(\epsilon^* \epsilon)$. Then by Prop. 3.11,

$$\|a\epsilon\|_\phi^2 = \langle a\epsilon, a\epsilon \rangle_\phi = \phi\big((a\epsilon)^* (a\epsilon)\big) \leq \|a\|_\mathcal{M}^2 \, \phi(\epsilon^* \epsilon) \approx 0,$$

therefore $ac_1 = ac_2 + a\epsilon \approx ac_2$, showing that $\widehat{ac_1} = \widehat{ac_2}$.

For $a \in \mathcal{M}$, clearly $\pi(a) : H \to H$ is a linear operator. Moreover, similar to the above calculation, for $\eta = \widehat{x} \in H$,

$$\|\pi(a)(\eta)\|_H^2 \approx \phi\big((ax)^* (ax)\big) \leq \|a\|_\mathcal{M}^2 \phi(x^* x) = \|a\|_\mathcal{M}^2 \|x\|_\phi^2 \approx \|a\|_\mathcal{M}^2 \|\eta\|_H^2,$$

so $\pi(a) \in \mathcal{B}(H)$ with $\|\pi(a)\| \leq \|a\|$.

To show the linearity of π, note that for $a, b \in \mathcal{M}, \alpha \in \mathbb{C}$ and $\eta = \widehat{x} \in H$,

$$\pi(a + \alpha b)(\eta) = \big((a + \alpha b)x\big)^\wedge = \widehat{ax} + \alpha \, \widehat{bx} = \pi(a)(\eta) + \alpha \left(\pi(b)(\eta) \right).$$

Also, for $a \in \mathcal{M}, \eta_1 = \widehat{x}_1, \eta_2 = \widehat{x}_2 \in H$,

$$\langle \eta_1, \pi(a^*)(\eta_2) \rangle_H \approx \langle x_1, a^* x_2 \rangle_\phi = \phi\big((a^* x_2)^* x_1\big)$$
$$= \phi\big(x_2^* (ax_1)\big) = \langle ax_1, x_2 \rangle_\phi \approx \langle \pi(a)(\eta_1), \eta_2 \rangle_H,$$

showing that $\pi(a^*) = \big(\pi(a)\big)^*$, the adjoint operator of $\pi(a)$.

Moreover, $\forall \widehat{x} \in H\,\big(\pi(1)(\widehat{x}) = \widehat{x}\big)$, *i.e.* $\pi(1) = 1$. Therefore π is a *homomorphism and so (π, H) is a GNS representation of $\mathcal{M}$.

Finally, define $\xi := \widehat{1}$, where 1 is the unit element in $\mathcal{M}$.

Then for each $a \in \mathcal{M}$, we have $\pi(a)(\xi) = \widehat{a \cdot 1} = \widehat{a}$, so

$$\phi(a) = \phi(1 \cdot a) = \langle a, 1 \rangle_\phi = \langle \widehat{a}, \widehat{1} \rangle_H = \langle \pi(a)(\xi), \xi \rangle_H.$$

Now we need to reduce the size of (π, H) in order to obtain a cyclic representation.

Let $X := \pi[\mathcal{M}](\widehat{\xi}) = \{\pi(x)(\widehat{1}) \,|\, x \in \mathcal{M}\} = \{\widehat{x} \,|\, x \in \mathcal{M}\}$. Clearly X is an inner-product subspace of H. Note that $\pi(a)(\widehat{x}) = \widehat{ax} \in X$ holds for every $a \in \mathcal{M}$ and $\widehat{x} \in X$, so X is invariant under $\pi[\mathcal{M}]$.

Therefore, if we replace H by $\bar{X}$ and π by its restriction on $\bar{X}$, then (π, H) is a cyclic GNS representation of $\mathcal{M}$ with cyclic element given by $\widehat{1} \in H$ satisfying $\forall x \in \mathcal{M}\,\big(\phi(x) = \langle \pi(x)(\widehat{1}), \widehat{1} \rangle_H\big)$. $\square$

Given a family $\{(\pi_i, H_i) \,|\, i \in I\}$ of GNS representations of $\mathcal{M}$, we define the direct sum $\bigoplus_{i \in I}(\pi_i, H_i) = (\pi, H)$ as follows:

First of all, $H := \bigoplus_{i \in I} H_i$ as given on p.159.

Then $\pi := \bigoplus_{i \in I} \pi_i : \mathcal{M} \to \mathcal{B}(H)$ is given by $\pi(a) = \big(\pi_i(a)\big)_{i \in I}$, where $a \in \mathcal{M}$. So $\pi(a)(\eta) = \{\pi_i(a)(\eta_i)\}_{i \in I}$ for every $\eta \in H$.

Let $a \in \mathcal{M}$. Note that by Thm. 3.15, for all $\eta \in H$, and each $i \in I$, $\|\pi_i(a)(\eta_i)\| \le \|\eta_i\|$, hence $\|\pi(a)(\eta)\|_H \le \|\eta\|_H$. Moreover, it is clear that $\pi(a) : H \to H$ is linear. So $\pi(a) \in \mathcal{B}(H)$.

It is not hard to check that π is a *homomorphism, *i.e.* $\bigoplus_{i \in I}(\pi_i, H_i)$ is a GNS representations of $\mathcal{M}$.

Example 3.17. Let $\mathcal{M} = C(\Omega)$ for some compact space Ω. Let μ_i, $i \in I$, be σ-additive complex Borel measures on Ω. For each $i \in I$ we let $H_i := \mathcal{B}\big(L_2(\mu_i)\big)$ and define $\pi_i : C(\Omega) \to H_i$ such that for each $f \in C(\Omega)$,

$$\pi_i(f) : g \mapsto fg, \quad \text{where } g \in L_2(\mu_i).$$

Then it is clear that $\bigoplus_{i \in I}(\pi_i, H_i)$ is a GNS representation of $\mathcal{M}$. $\square$

The following fundamental result shows that any unital C^*-algebra can be given a concrete representation as an operator algebra on a Hilbert space, *i.e.* any unital C^*-algebra embeds as a C^*-subalgebra of the algebra of operators on some Hilbert space. This can be viewed as a rough classification of unital C^*-algebras.

Theorem 3.20. (GNS Theorem) *Let $\mathcal{M}$ be a unital C^*-algebra.*
Then there is a Hilbert space H and a *isomorphism $\pi : \mathcal{M} \to \mathcal{B}(H)$.
(Note that π is isometric, by Thm. 3.15(ii).)

Proof. For each $\phi \in \mathbb{S}(\mathcal{M})$, let (π_ϕ, H_ϕ) denote the GNS representation
of $\mathcal{M}$ and $\xi_\phi \in H_\phi$ the cyclic element, as given in Thm. 3.19.
Define $(\pi, H) := \bigoplus_{\phi \in \mathbb{S}(\mathcal{M})} (\pi_\phi, H_\phi)$.
So $\pi : \mathcal{M} \to \mathcal{B}(H)$ is a *homomorphism.
For each $\phi \in \mathbb{S}(\mathcal{M})$, we have by Thm. 3.18 that $\phi(1) = 1$ and so

$$1 = \phi(1) = \big\langle \pi_\phi(1)(\xi_\phi), \xi_\phi \big\rangle = \langle \xi_\phi, \xi_\phi \rangle = \|\xi_\phi\|^2_{H_\phi},$$

i.e. $\xi_\phi \in S_{H_\phi}$.
Therefore it holds for each $a \in \mathcal{M}$ and each $\phi \in \mathbb{S}(\mathcal{M})$ that

$$\|\pi_\phi(a)\|^2_{\mathcal{B}(H_\phi)} \geq \|\pi_\phi(a)(\xi_\phi)\|^2_{H_\phi} = \big\langle \pi_\phi(a)(\xi_\phi), \pi_\phi(a)(\xi_\phi) \big\rangle_{H_\phi}$$

$$= \big\langle \pi_\phi(a)^* \pi_\phi(a)(\xi_\phi), \xi_\phi \big\rangle_{H_\phi} = \big\langle \pi_\phi(a^* a)(\xi_\phi), \xi_\phi \big\rangle_{H_\phi} = \phi(a^* a).$$

By Thm. 3.17(iii), $(a^* a) \in \mathcal{M}^+$, so it follows from Prop. 3.16 for some
$\phi \in \mathbb{S}(\mathcal{M})$ that

$$\phi(a^* a) = \|a^* a\| = \|a\|^2.$$

Putting all these together, we see for any $a \in \mathcal{M}$ that

$$\|\pi(a)\|_{\mathcal{B}(H)} \geq \sup_{\phi \in \mathbb{S}(\mathcal{M})} \|\pi_\phi(a)\|_{\mathcal{B}(H_\phi)} \geq \|a\|.$$

In particular, π is injective, therefore π is a *isomorphism.
(Since *isomorphisms are isometries (Thm. 3.15(ii)), we actually get
$\|\pi(a)\|_{\mathcal{B}(H)} = \|a\|$ for any $a \in \mathcal{M}$ in the above.) $\square$

So the GNS Theorem (Thm. 3.20) says that every unital C^*-algebra has
a faithful GNS representation as a unital C^*-subalgebra of some $\mathcal{B}(H)$.
The representation $\bigoplus_{\phi \in \mathbb{S}(\mathcal{M})} (\pi_\phi, H_\phi)$ in the proof of Thm. 3.20 is referred to as the ***universal representation*** of $\mathcal{M}$.
The terminology is justified by the following which is left as an exercise.

Proposition 3.17. *Let (π, H) be the universal representation of a unital*
C^*-algebra $\mathcal{M}$. Then for any GNS representation $(\tilde{\pi}, \tilde{H})$ of $\mathcal{M}$, there is a
*homomorphism $\varrho : \pi[\mathcal{M}] \to \tilde{\pi}[\mathcal{M}]$ such that $\tilde{\pi} = \varrho \circ \pi$. $\square$

The universal representations used in the proof of Thm. 3.20 is in fact
unique in a unitary sense. We now make precise of this notion.

We say that two GNS representations (π_1, H_1) and (π_2, H_2) of $\mathcal{M}$ are **unitarily equivalent** if there is a Hilbert space isomorphism $U : H_1 \to H_2$ (*i.e.* a surjective linear isometry) such that

$$\forall x \in \mathcal{M} \left(\pi_1(x) = U^{-1} \circ \pi_2(x) \circ U \right).$$

The proof of the following is left as an exercise.

Proposition 3.18. *Let* (π, H) *be a GNS representation of a unital C^*-algebra $\mathcal{M}$. Then there is* $\{(\pi_i, H_i) \mid i \in I\}$, *where each* (π_i, H_i) *is a cyclic GNS representation of $\mathcal{M}$, such that* $\bigoplus_{i \in I}(\pi_i, H_i)$ *and* (π, H) *are unitarily equivalent.*   □

3.2.4   Notes and exercises

A self-adjoint element of a C^*-algebra is also called **Hermitian**, as in the special case of square matrices over $\mathbb{C}$, such matrices are also commonly called Hermitian matrices.

Both Lem. 3.3 and Thm. 3.13 are due to Gelfand. Because of Lem. 3.3, hom($\mathcal{M}$) is identifiable with $\mathfrak{M}(\mathcal{M})$, hence hom($\mathcal{M}$) is also called the **maximal ideal space** of $\mathcal{M}$.

hom($\mathcal{M}$) is also called the spectrum (or the Gelfand spectrum) of $\mathcal{M}$.

Because of Thm. 3.14, commutative C^*-algebras correspond to compact spaces, hence one can view the classical topology theory as theory of commutative C^*-algebras. By the same token, the theory of general C^*-algebras can be viewed as the theory of **noncommutative topology**.

By combining the Riesz Representation Theorem (Thm. 2.14) and Thm. 3.14, one sees that the dual of a commutative unital C^*-algebra is identifiable with $\mathbb{M}(\Omega)$, the space of σ-additive complex Borel measures on some compact space Ω.

The Stone-Weierstrass Theorem (Thm. 3.11), as a generalization of the Weierstrass Approximation Theorem on uniform approximation of continuous functions in $C([0,1])$ by polynomials, was first proved by M.H. Stone. For a proof due to L. de Branges which involve the Riesz Representation Theorem for $C(\Omega)'$, the Krein-Milman Theorem and Alaoglu's Theorem, see [Conway (1990)].

Quantum physics can be given C^*-algebra formulations. Then states on algebras correspond to physical interpretation of states in a quantum system, *i.e.* connections between observables and measurements. Most physical states are statistical mixture of other states, those not of this type are pure states.

In most topology textbooks, the Tietze Extension Theorem (Cor. 3.13) is proved by a more direct construction using the Urysohn's Lemma. The proof here via the Stone-Weierstrass Theorem follows [Reed and Simon (1980)].

The important rôle played by the positive elements cannot be exaggerated, as a theorem in [Sherman (1951)] shows that a unital C^*-algebra $\mathcal{M}$ is commutative iff $\mathcal{M}^+$ forms a *lattice* (*i.e.* a partial order with any two elements having a unique supremum and a unique infimum in $\mathcal{M}^+$.) We will see later that positive elements in von Neumann algebras plays a key rôle in the form of projections.

EXERCISES

(1) Give an example of a nonunital Banach algebra $\mathcal{M}$ with a closed proper ideal I such that $\mathcal{M}/I$ is a unital Banach algebra.

(2) Complete the proof of the Tietze Extension Theorem (Cor. 3.13).

(3) Determine which of the converse of the assertions in Thm. 3.9 could fail for a non-normal element in a unital C^*-algebra.

(4) Let H be a complex Hilbert space and $a \in \mathcal{B}(H)$. Show that a is a partial isometry iff $\forall \xi \in \text{Ker}(a)^\perp \left(\|a(\xi)\| = \|\xi\| \right)$.

(5) Let $\mathcal{M}$ be a unital C^*-algebra and $a, a_n, n \in \mathbb{N}$, be normal elements in $\mathcal{M}$ such that $a_n \to a$. Let $f \in C(\sigma(a))$. Show that $f(a_n) \to f(a)$ in $\mathcal{M}$.

(6) Let $\mathcal{M}$ be a unital C^*-algebra and $a \in \mathcal{M}$ be normal. Show that the mapping $C\big(\sigma(a)\big) \ni f \mapsto f(a) \in \mathcal{M}_a$ is a *isomorphism and is unique among those extending the Riesz functional calculus.

(7) Give a direct proof of Cor. 3.19.

(8) Complete the proof of Prop. 3.13. Can the result be strengthened by require the unitary element u be of the form e^{ic}?

(9) Show that for commutative unital Banach algebras, the Gelfand transform need not be surjective.

(10) Let $\mathcal{M}$ be a unital C^*-algebra. Show that if $\pi : \mathcal{M} \to \mathcal{B}(H)$ is a linear, $\forall x \in \mathcal{M} \left(\pi(x^*) = (\pi(x))^* \right)$ and $\pi[\mathcal{M}][H]$ is dense in H, then $\pi(1) = 1$.

(11) A GNS representation (π, H) of a unital C^*-algebra $\mathcal{M}$ is *irreducible* if H has no proper closed $\pi[\mathcal{M}]$-invariant subspace. Given a positive functional $\phi \in \mathcal{M}'$, show that the GNS representation (π, H) corresponding to ϕ in Thm. 3.19 is irreducible iff ϕ is a pure state.

(12) Show that in the GNS Theorem (Thm. 3.20), if $\mathcal{M}$ is separable, then the Hilbert space H can be chosen to be separable.

(13) Prove Prop. 3.17 and Prop. 3.18.

(14) Show that a C^*-algebra is commutative iff all elements are normal.

3.3   The Nonstandard Hull of a C^*-Algebra

A C^*-algebra is just the slim silhouette of its nonstandard hull.

This section is about the nonstandard hull of an internal unital C^*-algebra and how important properties get transferred between them.

Given an internal unital C^*-algebra $\mathcal{M}$, the nonstandard hull $\widehat{\mathcal{M}}$ constructed in § 3.1.3 is therefore a unital Banach algebra.

For each $a \in \mathrm{Fin}(\mathcal{M})$ we define $(\widehat{a})^*$ by $\widehat{a^*}$. This is well-defined, as $\|a\| = \|a^*\|$ and hence also $a \approx 0$ iff $a^* \approx 0$. It is straightforward to see that $\widehat{\mathcal{M}}$ becomes a unital C^*-algebra under this definition of involution. Clearly if $\mathcal{M}$ is commutative, then $\widehat{\mathcal{M}}$ is commutative. Moreover, for unital C^*-algebra $\mathcal{M}$, the involution on $^*\widehat{\mathcal{M}}$ extends the corresponding one on $\mathcal{M}$. So we have:

Proposition 3.19.

(i) *If $\mathcal{M}$ be an internal unital C^*-algebra, then $\widehat{\mathcal{M}}$ forms a unital C^*-algebra. Moreover, if $\mathcal{M}$ is commutative, so is $\widehat{\mathcal{M}}$.*

(ii) *If $\mathcal{M}$ is a unital C^*-algebra, $\mathcal{M}$ is a unital C^*-subalgebra of $^*\widehat{\mathcal{M}}$.*   □

3.3.1   Basic properties

Following the convention on p.195, many $*$ signs are suppressed. For example, we write $\widehat{M}$ and $\mathcal{B}(\widehat{H})$ instead of the more precise forms $^*\widehat{\mathcal{M}}$ and $^*\mathcal{B}(\widehat{H})$ used in previous sections.

If a unital C^*-algebra $\mathcal{M}$ is a unital C^*-subalgebra of $\mathcal{B}(H)$ for some Hilbert space H, we have $\widehat{\mathcal{M}} \subset \widehat{\mathcal{B}(H)} \subset \mathcal{B}(\widehat{H})$ as unital C^*-subalgebras. Here the Hilbert space nonstandard hull $\widehat{H}$ is as given in Prop. 2.36 and the last inclusion follows from the identification in Prop. 2.16(ii). This identification can be generalized for GNS representations in the following.

Proposition 3.20. *Let $\mathcal{M}$ be an internal unital C^*-algebra and (π, H) be an internal GNS representation of $\mathcal{M}$.*

Define the mapping $\widehat{\pi} : \widehat{\mathcal{M}} \to \widehat{\mathcal{B}(H)} \subset \mathcal{B}(\widehat{H})$ by

$$\widehat{\pi}(\widehat{a})(\widehat{\eta}) := \widehat{\pi(a)(\eta)}, \ a \in \mathrm{Fin}(\mathcal{M}), \ \eta \in \mathrm{Fin}(H).$$

Then $(\widehat{\pi}, \widehat{H})$ is GNS representation of $\widehat{\mathcal{M}}$.

Moreover, if π is faithful, then $\widehat{\pi}$ is faithful.

Proof. By Thm. 3.15(i), $\forall x \in \mathcal{M} \left(\|\pi(x)\|_{\mathcal{B}(H)} \leq \|x\|_{\mathcal{M}} \right)$, so it follows that the mapping $\widehat{\pi}$ is well-defined, $\widehat{\pi}(1) = 1$, $\widehat{\pi} \in \mathcal{B}\big(\widehat{\mathcal{M}}, \widehat{\mathcal{B}(H)}\big)$ and is a *homomorphism, *i.e.* $\big(\widehat{\pi}, \widehat{H}\big)$ is GNS representation of $\widehat{\mathcal{M}}$.

If π is faithful, then π is an isometry by Thm. 3.15(ii). So we have for each $a \in \mathrm{Fin}(\mathcal{M})$ that

$$\|\widehat{\pi}(\widehat{a})\| \approx \sup_{\eta \in S_H} |\pi(a)(\eta)| = \|\pi(a)\| = \|a\| \approx \|\widehat{a}\|,$$

i.e. $\widehat{\pi}$ is an isometry, hence injective, therefore it is faithful. $\qquad \square$

In particular, the GNS Theorem for an internal unital C^*-algebra has a correspondence with that for the nonstandard hull.

As a straightforward consequence by the transfer, we get the following:

Corollary 3.22. *Let $\mathcal{M}$ be a unital C^*-algebra and (π, H) be a GNS representation of $\mathcal{M}$. Then $\big(\widehat{\pi}, \widehat{H}\big)$ is GNS representation of $\widehat{\mathcal{M}}$. If π is faithful, so is $\widehat{\pi}$.* $\qquad \square$

The following is an immediate consequence of the Gelfand transform. This result was first proved by Henson from a general result on the equivalence of Banach spaces having isomorphic nonstandard hulls. See [Henson (1988)] Thm. 5.1.

Theorem 3.21. *Let X be a Tychonoff space, then there is a compact space Ω such that $\widehat{C_{\mathrm{b}}(X)}$ is *isomorphic to $C_{\mathrm{b}}(\Omega)$.*

Proof. By Prop. 3.19(i), $C_{\mathrm{b}}(X)$ is a commutative unital C^*-algebra. So the result follows from Thm. 3.14 by taking $\Omega = \mathrm{hom}\big(C_{\mathrm{b}}(X)\big)$. $\qquad \square$

Now we would like to make a comparison for various notions about elements in an internal C^*-algebra and their correspondents in the nonstandard hull.

Proposition 3.21. *Let $\mathcal{M}$ be an internal unital C^*-algebra.*

(i) *For any $a \in \mathrm{Fin}(\mathcal{M})$, we have $\sigma(\widehat{a}) \supset {}^{\circ}\big[\sigma(a)\big]$.*
(ii) *If $a \in \mathrm{Fin}(\mathcal{M})$ is normal, we have $\sigma(\widehat{a}) = {}^{\circ}\big[\sigma(a)\big]$.*

Proof. (i): Let $\lambda \in \sigma(a)$. Since $\rho(a) \leq \|a\|$, $\lambda \in \mathrm{Fin}({}^*\mathbb{C})$ and $(a - \lambda) \in \mathrm{Fin}(\mathcal{M})$.

But, as $\lambda \in \sigma(a)$, we have in particular $(a - \lambda) \notin \big(\mathrm{Fin}(\mathcal{M})\big)^{-1}$, so Cor. 3.11 implies that $(\widehat{a} - {}^{\circ}\lambda) = \widehat{a - \lambda} \notin \big(\widehat{\mathcal{M}}\big)^{-1}$, *i.e.* ${}^{\circ}\lambda \in \sigma(\widehat{a})$.

(ii): We only need to prove $\sigma(\widehat{a}) \subset {}^{\circ}\big[\sigma(a)\big]$.

Suppose there is some $\lambda \in \sigma(\widehat{a}) \setminus \big({}^{\circ}\big[\sigma(a)\big]\big)$, so $\mathrm{dist}(\lambda, \sigma(a)) \not\approx 0$. Note in particular that $\lambda \notin \sigma(a)$. Then by Cor. 3.8, we get

$$\rho\big((a - \lambda)^{-1}\big)^{-1} = \mathrm{dist}(\lambda, \sigma(a)) \not\approx 0.$$

Hence $\rho\big((a - \lambda)^{-1}\big)$ must be finite. By remarks on p.211, $(a - \lambda)^{-1}$ is normal, hence

$$\big\|(a - \lambda)^{-1}\big\| = \rho\big((a - \lambda)^{-1}\big)$$

by Thm. 3.9(vii). Therefore $(a - \lambda) \in \big(\mathrm{Fin}(\mathcal{M})\big)^{-1}$. Since $(a - \lambda) \in \mathrm{Fin}(\mathcal{M})$, we have by Cor. 3.11 that

$$(\widehat{a} - \lambda) = \widehat{(a - \lambda)} \in \big(\widehat{\mathcal{M}}\big)^{-1},$$

showing that $\lambda \notin \sigma(\widehat{a})$, a contradiction.

Therefore $\sigma(\widehat{a}) \subset {}^{\circ}\big[\sigma(a)\big]$. $\square$

Normality cannot be dropped from Prop. 3.21(ii). See Notes at the end of this section.

Recall the use of $\mathrm{Fin}(\cdot)$ mentioned on p.80.

Theorem 3.22. *Let $\mathcal{M}$ be an internal unital C^*-algebra.*

(i) $\mathrm{Re}(\widehat{\mathcal{M}}) = \big(\mathrm{Fin}(\mathrm{Re}(\mathcal{M}))\big)^{\wedge}.$

(ii) $\mathrm{Fin}(\mathcal{M}^+) = \big[\mathrm{Fin}(\mathrm{Re}(\mathcal{M}))\big]^2.$

(iii) $\mathrm{Fin}(\mathcal{M}^+) = \{y^* y \,|\, y \in \mathrm{Fin}(\mathcal{M})\}.$

(iv) $\big(\widehat{\mathcal{M}}\big)^+ = \big(\mathrm{Fin}(\mathcal{M}^+)\big)^{\wedge}.$

(v) $\mathcal{U}(\widehat{\mathcal{M}}) = (\mathcal{U}(\mathcal{M}))^{\wedge}.$ (Note: $\mathcal{U}(\mathcal{M}) \subset \mathrm{Fin}(\mathcal{M}).$)

(vi) $\mathrm{Proj}(\widehat{\mathcal{M}}) = (\mathrm{Proj}(\mathcal{M}))^{\wedge}.$ (Note: $\mathrm{Proj}(\mathcal{M}) \subset \mathrm{Fin}(\mathcal{M}).$)

(vii) *Let $a \in \mathrm{Fin}(\mathcal{M})$. Then $\widehat{a}$ is a partial isometry iff there is a partial isometry $c \in \mathcal{M}$ such that $\widehat{a} = \widehat{c}$.*

(viii) *Let $a \in \mathrm{Fin}(\mathcal{M})$. Then $\widehat{a}$ is an isometry iff there is an isometry $c \in \mathcal{M}$ such that $\widehat{a} = \widehat{c}$.*

(ix) *Let $a \in \mathrm{Fin}(\mathcal{M})$. Then $\widehat{a}$ is a coisometry iff there is a coisometry $c \in \mathcal{M}$ such that $\widehat{a} = \widehat{c}$.*

(x) *Let $a_1, a_2 \in \mathrm{Fin}(\mathcal{M})$. Then $\widehat{a}_1 \leq \widehat{a}_2$ iff there are $c_1, c_2 \in \mathrm{Fin}(\mathcal{M})$ such that $c_1 \approx a_1$, $c_2 \approx a_2$ and $c_1 \leq c_2$.*

Moreover, if $\widehat{a}_1 \in \big(\widehat{\mathcal{M}}\big)^+$, then c_1, c_2 can be chosen from $\mathrm{Fin}(\mathcal{M}^+)$.

BANACH ALGEBRAS   245

Proof. In the following, we consider an arbitrary $a \in \mathrm{Fin}(\mathcal{M})$.

(i):

$$\widehat{a} \in \mathrm{Re}(\widehat{\mathcal{M}}) \quad \text{iff} \quad \widehat{a} = (\widehat{a})^* = \widehat{a^*} \quad \text{iff} \quad a \approx a^* \quad \text{iff} \quad a \approx \frac{a + a^*}{2} \left(= \mathrm{Re}(a) \right)$$

$$\text{iff} \quad a \approx c \quad \text{for some } c \in \mathrm{Fin}(\mathrm{Re}(\mathcal{M})).$$

(ii): ($\supset$) is trivial. For the other direction, if $a \in \mathrm{Fin}(\mathcal{M}^+)$, then $a = b^2$ for some $b \in \mathrm{Re}(\mathcal{M})$, so $b^* = b$ and $\|b\|^2 = \|b^* b\| = \|a\|$, therefore we have $b \in \mathrm{Fin}(\mathrm{Re}(\mathcal{M}))$.

(iii): If $a \in \mathrm{Fin}(\mathcal{M}^+)$, then there is $b \in \mathcal{M}$ such that $a = b^* b$ by Thm. 3.17(iii). Since $\|b\|^2 = \|b^* b\| = \|a\|$ is finite, we have $b \in \mathrm{Fin}(\mathcal{M})$. The other direction is clear.

(iv): follows from (i) and (ii):

$$\left(\widehat{\mathcal{M}}\right)^+ = \left[\mathrm{Re}(\widehat{\mathcal{M}})\right]^2 = \left[\left(\mathrm{Fin}(\mathrm{Re}(\mathcal{M}))\right)^\wedge\right]^2 = \left(\left[\mathrm{Fin}(\mathrm{Re}(\mathcal{M}))\right]^2\right)^\wedge$$

$$= \left(\mathrm{Fin}(\mathcal{M}^+)\right)^\wedge.$$

(v): Let $\widehat{a} \in \mathcal{U}(\widehat{\mathcal{M}})$. In particular, $\widehat{a} \in \left(\widehat{\mathcal{M}}\right)^{-1}$, so, by Cor. 3.11, we can assume that $a \in \left(\mathrm{Fin}(\mathcal{M})\right)^{-1}$. Then a has polar decomposition (Prop 3.13(ii)) as $|a| u$ for some $u \in \mathcal{U}(\mathcal{M})$.

From the Gelfand transform (Thm. 3.14) and the continuous functional calculus (Cor. 3.18), since $\widehat{a^* a} \approx \widehat{a}^* \widehat{a} = 1$, $|a| = \sqrt{a^* a}$ and $\sqrt{\cdot}$ is continuous, we have $|a| \approx 1$.

Hence $a \approx u$ and so $\widehat{a} = \widehat{u}$, therefore $\widehat{a} \in (\mathcal{U}(\mathcal{M}))^\wedge$. The other inclusion is trivial.

(vi): One inclusion is trivial. For the other, let $\widehat{a} \in \mathrm{Proj}(\widehat{\mathcal{M}})$. Since $\widehat{a} \in \left(\widehat{\mathcal{M}}\right)^+$ we can assume by (i) that $a \in \mathrm{Fin}(\mathcal{M}^+)$, hence $\sigma(a) \subset {}^*[0, \infty)$, by Thm. 3.16(iv). Moreover, a is normal.

By Thm. 3.16(vii) and Prop. 3.21(i), ${}^\circ\sigma(a) \subset \sigma(\widehat{a}) \subset \{0, 1\}$, so, by saturation, there is some $0 \approx \epsilon \in {}^*\mathbb{R}^+$ such that

$$\sigma(a) \subset \left([0, \epsilon] \cup [1 - \epsilon, 1 + \epsilon]\right).$$

Now let $f \in C(\sigma(a))$ be such that $f(z) = 0$ for all $z \in \left(\sigma(a) \cap [0, \epsilon]\right)$ and $f(z) = 1$ for all $z \in \left(\sigma(a) \cap [1 - \epsilon, 1 + \epsilon]\right)$. By a being normal and Cor. 3.18, $f(a) \in \mathcal{M}$ and $\sigma\left(f(a)\right) = f\left[\sigma(a)\right] \subset \{0, 1\}$, *i.e.* $f(a) \in \mathrm{Proj}(\mathcal{M})$.

Next let $g \in C(\sigma(a))$ be such that $g(z) = f(z) - z$ for all $z \in \sigma(a)$. Then by Cor. 3.18,

$$\|f(a) - a\|_{\mathcal{M}} = \|g(a)\|_{\mathcal{M}} = \|g\|_{C(\sigma(a))} \leq \epsilon \approx 0,$$

Hence $a \approx f(a)$ and so $\widehat{a} = \widehat{f(a)} \in (\mathrm{Proj}(\mathcal{M}))^\wedge$.

(vii): Clearly if $c \in \mathcal{M}$ is a partial isometry, then $\widehat{c}$ is a partial isometry in $\widehat{\mathcal{M}}$.

For the other implication, suppose $\widehat{a}$ is a partial isometry, so $\widehat{a}^* \widehat{a}$, i.e. $\widehat{a^* a}$, is a projection.

Fix $g \in C(\sigma(a^* a))$ so that $g(z) = 0$ for all $z \in \big(\sigma(a^* a) \cap [0,\epsilon]\big)$ and $g(z) = z^{-1/2}$ for all $z \in \big(\sigma(a^* a) \cap [1-\epsilon, 1+\epsilon]\big)$, where, as in (vi), for some fixed $0 \approx \epsilon \in {}^*\mathbb{R}^+$ we have $\sigma(a^* a) \subset \big([0,\epsilon] \cup [1-\epsilon, 1+\epsilon]\big)$.

Now let $f \in C(\sigma(a^* a))$ be $f(z) = z g^2(z)$, hence $f(z) = 0$ whenever $z \in \big(\sigma(a) \cap [0,\epsilon]\big)$ and $f(z) = 1$ whenever $z \in \big(\sigma(a) \cap [1-\epsilon, 1+\epsilon]\big)$, and so $f(a^* a) \in \mathrm{Proj}(\mathcal{M})$ as in (vi).

Now define $c = a g(a^* a)$. Then by $\big(g(a^* a)\big)^* = \bar{g}(a^* a) = g(a^* a)$ and the commutativity of $\mathcal{M}_{a^* a}$,

$$c^* c = \big(g(a^* a)\big)^* (a^* a) \big(g(a^* a)\big) = (a^* a) \big(g^2(a^* a)\big) = f(a^* a) \in \mathrm{Proj}(\mathcal{M}).$$

Therefore c is a partial isometry in $\mathcal{M}$.

Next note that ${}^\circ \sigma\big(a^* a - g(a^* a)\big) = \{0\}$, so $\sigma\big(\widehat{a^* a} - \widehat{g(a^* a)}\big) = \{0\}$ by Prop. 3.21(ii). i.e. $\widehat{a}^* \widehat{a} = \widehat{a^* a} = \widehat{g(a^* a)}$, hence

$$\widehat{c} = \widehat{a}\,\widehat{g(a^* a)} = \widehat{a}\,\widehat{a}^* \widehat{a} = \widehat{a}$$

by Thm. 3.9(viii) and $\widehat{a}$ being a partial isometry.

(viii): For the nontrivial direction, let $\widehat{a}^* \widehat{a} = 1$. In particular, $\overset{*}{a}$ is a partial isometry. Then it follows from (vii) for some partial isometry $c \in \mathcal{M}$ that $c \approx a$. Then $1 \approx a^* a \approx c^* c \in \mathrm{Proj}(\mathcal{M})$, therefore $c^* c = 1$ according to Cor. 3.12(ii), i.e. c is an isometry.

(ix): Same proof as in (viii) by interchanging a with a^*.

(x): For the nontrivial direction, suppose $\widehat{a}_1 \leq \widehat{a}_2$. Since $(\widehat{a}_2 - \widehat{a}_1) \in \big(\widehat{\mathcal{M}}\big)^+$, it follows from (iv) that $(a_2 - a_1) \approx c$ for some $c \in \mathrm{Fin}(\mathcal{M}^+)$. So we just let $c_1 = a_1$ and $c_2 = a_1 + c$. If $\widehat{a}_1 \in \big(\widehat{\mathcal{M}}\big)^+$, we get by (iv) $c_1 \in \mathrm{Fin}(\mathcal{M}^+)$ such that $c_1 \approx a_1$ and again let $c_2 = a_1 + c$. $\qquad\square$

There is a quite noticeable absence of a similar statement about normal elements in Thm. 3.22. This will be explained in the Notes.

Corollary 3.23. *Let $\mathcal{M}$ be an internal unital C^*-algebra. Suppose that $A \subset \mathrm{Fin}(\mathcal{M}^+)$ is finite and $b \in \mathrm{Fin}(\mathcal{M}^+)$ is such that $\forall x \in A \big(\widehat{b} \geq \widehat{x}\big)$.*

Then there is $c \in \mathrm{Fin}(\mathcal{M}^+)$ such that $c \approx b$ and $\forall x \in A \big(c \geq x\big)$.

Proof. If $A = \{a\}$, then $\widehat{b} \geq \widehat{a}$. Since $\widehat{b - a} = (\widehat{b} - \widehat{a}) \in \left(\widehat{\mathcal{M}}\right)^+$, there is by Thm. 3.22(iv) some $d \in \mathrm{Fin}(\mathcal{M}^+)$ such that $d \approx b - a$. Define $c = a + d$, so $c \in \mathcal{M}^+$ by Thm. 3.17(i) and $b \approx c \geq a$.

Now suppose $A = \{a_1, \ldots, a_{n+1}\}$ and assume inductively that there is $c_0 \in \mathrm{Fin}(\mathcal{M}^+)$ such that $b \approx c_0 \geq a_1, \ldots, a_n$.

Since $\widehat{c_0} = \widehat{b} \geq \widehat{a}_{n+1}$, by repeating the above argument, we get some $d \in \mathrm{Fin}(\mathcal{M}^+)$ such that $c_0 \approx b \approx (a_{n+1} + d) \geq a_{n+1}$. By Prop. 3.12, let $c \in \mathrm{Fin}(\mathcal{M}^+)$ such that $c \approx (a_{n+1} + d) \approx c_0$ with $c \geq (a_{n+1} + d)$ and $c \geq c_0$. Therefore we have $c \approx b$ and $c \geq a_1, \ldots, a_{n+1}$. $\square$

Theorem 3.23.   (The Nonstandard Extended Functional Calculus) *Let $\mathcal{M}$ be an internal unital C^*-algebra and $a \in \mathrm{Fin}(\mathcal{M})$ be normal.*

- *For every $F \in \mathrm{Fin}\big(C(\sigma(a)\big)$, $F(a) \in \mathrm{Fin}(\mathcal{M}_a)$. Hence $\widehat{F(a)} \in \widehat{\mathcal{M}}$.*
- *For every $f \in C(\sigma(\widehat{a}))$ there is an S-continuous $F \in \mathrm{Fin}\big(C(\sigma(a))\big)$ such that $^\circ F = f$.*
- *Let $F \in \mathrm{Fin}\big(C(\sigma(a)\big)$ be S-continuous. Then $^\circ F \in C(\sigma(\widehat{a}))$ and $\widehat{F(a)} = (^\circ F)(\widehat{a})$.*

Therefore $\{\widehat{F(a)} \mid F \in \mathrm{Fin}\big(C(\sigma(a))\big)\} \subset \widehat{\mathcal{M}_a}$ is an extension of the continuous functional calculus for $\widehat{a}$.

Proof.   Let $F \in \mathrm{Fin}\big(C(\sigma(a)\big)$, then by the continuous functional calculus (Cor. 3.18), $\|F(a)\|_{\mathcal{M}} = \|F\|_{C(\sigma(a))}$ and so $F(a) \in \mathrm{Fin}(\mathcal{M}_a)$.

Next let $f \in C(\sigma(\widehat{a}))$. Note that $\sigma(\widehat{a}) = {}^\circ\sigma(a)$, by Prop. 3.21. Then by the Tietze Extension Theorem (Cor. 3.13), we extend f to some continuous $g : D \to \mathbb{C}$, where $D \subset \mathbb{C}$ is compact such that $\sigma(a) \subset {}^*D$. Now let $F = {}^*g \restriction_{\sigma(a)}$, so $F \in C(\sigma(a))$ is S-continuous and $^\circ F = f$.

To prove the last statement, let $F \in \mathrm{Fin}\big(C(\sigma(a)\big)$ be S-continuous. Clearly $^\circ F \in C(\sigma(\widehat{a}))$, as $^\circ\sigma(a) = \sigma(\widehat{a})$. Let $f = {}^\circ F$.

To show $\widehat{F(a)} = f(\widehat{a})$, first note that $f(\widehat{a}) \in \widehat{\mathcal{M}}_{\widehat{a}} \subset \widehat{\mathcal{M}}_a$, so there is $b \in \mathrm{Fin}(\mathcal{M}_a)$ such that $b \approx f(\widehat{a})$. Hence it suffices to show $b \approx F(a)$.

Let $\sim$ denote either the Gelfand transform for $\widehat{\mathcal{M}}_{\widehat{a}}$ (as $\widehat{a}$ is normal) or the internal Gelfand transform for $\mathcal{M}_a$.

By the isometry property in Thm. 3.14, we only need to show

$$\forall \theta \in \mathrm{hom}(\mathcal{M}_a) \left(\widetilde{b}(\theta) \approx \widetilde{F(a)}(\theta)\right).$$

For $\theta \in \mathrm{hom}(\mathcal{M}_a)$, define $\widehat{\theta} : \widehat{\mathcal{M}}_a \to \mathbb{C}$ by $\widehat{\theta}(\widehat{x}) := {}^\circ\theta(x)$, $x \in \mathrm{Fin}(\mathcal{M}_a)$. It is easy to see that $\widehat{\theta}$ is well defined and $\widehat{\theta} \in \mathrm{hom}(\widehat{\mathcal{M}}_a)$.

Note that $\widehat{\theta} \restriction_{\widehat{\mathcal{M}_{\widehat{a}}}} \in \hom(\widehat{\mathcal{M}_{\widehat{a}}})$ whenever $\widehat{\theta} \in \hom(\widehat{\mathcal{M}_a})$. Finally, let $\theta \in \hom(\mathcal{M}_a)$, then

$$\widetilde{b}(\theta) = \theta(b) \approx \widehat{\theta}(\widehat{b}) = \widehat{\theta}(f(\widehat{a})) = \widetilde{f(\widehat{a})}(\widehat{\theta}) = f(\widetilde{\widehat{a}}(\widehat{\theta})) = f(\widehat{\theta}(\widehat{a}))$$

$$\approx F(\theta(a)) = F(\widetilde{a}(\theta)) = \widetilde{F(a)}(\theta),$$

as required. $\square$

In the case a is a normal element in a unital C^*-algebra $\mathcal{M}$, one can apply Thm.3.23 to $^*a$, which is normal by transfer, and $^*\mathcal{M}$ to obtain the extended functional calculus from $\widehat{F(^*a)} \in \widehat{^*\mathcal{M}}$, where $F \in \mathrm{Fin}\big(C(\sigma(^*a))\big)$.

3.3.2 Notes and exercises

The long list in Thm. 3.22 does not include a similar statement about normal elements. Of course, in an internal unital C^*-algebra $\mathcal{M}$, if an element $a \in \mathrm{Fin}(\mathcal{M})$ is normal, then $\widehat{a}$ remains normal, but the converse is false, as the following example shows.

Fix $N \in {}^*\mathbb{N} \setminus \mathbb{N}$. Let $\mathcal{M} = \mathcal{B}(\mathbb{C}^{2N})$, the internal algebra of $(2N \times 2N)$-matrices over $\mathbb{C}$. So $\mathcal{M}$ forms an internal unital C^*-algebra. Let $a \in \mathcal{M}$ be the $(2N \times 2N)$-matrix having the following subdiagonal entries

$$\frac{1}{N}, \frac{2}{N}, \dots, \frac{N-1}{N}, \frac{N}{N}, \frac{N-1}{N}, \dots, \frac{2}{N}, \frac{1}{N}$$

with entries of 0 elsewhere. *i.e.* $a = [a_{i,j}]_{1 \le i,j \le 2N}$ is such that

$$a_{n+1,n} = \begin{cases} \frac{n}{N}, & \text{if } 1 \le n \le N \\[2mm] \frac{2N-n}{N}, & \text{if } N+1 \le n \le 2N-1 \end{cases} \quad \text{and } a_{i,j} = 0 \text{ if } i-j \ne 1.$$

Let $[x,y] := (xy - yx)$ denote the **commutator** of elements x,y in a ring. Then $[a, a^*]$ is a diagonal matrix with the following diagonal entries

$$-\frac{1}{N^2}, \dots, -\frac{2n-1}{N^2}, \dots, -\frac{2N-1}{N^2}, \frac{2N-1}{N^2}, \dots, \frac{2n-1}{N^2}, \dots, \frac{1}{N^2}.$$

Therefore $[a, a^*] \ne 0$, *i.e.* a is not normal. However $\|[a, a^*]\| < 2N^{-1} \approx 0$, *i.e.* $\widehat{a}$ is a normal element in $\widehat{\mathcal{M}}$.

In this example, the only eigenvalue of a is 0. As $\mathcal{M} = \mathcal{B}(\mathbb{C}^{2N})$, this means that $\sigma(a) = \{0\}$. But $\|a\| = 1$, so $\|\widehat{a}\| = 1$. As $\widehat{a}$ is normal, we have $\rho(\widehat{a}) = \|\widehat{a}\| = 1$, therefore $^\circ\sigma(a) \subsetneq \sigma(\widehat{a})$, showing that Prop. 3.21(ii) cannot be improved.

Note also $\sigma(a) = \{0\}$ here shows that one direction in Thm. 3.16 (iii) and (iv) need not hold if the element is not normal.

Still with this example of the normal $\hat{a}$ such that $\hat{a} \neq \hat{c}$ for any normal $c \in \mathcal{M}$. Consider the decomposition $a = a_1 + ia_2$, where $a_1, a_2 \in \text{Fin}(\text{Re}(\mathcal{M}))$. Then $[a, a^*] = -2i[a_1, a_2]$. So $[a_1, a_2] \approx 0$, *i.e.* $\hat{a}_1\hat{a}_2 = \hat{a}_2\hat{a}_1$.

This gives an example of a pair of commuting $\hat{a}_1$ and $\hat{a}_2$ but there are no $c_1, c_2 \in \mathcal{M}$ such that $c_1 \approx a_1$, $c_2 \approx a_2$ and c_1, c_2 commute. For otherwise, let $c = c_1 + ic_2$, then $a \approx c$ and $[c, c^*] = 2i[c_1, c_2] = 0$, so c is normal, a contradiction.

Furthermore, the internal unital C^*-subalgebra $\mathcal{M}_a$ is noncommutative, while $\widehat{\mathcal{M}_{\hat{a}}}$ is commutative.

As an example similar to the above, consider $\mathcal{M} = {}^*\mathcal{B}(\ell_2)$ and a weighted unilateral shift operator $S \in \mathcal{M}$ (see also Example 3.12) defined as follows. Fix $N \in {}^*\mathbb{N} \setminus \mathbb{N}$. For every $\xi = \{\xi_n\}_{n \in {}^*\mathbb{N}} \in {}^*\ell_2$,

$$S(\xi)_0 := 0, \quad S(\xi)_n := \frac{\min\{n, N\}}{N}\xi_{n-1} \quad \text{if} \quad n \in {}^*\mathbb{N}^+.$$

Then $[S, S^*](\xi)_n = \begin{cases} -(2n+1)N^{-2}\xi_n & \text{if } 0 \leq n < N \\ 0 & \text{if } N \leq n \in {}^*\mathbb{N} \end{cases}$ and so $\hat{S}$ is normal. By §1.1 in [Davidson and Szarek (2001)], by comparing the Fredholm indices of operators close to S with the index of a normal operator, one can show that $\|S - T\| \geq 1$ for any normal $T \in \mathcal{M}$. therefore, there is no normal $T \in \mathcal{M}$ such that $\hat{T} = \hat{S}$.

But here ${}^\circ\sigma(S) = \sigma(\hat{S})$.

See [Conway (1991)] for more results on shift operators.

EXERCISES

(1) In the example mentioned in the above Notes, is $\widehat{\mathcal{M}_a}$ commutative? This is related to the converse of the second statement in Prop. 3.19(i).
(2) Check that ${}^\circ\sigma(S) = \sigma(\hat{S})$ in the example mentioned in the above Notes.
(3) Given an internal unital C^*-algebra $\mathcal{M}$, characterizes those $a \in \text{Fin}(\mathcal{M})$ such that $\hat{a}$ is normal in $\widehat{\mathcal{M}}$.
(4) Let $\mathcal{M}$ be an internal unital C^*-algebra and (π, H) be a GNS representation of $\widehat{\mathcal{M}}$. Is (π, H) unitarily equivalent to $(\hat{\pi}_0, \hat{H}_0)$ for some internal GNS representation (π_0, H_0) of $\mathcal{M}$?
(5) Prove or disprove: If (π, H) is an internal cyclic GNS representation of an internal unital C^*-algebra $\mathcal{M}$, then $(\hat{\pi}, \hat{H})$ is a cyclic GNS representation of $\widehat{\mathcal{M}}$.

3.4   Von Neumann Algebras

Bicommutant and bidual find peace and tranquility in the paradise of von Neumann algebras.

By the universal representation in the GNS Theorem (Thm. 3.20), every unital C^*-algebra can be identified with a unital C^*-subalgebra of $\mathcal{B}(H)$ for some Hilbert space H.

So we only work with such unital C^*-subalgebra in this section.

This section deals with an important class of C^*-algebras: the von Neumann algebras. A von Neumann algebra satisfies certain topological closure properties in $\mathcal{B}(H)$ and is algebraically characterized by being equal to its bicommutant and its bidual. The weak nonstandard hull is used to prove the bidual result. Spectral measures and spectral integrals are briefly discussed. The tracial nonstandard hull construction is introduced.

3.4.1   *Operator topologies and the bicommutant*

For a Hilbert space H, we first define two topologies on $\mathcal{B}(H)$ which are weaker than the norm topology in general.

For each pair $\xi, \eta \in H$, we let

$$\omega_{\xi,\eta} : \mathcal{B}(H) \to \mathbb{C}$$

denote the mapping $x \mapsto \langle x(\xi), \eta \rangle$. Clearly $\omega_{\xi,\eta}$ is linear.

Since $|\langle x(\xi), \eta \rangle| \leq \|x(\xi)\| \|\eta\| \leq \|x\| \|\xi\| \|\eta\|$, we have $\|\omega_{\xi,\eta}\| \leq \|\xi\| \|\eta\|$. Hence $\omega_{\xi,\eta} \in \mathcal{B}(H)'$.

Note that the mappings $\mathcal{B}(H) \ni x \mapsto |\langle x(\xi), \eta \rangle|$ are seminorms on $\mathcal{B}(H)$, which we denote by $p_{\xi,\eta}$.

On the other hand, for each $\xi \in H$, the mapping $\mathcal{B}(H) \ni x \mapsto \|x(\xi)\|$ is also a seminorm—denoted by p_ξ.

The **weak operator topology** (**WOT**) on $\mathcal{B}(H)$ is defined to be the topology generated by the seminorms $p_{\xi,\eta}$, $\xi, \eta \in H$.

While the **strong operator topology** (**SOT**) on $\mathcal{B}(H)$ is defined to be the topology generated by the seminorms p_ξ, $\xi \in H$.

So a net $\{a_i\}_{i \in I} \subset \mathcal{B}(H)$ converges to $a \in \mathcal{B}(H)$ in WOT if

$$\lim_{i \in I} \langle a_i(\xi), \eta \rangle = \langle a(\xi), \eta \rangle \quad \text{for all} \quad \xi, \eta \in H. \qquad (a_i \xrightarrow{\text{WOT}} a.)$$

Likewise, $\{a_i\}_{i \in I} \subset \mathcal{B}(H)$ converges to $a \in \mathcal{B}(H)$ in SOT if

$$\lim_{i \in I} a_i(\xi) = a(\xi) \quad \text{for all} \quad \xi \in H. \qquad (a_i \xrightarrow{\text{SOT}} a.)$$

Therefore norm convergence $\Rightarrow$ SOT-convergence $\Rightarrow$ WOT-convergence, *i.e.* the norm topology is stronger (*i.e.* finer) than SOT and SOT is stronger than WOT. Of course, if H is finite dimensional, all three are equivalent. We leave it as an exercise the check that in general, one is strictly stronger than the other.

In the following, we assume that $\mathcal{M}$ is a unital C^*-algebra and $\mathcal{B}(H)$ is obtained from the universal representation of $\mathcal{M}$ by the GNS Theorem (Thm. 3.20). Moreover we regard $\mathcal{M} \subset \mathcal{B}(H)$ as a unital C^*-subalgebra.

We say that $\mathcal{M}$ is a *von Neumann algebra* if $\mathcal{M} = \overline{\mathcal{M}}^{\text{WOT}}$, *i.e.* if $\mathcal{M}$ is WOT-closed in $\mathcal{B}(H)$.

In such case, we also call $\mathcal{M}$ a *von Neumann algebra on* H.

In particular, $\mathcal{B}(H)$ itself is a von Neumann algebra.

It turns out that the notion of von Neumann algebras can be given a purely algebraic characterization.

Generalizing the definition on p.163, for each nonempty $X \subset \mathcal{B}(H)$, the *commutant* of X is defined as

$$\{X\}' := \{y \in \mathcal{B}(H) \mid \forall x \in X \ (xy = yx)\}.$$

(This notation should not cause confusion with the similar notation for dual spaces.)

It is clear that $\{X\}'$ always contain 1 and is a subalgebra of $\mathcal{B}(H)$ for any nonempty $X \subset \mathcal{B}(H)$. It is a C^*-subalgebra if X is closed under involution.

Instead of $\{\{X\}'\}'$, ... one simply writes $\{X\}''$, $\{X\}'''$ *etc.*

With a moment of thought, one sees that $\{X\}' = \{X\}'''$.

We call $\{X\}''$ the *bicommutant* of X.

Note that $X \subset \{X\}''$ holds always.

Lemma 3.5. $\{\mathcal{M}\}'$ *is WOT-closed.*

Proof.   Let $a \in \mathcal{B}(H)$ be such that $a_i \xrightarrow{\text{WOT}} a$ for some net $\{a_i\}_{i \in I} \subset \{\mathcal{M}\}'$. We must show that $a \in \{\mathcal{M}\}'$.

Fix $c \in \mathcal{M}$. we need to check that $ac = ca$.

From the definition, $\forall \xi, \eta \in H \ (\langle a_i(\xi), \eta \rangle \to \langle a(\xi), \eta \rangle)$. Consequently we have the following for any $\xi, \eta \in H$:

$$\langle (a_i c)(\xi), \eta \rangle = \langle a_i\big(c(\xi)\big), \eta \rangle \to \langle a\big(c(\xi)\big), \eta \rangle = \langle (ac)(\xi), \eta \rangle,$$

i.e. $a_i c \xrightarrow{\text{WOT}} ac$.

On the the hand, for any $\xi, \eta \in H$,

$$\langle (c a_i)(\xi), \eta \rangle = \langle a_i(\xi), c^*(\eta) \rangle \to \langle a(\xi), c^*(\eta) \rangle = \langle (ca)(\xi), \eta \rangle,$$

252   *Nonstandard Methods in Functional Analysis*

i.e. $ca_i \xrightarrow{\text{SOT}} ca$.

Since $ca_i = a_i c$, $i \in I$, we get $ac = ca$.

As this holds for all $c \in \mathcal{M}$, we have $a \in \{\mathcal{M}\}'$. $\qquad\square$

Combining Lem. 3.5 with previous remarks, we see the following:

Corollary 3.24. $\{\mathcal{M}\}'$ *always forms a von Neumann subalgebra of* $\mathcal{B}(H)$ *and* $\mathcal{M}$ *extends to the von Neumann algebra* $\{\mathcal{M}\}'' \subset \mathcal{B}(H)$. $\qquad\square$

Theorem 3.24. (Von Neumann's Bicommutant Theorem) *The following are equivalent for a unital C^*-subalgebra $\mathcal{M} \subset B(H)$.*

(i) $\mathcal{M}$ *is WOT-closed. (i.e. $\mathcal{M}$ is a von Neumann algebra).*

(ii) $\mathcal{M}$ *is SOT-closed.*

(iii) $\mathcal{M} = \{\mathcal{M}\}''$.

In fact $\overline{\mathcal{M}}^{\text{WOT}} = \overline{\mathcal{M}}^{\text{SOT}} = \{\mathcal{M}\}''$ *holds.*

Proof. ((i) $\Rightarrow$ (ii)) is trivial as SOT is a stronger than WOT.

((iii) $\Rightarrow$ (i)) : If $\mathcal{M} = \{\mathcal{M}\}''$ then, by Lem. 3.5, $\mathcal{M}$ is WOT-closed.

((ii) $\Rightarrow$ (iii)) : We will show that $\mathcal{M}$ is SOT-dense in $\{\mathcal{M}\}''$. Then $\mathcal{M} = \overline{\mathcal{M}}^{\text{SOT}}$ implies, by Lem. 3.5, that $\mathcal{M} = \{\mathcal{M}\}''$.

So we let $a \in \{\mathcal{M}\}''$ and we need to show that $a \in \overline{\mathcal{M}}^{\text{SOT}}$. In other words, for any $\xi_1, \ldots, \xi_n \in H$, $n \in \mathbb{N}$, and $\epsilon \in \mathbb{R}^+$, we need to find $c \in \mathcal{M}$, which is dependent on ϵ and the $\xi_1, \ldots, \xi_n$, such that

$$\bigwedge_{i=1}^{n} \left(p_{\xi_i}(a - c) < \epsilon \right). \tag{3.5}$$

First consider the case $n = 1$. So we fix $\xi \in H$.

Define a closed subspace $X := \overline{\{x(\xi) \,|\, x \in \mathcal{M}\}} \subset H$. Let $q := \pi_X$, the orthogonal projection of H onto X. Thus $q \in \mathrm{Proj}(\mathcal{B}(H))$.

As $q[X] = \pi_X[X] = X$, we note that $\forall x \in \mathcal{M} \left(qxq = xq \right)$.

Consequently, for any $c \in \mathcal{M}$ we have

$$qc = (c^* q)^* = (qc^* q)^* = q(qc^*)^* = qcq = cq,$$

i.e. $q \in \{\mathcal{M}\}'$.

Hence $aq = qa$ and so $a(\xi) = aq(\xi) = qa(\xi) \in X$, which is the closure of $\{x(\xi) \,|\, x \in \mathcal{M}\}$. Therefore, for some $c \in \mathcal{M}$ we have $p_\xi(a - c) < \epsilon$.

Next consider the general case $\xi_1, \ldots, \xi_n \in H$, $n \in \mathbb{N}$.

We form the direct sum of the Hilbert space H with itself n times and write $H^{(n)} := \oplus_{i=1}^{n} H$. Elements in $H^{(n)}$ are represented as column vectors.

Then elements in $\mathcal{B}(H^{(n)})$ are identified with $n \times n$-matrices over $\mathcal{B}(H)$ and their action on $H^{(n)}$ corresponds to the matrix multiplication.

For $x \in \mathcal{B}(H)$ we let $\oplus^{(n)} x$ denote the diagonal matrix in $\mathcal{B}(H^{(n)})$ with diagonal entries given by x.

Let $\pi : \mathcal{B}(H) \to \mathcal{B}(H^{(n)})$ be given by $\pi(x) := \oplus^{(n)} x$.

Then $\mathcal{M}$ is *isomorphic to $\pi[\mathcal{M}]$, a unital C^*-subalgebra of $\mathcal{B}(H^{(n)})$.

First we claim that $\oplus^{(n)} a \in \{\pi[\mathcal{M}]\}''$. To see this, let any $b \in \{\pi[\mathcal{M}]\}'$. Write b as the matrix $[b_{ij}]_{1 \leq i,j \leq n}$. For $x \in \mathcal{M}$, by comparing the entries in $\left(\oplus^{(n)} x \right) b = b \left(\oplus^{(n)} x \right)$, we see that $x b_{ij} = b_{ij} x$ holds for all b_{ij}. Therefore b has all entries $b_{ij} \in \{\mathcal{M}\}'$. As a result, $a b_{ij} = b_{ij} a$ holds for all b_{ij}, and same comparison again shows that $\left(\oplus^{(n)} a \right) b = b \left(\oplus^{(n)} a \right)$. Since this holds for all $b \in \{\pi[\mathcal{M}]\}'$, we have $\oplus^{(n)} a \in \{\pi[\mathcal{M}]\}''$.

Repeat the previous argument for the case $n = 1$ but in the setting given by $\pi[\mathcal{M}] \subset \mathcal{B}(H^{(n)})$, $\oplus^{(n)} a \in \{\pi[\mathcal{M}]\}''$ and $\xi = (\xi_1, \ldots, \xi_n)^{\mathrm{T}} \in H^{(n)}$, we conclude that

$$p_\xi \left(\left(\oplus^{(n)} a \right) - \left(\oplus^{(n)} c \right) \right) < \epsilon \quad \text{holds for some} \quad \oplus^{(n)} c \in \pi[\mathcal{M}].$$

Therefore, from the definition of the direct sum of Hilbert spaces, we obtain (3.5).

The proof of ((ii) $\Rightarrow$ (iii)) shows that $\mathcal{M}$ is SOT-dense in $\{\mathcal{M}\}''$. By Lem. 3.5, $\{\mathcal{M}\}''$ is SOT-closed, and since $\mathcal{M} \subset \{\mathcal{M}\}''$, we have $\overline{\mathcal{M}}^{\mathrm{SOT}} = \{\mathcal{M}\}''$. On the other hand, trivially $\overline{\mathcal{M}}^{\mathrm{SOT}} \subset \overline{\mathcal{M}}^{\mathrm{WOT}}$ and also, by $\mathcal{M} \subset \{\mathcal{M}\}''$ and $\{\mathcal{M}\}''$ WOT-closed (Lem. 3.5), $\overline{\mathcal{M}}^{\mathrm{WOT}} \subset \{\mathcal{M}\}''$, so we conclude that

$$\overline{\mathcal{M}}^{\mathrm{SOT}} = \overline{\mathcal{M}}^{\mathrm{WOT}} = \{\mathcal{M}\}'' \qquad \square$$

With $\mathcal{M}$ identified with a unital C^*-subalgebra of $\mathcal{B}(H)$ through the universal representation, $\{\mathcal{M}\}''$, the von Neumann algebra generated by $\mathcal{M}$ in $\mathcal{B}(H)$, is called the **universal enveloping von Neumann algebra of $\mathcal{M}$**.

Example 3.18.

- As mentioned already, $\mathcal{B}(H)$ is a von Neumann algebra. In particular, $\mathbb{C}$ is also a von Neumann algebra.
- Let μ be a σ-additive complex Borel measure on a compact space and $\mathcal{M} := \{m_f \mid f \in L_\infty(\mu)\}$, where for $f \in L_\infty(\mu)$, we define $m_f \in B(L_2(\mu))$, a **multiplication operator**, to be the linear operator $L_2(\mu) \ni g \mapsto fg$. Then $\mathcal{M}$ is a unital C^*-subalgebra of $B(L_2(\mu))$. Moreover, we leave it as an exercise to check that $\mathcal{M}$ forms a commutative von Neumann algebra. $\qquad \square$

3.4.2   Nonstandard hulls vs. von Neumann algebras

In this subsection we show that other than unimportant cases, the norm-nonstandard hull of an internal unital C^*-algebra is never a von Neumann algebra. However, there is a modified construction, called the *tracial non-standard hull*, which is applicable to certain von Neumann algebras and produces von Neumann algebras.

However, a probably better solution is to expand the class of von Neumann algebras so that the class is stable under the norm-nonstandard hull construction—this will be done in §3.5.2.

Proposition 3.22. *Let $\mathcal{M}$ be an internal unital C^*-algebra containing nonzero $\{p_n\}_{n \in \mathbb{N}} \subset \mathrm{Proj}(\mathcal{M})$ so that $p_n p_m = 0$ whenever $n \neq m$. (i.e. the projections are **mutually orthogonal**.)*

Then the norm-nonstandard hull $\widehat{M}$ fails to be a von Neumann algebra.

Proof.   Regard $\mathcal{M} \subset \mathcal{B}(H)$ for some internal Hilbert space H given by the universal representation.

Extend $\{p_n\}_{n \in \mathbb{N}}$ to an internal sequence of nonzero mutually orthogonal projections $\{p_n\}_{n \leq N} \subset \mathrm{Proj}(\mathcal{M})$ for some $N \in {}^*\mathbb{N} \setminus \mathbb{N}$.

Since $\left\| \sum_{i=0}^n p_i \right\|^2 = \left\| \left(\sum_{i=0}^n p_i \right) \overline{\left(\sum_{i=0}^n p_i \right)} \right\| = \left\| \sum_{i=0}^n p_i \right\|$ holds for any $n \leq N$ by mutual orthogonality, we have

$$\forall n \leq N \left(\left\| \sum_{i=0}^n p_i \right\| = 1 \right). \tag{3.6}$$

By Thm. 3.22(vi), $\{\widehat{p}_n\}_{n \leq N} \subset \mathrm{Proj}(\widehat{\mathcal{M}})$.
Define $c : \widehat{H} \to \widehat{H}$ by

$$\widehat{\xi} \mapsto c(\widehat{\xi}) := \lim_{n \to \infty} \sum_{i=0}^n \widehat{p_i(\xi)} = \lim_{n \to \infty} \sum_{i=0}^n \widehat{p}_i(\widehat{\xi}), \quad \text{where} \quad \xi \in \mathrm{Fin}(H).$$

From (3.6), it is clear that the above sequence is norm-convergent in $\widehat{H}$, c is well-defined and $c \in S_{\mathcal{B}(\widehat{H})}$.

Note that given any $\epsilon \in \mathbb{R}^+$ and $\xi, \ldots, \xi_n \in \mathrm{Fin}(H)$, $n \in \mathbb{N}$, we have

$$\left\| c(\widehat{\xi}_i) - \left(\sum_{i=0}^m \widehat{p}_i(\widehat{\xi}) \right) \right\| < \epsilon$$

for all large $m \in \mathbb{N}$.

Therefore we have $c \in \overline{\mathcal{M}^{\mathrm{SOT}}}$

Consequently the desired conclusion follows from the von Neumann's Bicommutant Theorem (Thm. 3.24) if we can show that $c \notin \widehat{\mathcal{M}}$.

Suppose otherwise, $c = \hat{a}$ for some $a \in \mathrm{Fin}(\mathcal{M})$. Consider

$$A := \big\{ n < N \mid \forall \xi \in S_{p_n[H]} \left(\|a(\xi) - \xi\| < 1/2 \right) \big\}.$$

Note that A is an internal set.

If $n \in \mathbb{N}$ and $\xi \in S_{p_n[H]}$, we have $c(\hat{\xi}) = \hat{\xi}$, so $a(\xi) \approx \xi$, in particular $n \in A$.

On the other hand, for infinite $n < N$ and $\xi \in S_{p_n[H]}$, we have $c(\hat{\xi}) = 0$, so $\|a(\xi) - \xi\| \approx 1$, and thus $n \notin A$.

In other words, $A = \mathbb{N}$, which is impossible as A is internal. $\qquad\square$

Similar result holds when the conditions on the projections are imposed on the nonstandard hull.

Corollary 3.25. *Let $\mathcal{M}$ be an internal unital C^*-algebra such that $\mathrm{Proj}(\widehat{\mathcal{M}})$ contains an infinite family of mutually orthogonal projections. Then $\widehat{\mathcal{M}}$ is not a von Neumann algebra.*

Proof. Again, regard $\mathcal{M} \subset \mathcal{B}(H)$, as given by the universal representation.

Let $\{\widehat{p_n}\}_{n \in \mathbb{N}} \subset \mathrm{Proj}(\widehat{\mathcal{M}})$ be mutually orthogonal. By Thm. 3.22(vi), we can assume that $\{p_n\}_{n \in \mathbb{N}} \subset \mathrm{Proj}(\mathcal{M})$.

Recall that for a closed subspace $Y \subset X$, we use π_Y to denote the orthogonal projection onto Y.

Now for $n \in \mathbb{N}^+$, we replace p_n by

$$p_n \cdot \pi_{(p_0[H] \oplus \cdots \oplus p_{n-1}[H])^\perp}.$$

Then clearly $\{p_n\}_{n \in \mathbb{N}}$ is a infinite family of mutually orthogonal projections in $\mathcal{M}$, hence Prop. 3.22 applies. $\qquad\square$

Corollary 3.26. *Let H be a complex Hilbert space.*
Then $\widehat{\mathcal{B}(H)}$ is a von Neumann algebra iff H is finite dimensional.

Proof. For the nontrivial direction, suppose $\widehat{\mathcal{B}(H)}$ is a von Neumann algebra. Then by Prop. 3.22, $\mathcal{B}(H)$ contains no infinite family of mutually orthogonal projections, so H must be finite dimensional. $\qquad\square$

In particular, $\widehat{\mathcal{B}(\ell_2)}$ is not a von Neumann algebra. Many von Neumann algebras contain infinitely many mutually orthogonal projections, so they all fail to have von Neumann norm-nonstandard hulls.

More on projections will be dealt with in §3.5.1.

Now we describe the ***tracial nonstandard hull*** construction.

Let $\mathcal{M}$ be a unital C^*-algebra. By a **tracial state** on $\mathcal{M}$ we mean some $\tau \in \mathbb{S}(\mathcal{M})$ such that

$$\forall x \in \mathcal{M} \left(\tau(x^* x) = \tau(x\, x^*) \right).$$

So the trace Tr on matrices in Example 3.15 can be normalized to form a tracial state. Of course, if $\mathcal{M}$ is commutative, every state is a tracial state.

Fix an internal unital C^*-algebra $\mathcal{M}$ which admits an internal tracial state τ on it. Observe that by Thm. 3.17, $\tau(x^* x) \in {}^*[0, \infty)$ holds for all $x \in \mathcal{M}$. As in the GNS construction (Thm. 3.19),

$$\mathcal{M} \ni x \mapsto \left(\tau(x^* x) \right)^{1/2} \in {}^*[0, \infty)$$

defines an internal seminorm on $\mathcal{M}$ denoted by $\|\cdot\|_\tau$.

Define an equivalence relation on $\mathcal{M}$ by letting $x \approx_\tau y$, $x, y \in \mathcal{M}$, whenever $\|x - y\|_\tau \approx 0$. Then let $\widehat{x}^\tau$ denote the equivalence classes *w.r.t.* $\approx_\tau$ and define $\widehat{\mathcal{M}}^\tau := \left\{ \widehat{x}^\tau \mid x \in \mathrm{Fin}(\mathcal{M}) \right\}$.

Here $\mathrm{Fin}(\mathcal{M})$ refers to the finite part *w.r.t.* the internal norm on $\mathcal{M}$. Note that, by Cor. 3.20, for any $a \in \mathcal{M}$, $\tau(a^* a) \leq \|a^* a\|\, \tau(1) = \|a\|^2$, hence $\|a\|_\tau \leq \|a\|$ and $a \approx 0$ implies that $a \approx_\tau 0$. Therefore, the idea is that, by using $\approx_\tau$ instead of $\approx$, a smaller nonstandard hull is produced and hopefully it would be of the right size to become a von Neumann algebra.

For $\widehat{x}^\tau, \widehat{y}^\tau \in \widehat{\mathcal{M}}^\tau$ and $\alpha \in \mathbb{C}$, we define

$$\widehat{x}^\tau + \alpha\, \widehat{y}^\tau := \widehat{(x + \alpha y)}^\tau \quad \text{and} \quad \left(\widehat{x}^\tau \right)^* := \widehat{(x^*)}^\tau$$

and $\|\widehat{x}^\tau\| := \inf\limits_{y \approx_\tau x} {}^\circ\|y\|$ from the internal norm on $x \in \mathrm{Fin}(\mathcal{M})$.

It is an easy exercise to prove the following:

Lemma 3.6. *Suppose $\mathcal{M}$ is an internal unital C^*-algebra having an internal tracial state τ. Then $\widehat{\mathcal{M}}^\tau$ forms a unital C^*-algebra under the above operations and norm.* $\qquad\square$

We comment that by Prop. 3.11, it holds for any $x, y \in \mathcal{M}$ that

$$(xy)^* (xy) \leq \|x\|^2 (y^* y).$$

Then by τ being a tracial state, one gets

$$\|xy\|_\tau \leq \min \left\{ \|x\|\, \|y\|_\tau, \|y\|\, \|x\|_\tau \right\} \tag{3.7}$$

and this is needed in the proof of the above.

It turns out τ gives rise to a tracial state on $\widehat{\mathcal{M}}^\tau$: this is done by defining $\widehat{\tau}(\widehat{x}^\tau) := {}^\circ\tau(x)$, $x \in \mathrm{Fin}(\mathcal{M})$. It is straightforward to check that $\widehat{\tau}$ forms a tracial state on $\widehat{\mathcal{M}}^\tau$. Moreover, by the construction of on $\widehat{\mathcal{M}}^\tau$, it follows immediately that $\widehat{\tau}$ is **faithful**, *i.e.* injective.

Theorem 3.25. *Suppose $\mathcal{M}$ is an internal unital C^*-algebra having an internal tracial state τ. Then $\widehat{\mathcal{M}}^\tau$ is a von Neumann algebra.*

Proof. Let (π, H) be the internal cyclic GNS representation of $\mathcal{M}$ corresponding to τ given by Thm. 3.19.

Note that $\mathcal{M}$ corresponds to a *dense subset of H. Moreover, because $\forall x \in \mathcal{M} \left(\|x\|_\tau \leq \|x\| \right)$, we make the identification $\mathrm{Fin}(\mathcal{M}) \subset \mathrm{Fin}(H)$.

Let $\widehat{H}$ be the norm-nonstandard hull construction of the Hilbert space H. Then $K := \widehat{\mathrm{Fin}(\mathcal{M})}$ is a Hilbert space, a closed subspace of $\widehat{H}$.

Define $\widehat{\pi} : \widehat{\mathcal{M}}^\tau \to \mathcal{B}(K)$ by $\widehat{\pi}(\widehat{x}^\tau) := {}^\circ\pi(x)\!\restriction_K$, $x \in \mathrm{Fin}(\mathcal{M})$.

Then it is readily seen that $\left(\widehat{\pi}, K\right)$ is a cyclic GNS representation of $\widehat{\mathcal{M}}^\tau$ corresponding to the tracial state $\widehat{\tau}$. In particular, it is faithful, *i.e.* $\widehat{\pi}$ is an isometric *isomorphism into $\mathcal{B}(K)$.

Suppose $\left\{\widehat{a_i^\tau}\right\}_{i \in I} \subset \widehat{\mathcal{M}}^\tau$ is a WOT-convergent net. By the Uniform Boundedness Principle (Thm. 2.4), $\left\{\widehat{a_i^\tau}\right\}_{i \in I}$ is bounded in norm. So we can assume that $\left\{\widehat{a_i^\tau}\right\}_{i \in I} \subset \bar{B}_{\widehat{\mathcal{M}}^\tau}$. Moreover, we could let $\{a_i\}_{i \in I} \subset \bar{B}_\mathcal{M}$.

For any $c \in \mathrm{Fin}(\mathcal{M})$ and $\xi, \eta \in \mathrm{Fin}(\mathcal{M}) \subset H$, we have

$$\langle \widehat{\pi}(\widehat{c}^\tau)(\widehat{\xi}), \widehat{\eta}\rangle_K \approx \langle \pi(c)(\xi), \eta\rangle_H \leq \|c\xi\|_\tau \|\eta\|_\tau \leq \|\xi\|_\mathcal{M} \|\eta\|_\tau \|c\|_\tau$$

$$\approx {}^\circ \|\xi\|_\mathcal{M} \|\widehat{\eta}\|_K \sqrt{\widehat{\tau}((\widehat{c}^\tau)^* (\widehat{c}^\tau))},$$

where (3.7) is use in the last inequality. Observe that we get an equality between the two ends in the above when $\xi = \eta = 1$.

Hence, to show that $\left\{\widehat{a_i^\tau}\right\}_{i \in I} \subset \widehat{\mathcal{M}}^\tau$ WOT-converges in $\widehat{\mathcal{M}}^\tau$, it suffices to show that $\lim\limits_{i,j \to \infty} \widehat{\tau}\!\left(((\widehat{(a_i - a_j)^\tau})^* ((\widehat{a_i - a_j)^\tau}))\right) = 0$, *i.e.*

$$\lim\limits_{i,j \to \infty} {}^\circ\!\left(\tau((a_i - a_j)^* (a_i - a_j))\right) = 0.$$

So we can assume that I is countable and $\{a_i\}_{i \in I}$ extends to some internal $\{a_i\}_{i \in J} \subset \bar{B}_\mathcal{M}$. Let $a := a_j$ for any choice of $j \in J$ such that $\forall i \in I \, (j \geq i)$, which exists by saturation. Then $a \in \bar{B}_\mathcal{M}$, so $\widehat{a}^\tau \in \widehat{\mathcal{M}}^\tau$. By saturation, we have $\lim\limits_{i \in I} {}^\circ\!\left(\tau((a - a_i)^* (a - a_i))\right) = 0$.

Therefore $\mathrm{WOT} - \lim\limits_{i \in I} \widehat{a_i^\tau} = \widehat{a}^\tau \in \widehat{\mathcal{M}}^\tau$.

So we conclude that $\widehat{\mathcal{M}}^\tau$ is WOT-closed and hence forms a von Neumann algebra. $\qquad\square$

Observe that if τ is a faithful tracial state on a unital C^*-algebra $\mathcal{M}$, then the above gives an isometric *isomorphism of $\mathcal{M}$ into $\widehat{{}^*\mathcal{M}^\tau}$. But in general the latter is much larger than the universal enveloping von Neumann algebra of $\mathcal{M}$.

3.4.3   *Weak nonstandard hulls and biduals*

We continue to work with a unital C^*-algebra $\mathcal{M}$ identified with its image in $\mathcal{B}(H)$ given by the universal representation. Beware that although this identification is helpful for the clarity of presentation, it may cause technical confusion at times.

The main result here is to use the weak nonstandard hull construction (see p.123) to show that the universal enveloping von Neumann algebra $\{\mathcal{M}\}''$ is isometrically isomorphic to the bidual $\mathcal{M}''$.

Note that this does not mean that a von Neumann algebra $\mathcal{M}$ is a reflexive Banach space, as the canonical evaluation map given in Prop. 2.23 needs not be preserved by the identification of $\mathcal{M}$ as a copy in $\mathcal{B}(H)$.

Since $\mathcal{M} \subset \mathcal{B}(H)$, for the $\omega_{\xi,\eta}$ defined on p.251, by a slight abuse of notation, the restriction $\omega_{\xi,\eta} \upharpoonright_{\mathcal{M}}$ is still denoted by $\omega_{\xi,\eta}$ and we also write $\omega_{\xi,\eta} \in \mathcal{M}'$.

Proposition 3.23. *Each $\phi \in \mathcal{M}'$ is a linear combination of some $\omega_{\xi,\xi}$ for some $\xi \in H$.*

Proof.   By Prop. 3.14, ϕ is a linear combination of positive functionals in $\mathcal{M}'$, hence is a linear combination of states.

By the GNS construction (Thm. 3.19) and Thm. 3.20, each $\theta \in \mathbb{S}(\mathcal{M})$ is the same as the mapping $\mathcal{M} \ni x \mapsto \langle x(\xi), \xi \rangle$ for some $\xi \in H$. $\square$

Because of Prop. 3.23, each $\phi \in \mathcal{M}'$ is also identified with a bounded linear functional on $\mathcal{B}(H)$ given by the corresponding $\omega_{\xi,\xi}$'s when the latter are regarded as bounded linear functional on $\mathcal{B}(H)$. For each $\phi \in \mathcal{M}'$ we still use the same symbol ϕ to denote the corresponding element in $\mathcal{B}(H)'$.

Note that by Prop. 3.23, the weak topology on $\mathcal{M}'$ is generated by the seminorms $p_{\xi,\xi}$, $\xi \in H$. Consequently, for $a \in {}^*\mathcal{M}$, we have

$$\forall \xi \in H \left(\omega_{\xi,\xi}(a) \approx 0 \right) \quad \text{iff} \quad a \approx_w 0 \quad \text{iff} \quad \forall \xi, \eta \in H \left(\omega_{\xi,\eta}(a) \approx 0 \right).$$

In the following, $\widehat{{}^*\mathcal{M}^w}$ denote the weak nonstandard hull of $\mathcal{M}$ (as a Banach space) and $\widehat{a}$ its elements, where $a \in \mathrm{Fin}_w({}^*\mathcal{M})$.

We regard $\mathcal{M}'' = \widehat{{}^*\mathcal{M}^w}$ by Thm. 2.13. Then an embedding

$$\pi : \mathcal{M}'' \to \mathcal{B}(H)$$

is defined as follows.

Fix $a \in \mathrm{Fin_w}(\,^*\mathcal{M})$. For $\xi \in H$, the mapping

$$H \ni \zeta \mapsto {}^\circ\langle \zeta, a(\xi) \rangle \in \mathbb{C}$$

is bounded linear functional on H, so by the Riesz Representation Theorem
(Thm. 2.28), there is a unique $\eta \in H$ such that $\langle \zeta, \eta \rangle = {}^\circ\langle \zeta, a(\xi) \rangle$.

If $c \in \,^*\mathcal{M}$ and $c \approx_w a$, then $\langle \zeta, c(\xi) \rangle \approx \langle \zeta, a(\xi) \rangle$.

So the mapping $\pi(\hat{a}) : H \to \mathbb{C}$ taking the above $\xi \in H$ to the unique
$\eta \in H$ is well-defined. *i.e.* for any $\zeta \in H$,

$$\langle \zeta, \pi(\hat{a})(\xi) \rangle = {}^\circ\langle \zeta, a(\xi) \rangle = \overline{{}^\circ\langle a(\xi), \zeta \rangle} = \overline{{}^\circ \omega_{\xi,\zeta}(a)}. \tag{3.8}$$

From (3.8), one sees that $\pi(\hat{a})$ is a bounded linear operator on H, *i.e.*
$\pi(\hat{a}) \in \mathcal{B}(H)$. Moreover, one also sees that π is injective: Suppose $a, c \in$
$\mathrm{Fin_w}(\,^*\mathcal{M})$ are such that $\pi(\hat{a}) = \pi(\hat{c})$. Then for any $\xi, \zeta \in H$,

$$\overline{\omega_{\xi,\zeta}(a)} \approx \langle \zeta, \pi(\hat{a})(\xi) \rangle = \langle \zeta, \pi(\hat{c})(\xi) \rangle \approx \overline{\omega_{\xi,\zeta}(c)}$$

so we have $a \approx_w c$ by Prop. 3.23, *i.e.* $\hat{a} = \hat{c}$.

Note also that by (3.8) we have:

$$\forall \xi, \zeta \in H \left(\omega_{\xi,\zeta}\big(\pi(\hat{a})\big) = \langle \pi(\hat{a})(\xi), \zeta \rangle = \overline{\langle \zeta, \pi(\hat{a})(\xi) \rangle} \approx \omega_{\xi,\zeta}(a) \right). \tag{3.9}$$

Theorem 3.26. *Let $\mathcal{M}$ be an unital C^*-algebra identified with its copy in
$\mathcal{B}(H)$ given by the universal representation.*

The there is an embedding $\pi : \mathcal{M}'' \to \mathcal{B}(H)$ satisfying the following:

(i) *π is linear and injective.*

(ii) *π is the identity mapping on $\mathcal{M}$. (As identified in $\mathcal{B}(H)$.)*

(iii) *π is an isometry.*

(iv) *$\pi[\mathcal{M}'']$ forms a von Neumann subalgebra of $\mathcal{B}(H)$.*

Moreover, $\pi[\mathcal{M}''] = \overline{\mathcal{M}}^{\text{WOT}}$.

Proof. Let $\pi : \mathcal{M}'' \to \mathcal{B}(H)$ be defined in the above discussion.

(i): π is clearly linear and we have just shown that it is injective.

(ii): For $a \in \mathcal{M} \subset \mathcal{B}(H)$, we have by (3.9) for all $\xi, \zeta \in H$ that
$\omega_{\xi,\zeta}\big(\pi(\hat{a})\big) \approx \omega_{\xi,\zeta}(a)$, hence $\omega_{\xi,\zeta}\big(\pi(\hat{a})\big) = \omega_{\xi,\zeta}(a)$ since both are in $\mathbb{C}$.
Therefore $\pi(a) = a$.

(iii): Let $a \in \mathrm{Fin_w}(\,^*\mathcal{M})$ and $r := \|\pi(\hat{a})\|_{\mathcal{B}(H)}$. Then

$$r = \sup_{\xi,\zeta \in S_H} \langle \xi, \pi(\hat{a})(\zeta) \rangle = \sup_{\xi,\zeta \in S_H} {}^\circ\langle \xi, a(\zeta) \rangle \tag{3.10}$$

$$= \sup_{\xi,\zeta \in S_H} {}^\circ |\omega_{\zeta,\xi}(a)| \leq \|\hat{a}\|_{\mathcal{M}''}.$$

For each internal subspace $K \subset {}^*H$, we define $\Theta(K) := \sup\limits_{\xi, \zeta \in S_K} |\omega_{\xi,\zeta}(a)|$.

Then by (3.10), for any $\epsilon \in \mathbb{R}^+$ and finite dimensional subspace $H_0 \subset H$, there is an internal subspace $K \subset {}^*H$ such that $H_0 \subset K \subset {}^*H$ and $|\Theta(K) - r| < \epsilon$. Hence, by saturation, we can fix some internal subspace $K \subset {}^*H$ such that $H \subset K$ and $\Theta(K) \approx r$.

Let $p \in {}^*\mathcal{B}(H)$ be the projection of ${}^*H$ onto K. Then for all $\xi, \zeta \in H$, we have $\omega_{\xi,\zeta}(a) = \omega_{\xi,\zeta}(ap)$, hence $\phi(a) = \phi(ap)$ for all $\phi \in \mathcal{M}'$, by Prop. 3.23. Note that here ϕ is simultaneously regarded as an element in $\mathcal{B}(H)'$ via the canonical linear combination of positive functionals.

Therefore

$$\|\widehat{a}\|_{\mathcal{M}''} = \sup\limits_{\phi \in S_{\mathcal{M}'}} {}^\circ|\phi(a)| = \sup\limits_{\phi \in S_{\mathcal{M}'}} {}^\circ|\phi(ap)| \lesssim \|ap\|_{{}^*\mathcal{B}(H)}.$$

Observing that $\|ap\|_{{}^*\mathcal{B}(H)} = \Theta(K)$, we therefore have

$$\|\widehat{a}\|_{X''} \lesssim \|ap\|_{{}^*\mathcal{B}(H)} = \Theta(K) \approx r. \tag{3.11}$$

Therefore (iii) follows from (3.10) and (3.11).

(iv): First show that $\pi[\mathcal{M}'']$ is closed under product. *i.e.* for any given $a, b \in \mathrm{Fin}_w({}^*\mathcal{M})$, we need to show $\pi(\widehat{a})\pi(\widehat{b}) \in \pi[\mathcal{M}'']$.

Consider a mapping that takes $\omega_{\xi,\zeta}$, where $\xi, \zeta \in H$, to $\langle \xi, \pi(\widehat{a})\pi(\widehat{b})(\zeta) \rangle$. Then by Prop. 3.23, this mapping extends to a linear mapping $\mathcal{M}' \to \mathbb{C}$. It is easy to see that this mapping is bounded, hence belong to $\mathcal{M}''$, therefore it is given by $\widehat{c}^*$ for some $c \in \mathrm{Fin}_w({}^*\mathcal{M})$. As a result, for all $\xi, \zeta \in H$,

$$\langle \xi, \pi(\widehat{a})\pi(\widehat{b})(\zeta) \rangle = \widehat{c}^*(\omega_{\xi,\zeta}) \approx \omega_{\xi,\zeta}(c^*) = \langle c^*(\xi), \zeta \rangle = \langle \xi, c(\zeta) \rangle \rangle \approx \langle \xi, \pi(\widehat{c})(\zeta) \rangle,$$

therefore $\pi(\widehat{a})\pi(\widehat{b}) = \pi(\widehat{c}) \in \pi[\mathcal{M}'']$.

Next show that $\pi[\mathcal{M}'']$ is closed under involution.

Let $a \in \mathrm{Fin}_w({}^*\mathcal{M})$. Then for any $\xi, \zeta \in H$,

$$\left\langle \xi, \left(\pi(\widehat{a})\right)^*(\zeta) \right\rangle = \langle \pi(\widehat{a})(\xi), \zeta \rangle = \overline{\langle \zeta, \pi(\widehat{a})(\xi) \rangle} \approx \overline{\langle \zeta, a(\xi) \rangle}$$
$$= \langle a(\xi), \zeta \rangle = \langle \xi, a^*(\zeta) \rangle \approx \langle \xi, \pi(\widehat{a^*})(\zeta) \rangle.$$

therefore $\left(\pi(\widehat{a})\right)^* = \pi(\widehat{a^*}) \in \pi[\mathcal{M}'']$.

Finally show that $\pi[\mathcal{M}'']$ is WOT-closed.

We do this by showing that $\pi[\mathcal{M}''] = \overline{\mathcal{M}}^{\mathrm{WOT}}$. So let $c \in \mathcal{B}(H)$ such that $c \in \overline{\mathcal{M}}^{\mathrm{WOT}}$. Then for any $\epsilon \in \mathbb{R}^+$ and $\xi_1, \ldots, \xi_n, \zeta_1, \ldots, \zeta_n \in H$, $n \in \mathbb{N}$, there is $x \in \mathcal{M}$ such that $\bigwedge\limits_{i=1}^{n} |\omega_{\xi_i,\zeta_i}(c - x)| < \epsilon$.

By saturation, there is $a \in {}^*\mathcal{M}$ such that $\omega_{\xi,\zeta}(a) \approx \omega_{\xi,\zeta}(c)$ for all $\xi, \zeta \in H$. Note in particular that $\|a\|_w \approx \|c\|_{\mathcal{B}(H)}$, so $a \in \mathrm{Fin}_w({}^*\mathcal{M})$.

Consequently, $\forall \xi, \zeta \in H \left(\langle c(\xi), \zeta \rangle = \langle \pi(\widehat{a})(\xi), \zeta \rangle \right)$.

Therefore $c = \pi(\widehat{a}) \in \pi[\mathcal{M}'']$. $\square$

In particular, Thm. 3.26 says that the weak nonstandard hull of a unital C^*-algebra always forms a von Neumann algebra.

To rephrase the results in Thm. 3.26, we state the following:

Corollary 3.27. (Sherman-Takeda Theorem) *Let $\mathcal{M}$ be a unital C^*-algebra. Then the bidual $\mathcal{M}''$ forms a unital C^*-algebra so that the von Neumann algebra generated by $\mathcal{M}$ in its universal representation is isometrically *isomorphic to $\mathcal{M}''$.* $\square$

So, roughly speaking, a von Neumann algebra $\mathcal{M}$ is a unital C^*-algebra coinciding with its bicommutant as well as coinciding with its bidual, *i.e.*

$$\text{`` } \mathcal{M} = \{\mathcal{M}\}'' = \mathcal{M}'' \text{ ''}.$$

Consequently a von Neumann algebra $\mathcal{M}$ always has a predual given by $\mathcal{M}'$.

The following shows that for a normal element a in a unital C^*-algebra, we can extend the continuous functional calculus to Borel functional calculus by identifying in the generated von Neumann algebra an element naturally corresponding to $f(a)$, for each Borel function f on $\sigma(a)$.

Recall $B(\Omega)$, the unital C^*-algebra of bounded Borel functions, given in Example 3.11.

Corollary 3.28. (The Borel Functional Calculus) *Let $\mathcal{M}$ be a unital C^*-algebra identified with its universal representation in $\mathcal{B}(H)$. Let $\tilde{\mathcal{M}}$ denote the von Neumann algebra generated by $\mathcal{M}$ in $\mathcal{B}(H)$.*

Let $a \in \mathcal{M}$ be a normal element and $\Omega := \sigma(a)$.

*Then there is $\pi : B(\Omega) \to \tilde{\mathcal{M}}$ which is a *isomorphism into $\tilde{\mathcal{M}}$ such that $\pi \restriction_{C(\Omega)}$ coincides with the continuous functional calculus given in Cor. 3.18.*

Proof. By von Neumann's Bicommutant Theorem (Thm. 3.24) and Thm. 3.26, we make the identification

$$\tilde{\mathcal{M}} = \overline{\mathcal{M}^{\mathrm{WOT}}} = \mathcal{M}'' \subset \mathcal{B}(H).$$

By the Riesz Representation Theorem (Thm. 2.14), we identify $C(\Omega)'$ with $\mathbb{M}(\Omega)$ (the space of σ-additive complex Borel measures on Ω), hence we also identify $\overline{C(\Omega)^{\mathrm{WOT}}} = C(\Omega)''$ with $\mathbb{M}(\Omega)'$. In particular, $\mathbb{M}(\Omega)' \subset \tilde{\mathcal{M}}$.

Then we only need to define $\pi : B(\Omega) \to \mathbb{M}(\Omega)'$.

For each $f \in B(\Omega)$, let $\pi(f)$ be the following mapping:

$$\pi(f) : \mu \mapsto \int_{\Omega} f d\mu, \quad \text{where} \quad \mu \in \mathbb{M}(\Omega).$$

It can be seen that $\pi(f) \in \mathbb{M}(\Omega)'$ and π corresponds to a *isomorphism of $B(\Omega)$ into $\tilde{\mathcal{M}}$.

Note that when $f \in C(\Omega)$, $\pi(f)$ is just the evaluation mapping from the canonical embedding of $C(\Omega)$ into $C(\Omega)''$. From this, it can be seen that $\pi \restriction_{C(\Omega)}$ coincides with the continuous functional calculus.   □

In the above, we simply regard $f(a)$ as the element in $\tilde{\mathcal{M}}$ given by the *isomorphism.

So we see that a von Neumann algebra is rich enough to contain all Borel functional calculus of its normal elements.

As a corollary to Cor. 3.28, by a proof similar to Thm. 3.23, we obtain the following:

Corollary 3.29.   (The Nonstandard Borel Functional Calculus) *Let $\mathcal{M}$ be a unital C^*-algebra and $\tilde{\mathcal{M}}$ be its enveloping von Neumann algebra. Let $\widehat{\tilde{\mathcal{M}}}$ denote the norm-nonstandard hull of $\tilde{\mathcal{M}}$. Suppose $a \in \mathcal{M}$ is normal.*

Then $F(a) \in \widehat{\tilde{\mathcal{M}}}$ is naturally defined for each $F \in \mathrm{Fin}\big(\,^*B(\sigma(a))\big)$. Moreover, when $^\circ F = f \in B(\sigma(a))$, it agrees with $f(a)$ given by the Borel functional calculus.   □

Cor. 3.29 appears to provide the ultimate functional calculus a normal element could possibly get.

We end this subsection by an application of the Borel functional calculus.

Example 3.19.   (Spectral measure and spectral integral.)

Let $\mathcal{M}$ be a unital C^*-algebra identified with its universal representation in $\mathcal{B}(H)$ and let $a \in \mathcal{M}$ be normal.

So $f(a) \in \mathcal{B}(H)$ is defined by Cor. 3.28 for each $f \in B\big(\sigma(a)\big)$.

Let $\mathbb{B}$ denote the collection of Borel subsets of $\sigma(a)$. Since for each $A \in \mathbb{B}$, the characteristic function $\chi_A \in B\big(\sigma(a)\big)$, we define:

$$p : \mathbb{B} \to \mathrm{Proj}(\mathcal{B}(H)) \quad \text{by} \quad p(A) = \chi_A(a).$$

It is clear that $\chi_A(a) \in \mathrm{Proj}(\mathcal{B}(H))$, as $\chi_A = \bar{\chi}_A = \chi_A^2$.

Note that, by Cor. 3.28, if $\mathcal{M}$ is a von Neumann algebra, then the $\chi_A(a) \in \mathrm{Proj}(\mathcal{M})$, hence $p : \mathbb{B} \to \mathrm{Proj}(\mathcal{M})$.

Note also that $p(\emptyset) = 0$ and $p(\sigma(a)) = 1$. Moreover, it is σ-additive in the sense that if $\{A_n\}_{n\in\mathbb{N}} \subset \sigma(a)$ is a countable family of disjoint subsets, then it is not hard to verify that

$$p\Big(\bigcup_{n\in\mathbb{N}} A_n\Big) = \Big(\text{SOT}-\lim_{n\to\infty} \sum_{m=0}^{n} p(A_m)\Big) \in \text{Proj}(\mathcal{B}(H)).$$

Therefore, p can be regarded as a vector measure, taking values in the Banach space $\mathcal{B}(H)$.

This vector measure p is called the **spectral measure** of a.

Note that for each pair $\xi, \zeta \in H$, we have a σ-additive complex function on $\mathbb{B}$ given by $\omega_{\xi,\zeta} \circ p$.

Furthermore, for any $f \in B(\sigma(a))$ and $\xi \in H$, the mapping

$$H \ni \zeta \mapsto \int_{\sigma(a)} f(x)\, d(\omega_{\xi,\zeta} \circ p)(x) \in \mathbb{C}$$

is a bounded linear functional on H, hence, by the Riesz Representation Theorem (Thm. 2.28), there is a unique $\eta \in H$ such that

$$\forall \zeta \in H \left(\langle \eta, \zeta \rangle = \int_{\sigma(a)} f(x)\, d(\omega_{\xi,\zeta} \circ p)(x) \right).$$

We define $\pi : B(\sigma(a)) \to \mathcal{B}(H)$ by letting $\pi(f)$ take $\xi \in H$ to the above η.

Also, one can check that (π, H) is a GNS representation of $B(\sigma(a))$.

For each $f \in B(\sigma(a))$, $\pi(f)$ is called the **spectral integral** of f *w.r.t.* p and is denoted by

$$\int f\, dp.$$

Note that $f(a) = \int f\, dp$, as it is clear from the definition that this holds for f being the characteristic function of a Borel subset of $\sigma(a)$.

In particular, we get $a = \int x\, dp(x)$. This can be viewed as the analog of the diagonalization of a normal matrix.

In the case when a Hilbert space K is given and a normal operator $N \in \mathcal{B}(K)$ is under consideration, one applies the above to $\mathcal{M} := \mathcal{B}(K)$ and retrieves the $\int f\, dp$, i.e. $f(a)$, where $f \in B(\sigma(a))$, from the above $\mathcal{B}(H)$ to $\mathcal{B}(K)$ by using Prop. 3.17 with a GNS representation of $\mathcal{M}$ provided by the identity mapping on itself. In other words, every normal operator admits the spectral measure and spectral integrals.

The spectral integral representation is a powerful tool in the study of operator theory. $\square$

3.4.4   *Notes and exercises*

Von Neumann algebras were first called ***rings of operators*** and were intensively studied and developed by von Neumann and Murray in the 1940's.

Comparing with general C^*-algebras, von Neumann algebras form closer link to the underlying Hilbert space, and, as such, they appear to be more applicable in modeling observables in quantum physics.

Let $\mathcal{M}$ be a commutative von Neumann algebra and $\phi \in \mathcal{M}' \setminus \{0\}$ be positive. Then $\mathcal{M}$ has a GNS representation (π, H) with a cyclic element $\xi \in H$ corresponding to ϕ as in the GNS construction (Thm. 3.19).

Let $\tilde{\mathcal{M}} := \pi[\mathcal{M}]$, a commutative unital C^*-subalgebra of $\mathcal{B}(H)$.

Let $\tilde{\phi} : \tilde{\mathcal{M}} \to \mathbb{C}$ be given by $\pi(a) \mapsto \langle \pi(a)(\xi), \xi \rangle$, $a \in \mathcal{M}$, then it can be seen that $\tilde{\phi} \in \tilde{\mathcal{M}}'$ and is positive.

By the Gelfand transform (Thm. 3.14), there is some compact space Ω and $\gamma : \tilde{\mathcal{M}} \to C(\Omega)$, a *isomorphism onto $C(\Omega)$ given by the Gelfand transform. Thus $(\tilde{\phi} \circ \gamma^{-1}) \in C(\Omega)'$, so, by the Riesz Representation Theorem (Thm. 2.14), there is a σ-additive complex Borel measures μ on Ω such that

$$\forall f \in C(\Omega) \left(\tilde{\phi}(\gamma^{-1}(f)) = \int_\Omega f \, d\mu \right).$$

Since $\tilde{\phi}$ is positive, μ can be taken to be a positive measure.

For $f \in L_\infty(\mu)$, let $m_f \in \mathcal{B}(L_2(\Omega))$ denote the multiplication operator in Example 3.18. Note that $C(\Omega) \subset L_\infty(\mu)$ by identifying elements with their equivalence classes.

Now define $\nu : \mathcal{M} \to \mathcal{B}(L_2(\mu))$ by $\nu(a) := m_{(\gamma \circ \pi)(a)}$. It is not hard to check that $(\nu, L_2(\mu))$ is a GNS representation of $\mathcal{M}$.

We claim that (π, H) and $(\nu, L_2(\mu))$ are unitarily equivalent.

For this we define a Hilbert space isomorphism $U : H \to L_2(\mu)$. By ξ being cyclic, it suffices to define U on the dense subset $\{a(\xi) \mid a \in \tilde{\mathcal{M}}\}$ of H. We let $U(a(\xi)) := \gamma(a)$, $a \in \tilde{\mathcal{M}}$.

Clearly U is linear.

For isometry, let $a \in \tilde{\mathcal{M}}$, then

$$\|a(\xi)\|_H^2 = \langle a(\xi), a(\xi) \rangle_H = \langle (a^* a)(\xi), \xi \rangle_H = \int_\Omega \gamma(a^* a) \, d\mu$$

$$= \int_\Omega \gamma(a^*) \, \gamma(a) \, d\mu = \int_\Omega \overline{\gamma(a)} \, \gamma(a) \, d\mu = \int_\Omega |\gamma(a)|^2 \, d\mu.$$

As $C(\Omega)$ is a dense subset of $L_2(\mu)$ under the L_2-norm, U is a surjective, hence a Hilbert space isomorphism.

So we have the unitary equivalence of (π, H) and $(\nu, L_2(\mu))$ given by $\nu(\cdot) = (U \circ \pi(\cdot) \circ U^{-1})$.

Moreover, it can be verified that the mapping

$$\mathcal{B}(H) \ni a \mapsto (U \circ a \circ U^{-1}) \in \mathcal{B}(L_2(\mu))$$

is a homeomorphism $w.r.t.$ the SOT-topology, so $\nu[\mathcal{M}]$ is a von Neumann subalgebra of $\mathcal{B}(L_2(\mu))$. Furthermore, $\{m_f \mid f \in C(\Omega)\}$ is SOT-dense in $L_\infty(\Omega)$, so we conclude that $\nu[\mathcal{M}]$ is isometrically *isomorphic to $L_\infty(\mu)$.

For each $\phi \in \mathcal{M}' \setminus \{0\}$ and cyclic GNS representation (π_ϕ, H_ϕ) of $\mathcal{M}$ corresponding to ϕ, we let the unitarily equivalent GNS representation $(\nu_\phi, L_2(\mu_\phi))$ and the positive measure μ_ϕ on the compact space Ω_ϕ be given as above, resulting a *isomorphic copy $L_\infty(\mu_\phi)$ of $\nu_\phi[\mathcal{M}]$.

Finally, consider $(\nu_\phi, L_2(\mu_\phi))$, where $\phi \in \mathbb{S}(\mathcal{M})$. So μ_ϕ is a probability measure.

Define $(\nu, H) := \bigoplus_{\phi \in \mathbb{S}(\mathcal{M})} (\nu_\phi, L_2(\mu_\phi))$. We let $\bigoplus_{\phi \in \mathbb{S}(\mathcal{M})} L_\infty(\mu_\phi)$ be given by the L_∞-direct sum, then it forms a von Neumann algebra. Then as in the proof the GNS Theorem (Thm. 3.20), we conclude that $\mathcal{M}$ is *isomorphic to $\bigoplus_{\phi \in \mathbb{S}(\mathcal{M})} L_\infty(\mu_\phi)$ as a von Neumann algebra.

It is in this sense that classical measure theory can be viewed as the theory of commutative von Neumann algebras. As suggested by the term *noncommutative topology theory* (p.240), one regards the theory of general von Neumann algebra as the **noncommutative measure theory**.

For more on commutative von Neumann algebra, see III.1 in [Takesaki (2002)].

The tracial nonstandard hull construction (Thm. 3.25) is of limited usage only, as some von Neumann algebras fail to admit any tracial states. It is mainly used for a finite von Neumann algebra (to be defined on p.271), which has plenty tracial states, or a von Neumann algebra from a class called type II$_1$, which admits a unique tracial state. See Ch. V. in [Takesaki (2002)] for more on this. For an application of the tracial nonstandard hull construction to hyperfinite dimensional matrix algebras, *cf.* [Hinokuma and Ozawa (1993)].

There is a ultraproduct version of the tracial nonstandard hull construction known among practitioners as the tracial ultraproduct. It has been used at least since [McDuff (1970)], although rarely put in clear details. See [Brown and Ozawa (2008)] for this approach and applications.

EXERCISES

(1) Show by examples that in general the norm topology is strictly stronger than the SOT and the SOT is strictly stronger than WOT.

(2) Show that if the Hilbert space H is infinite dimensional, none of WOT nor SOT are metrizable.

(3) Show that if the Hilbert space H is separable, then the WOT on $\bar{B}_{\mathcal{B}(H)}$ is metrizable. The same conclusion holds for SOT.

(4) Let H be a Hilbert space and $C \subset \bar{B}_{\mathcal{B}(H)}$ be convex. Show that $\overline{C^{\text{WOT}}} = \overline{C^{\text{SOT}}}$.

(5) Verify that the $\mathcal{M}$ given in Example 3.18 is a commutative von Neumann algebra.

(6) Complete all details in the proof of Cor. 3.28.

(7) On a fixed a Hilbert space, among all normal operators, characterize those compact ones using the spectral measures.

(8) Prove Lem. 3.6.

3.5   Some Applications of Projections

Projections not only constitute a von Neumann algebra's soul, they also constitute the von Neumann algebra's reincarnation.

As we have seen already normal elements contribute the most manageable part of a unital C^*-algebra, as they retain enough commutativity to make rich functional calculus and other desirable properties available. On the other hand, they are the ones that admit the spectral integral representation (Example 3.19), which in a sense says that a normal element is an "infinite linear combination of projections". Therefore projections play an important rôle in C^*-algebras.

Projections have been dealt with on a number of occasions: On p.103, in §3.2.1 (p.156: Thm. 2.29, Prop. 2.38, Prop. 2.39 and Cor. 2.31), on p.167 (Prop. 2.47), in §3.3.1 and §3.4.2.

In a sense, what makes von Neumann algebras special is the fact that they are enriched with projections, as this is evident from Example 3.19 and Ex. 3 on p.275.

However, the collection of projections could be rather trivial in a general C^*-algebra, see Ex. 1 on p.275.

In this section we consider projections in the norm-nonstandard hull of a internal C^*-algebra.

3.5.1   *Infinite C^*-algebras*

In this subsection, $\mathcal{M}$ always stands for an internal C^*-algebra.

Recall the partial order $\leq$ on $\mathcal{M}^+$ given on p.230.

Since $\mathrm{Proj}(\mathcal{M}) \subset \mathcal{M}^+$, so, by Thm. 3.9(v) that if $p \in \mathrm{Proj}(\mathcal{M})$ then $(1 - p) \in \mathrm{Proj}(\mathcal{M})$, it follows that 1 is the maximal element and 0 the minimal element in $\big(\mathrm{Proj}(\mathcal{M}), \leq \big)$.

For $p, q \in \mathrm{Proj}(\mathcal{M})$, if $p \leq q$, we also say that p is a **subprojection** of q. This is a natural terminology, as one regard p, q as orthogonal projections in $\mathcal{B}(H)$ given by the universal representation of $\mathcal{M}$, then $p \leq q$ is the same as $p[H] \subset q[H]$.

Proposition 3.24. *Let $p, q \in \mathrm{Proj}(\mathcal{M})$. Then $p \leq q$ iff $p = pq = qp$.*

Proof.   Suppose $p \leq q$. Let $a \in \mathcal{M}^+$ be such that $q = p + a$, then

$$p + a = q = q^2 = p + pa + ap + a^2, \quad i.e. \quad a = pa + ap + a^2,$$
$$i.e. \quad pa = pa + pap + pa^2, \qquad\qquad i.e. \quad pa^2 = -pap,$$
$$i.e. \quad 0 \leq pa^2p = -pap \leq 0, \qquad\qquad \text{hence} \quad pa^2p = 0,$$

where Prop. 3.11(ii) and Thm. 3.17(i) are used for the last inequality.

Consequently, $(ap)^*(ap) = pa^2p = 0$ and $(pa)(pa)^* = pa^2p = 0$, therefore $ap = 0 = pa$ and

$$qp = p + ap = p \quad \text{and} \quad pq = p + pa = p$$

as required.

For the converse, from assumption that $p = pq = qp$ we have $(q - p)^2 = q - p$, i.e. $(q - p) \in \text{Proj}(\mathcal{M}) \subset \mathcal{M}^+$, so $p \leq q$.   □

Given $p, q \in \text{Proj}(\mathcal{M})$, we define the following:

$$(p \wedge q) := pq, \quad (p \vee q) := (p + q - pq), \quad (p \smallsetminus q) := (p - pq).$$

It is clear that if $pq = qp$, then all $(p \wedge q), (p \vee q), (p \smallsetminus q) \in \text{Proj}(\mathcal{M})$. Moreover, the following is easy to check:

Proposition 3.25. *Let $\mathbb{B} \subset \text{Proj}(\mathcal{M})$ be such that $0, 1 \in \mathbb{B}$ and $\mathbb{B}$ commutes with itself, i.e. $\forall p, q \in \mathbb{B}\,(pq = qp)$. Suppose $\mathbb{B}$ is closed under $\wedge, \vee$ and $\smallsetminus$. Then $(\mathbb{B}, 0, 1, \wedge, \vee, \smallsetminus)$ forms a Boolean algebra.*
Moreover, $p \leq q$ iff $p = (p \wedge q)$, where $p, q \in \mathbb{B}$.   □

Example 3.20. Let Ω be a compact space with a σ-additive complex Borel measure μ on $\mathcal{B}(\Omega)$, the Borel subsets of Ω. Let $\mathcal{M} := \{m_f \mid f \in L_\infty(\Omega)\}$ be as in Example 3.18.

Consider $\mathbb{B} := \{m_f \mid f = \chi_A \text{ where } A \in \mathcal{B}(\Omega)\}$. Then $\mathbb{B} \subset \text{Proj}(\mathcal{M})$ and forms a Boolean algebra—in fact a σ-algebra.

Note that for $A_1, A_2 \in \mathcal{B}(\Omega)$, $\mu(A_1 \triangle A_2) = 0$ iff $m_{\chi_{A_1}} = m_{\chi_{A_2}}$.

It is easily seen that $\mathbb{B}$ is isomorphic to the measure algebra given by μ.

See also Problem 23 on p.301.   □

Given $p, q \in \text{Proj}(\mathcal{M})$, if there is a partial isometry $c \in \mathcal{M}$ satisfying that $p = c^*c$ and $q = cc^*$, then we write $p \sim q$ and say that the projections p, q are **Murray-von Neumann equivalent**. This is an important equivalence relation, especially in the context of von Neumann algebras.

BANACH ALGEBRAS 269

Note that if $p \in \text{Fin}(\mathcal{M})$ is such that $p \sim 0$, then there is a partial isometry $a \in \mathcal{M}$ with $a^* a = p$ and $aa^* = 0$, but the latter implies $\|a\|^2 = 0$, i.e. $a = 0$ and thus $p = 0$. Hence $p \sim 0$ iff $p = 0$.

Another common equivalence relation is the following: $p, q \in \text{Proj}(\mathcal{M})$ are **unitarily equivalent** if there is $u \in \mathcal{U}(\mathcal{M})$ such that $upu^* = q$. Note this implies Murray-von Neumann equivalence: If $upu^* = q$, then we have

$$(up)(up)^* = q \quad \text{and} \quad (up)^*(up) = pu^* up = p^2 = p, \quad i.e. \ p \sim q.$$

Other than the ordering $\leq$ on $\text{Proj}(\mathcal{M})$, we mention another partial ordering which is finer than $\leq$: Let $p, q \in \text{Proj}(\mathcal{M})$. We define

$$p \precsim q \quad \text{iff} \quad p = (c^* c) \leq q \quad \text{for some partial isometry } c \in \mathcal{M}.$$

If $p \precsim q$, we say that p is **subordinate to** q.

In general, $p \precsim q \precsim p$ does not imply $p \sim q$. However, we leave it as an exercise to show the following:

Proposition 3.26. *Let $\mathcal{M}$ be a von Neumann algebra.*
Then for any $p, q \in \text{Proj}(\mathcal{M})$, if $p \precsim q \precsim p$ then $p \sim q$. $\qquad\square$

The following is proved from some technical calculation results which we refer the readers to [Blackadar (2006)].

Lemma 3.7. *Suppose that $\mathcal{M}$ is an internal unital C^*-algebra.*

(i) *Let $p, q \in \text{Proj}(\mathcal{M})$ such that $p \approx q$. Then $p \sim q$.*
(ii) *Let $a, b \in \text{Proj}(\widehat{\mathcal{M}})$. Then*
   $a \sim b$ iff there are $p, q \in \text{Proj}(\mathcal{M})$, $\widehat{p} = a$, $\widehat{q} = b$ and $p \sim q$.
(iii) *Let $a, b \in \text{Proj}(\widehat{\mathcal{M}})$. Then*
   $a \leq b$ iff there are $p, q \in \text{Proj}(\mathcal{M})$, $\widehat{p} = a$, $\widehat{q} = b$ and $p \leq q$.

Proof. (i): Apply II.3.3.4 in [Blackadar (2006)], which is a more general result.

(ii): If $a \sim b$, then there is a partial isometry $\widehat{a} \in \widehat{\mathcal{M}}$ such that $\widehat{a^* a} = (\widehat{a})^* \widehat{a} = a$ and $\widehat{aa^*} = \widehat{a}(\widehat{a})^* = b$.

By Thm. 3.22(vii), we can assume that a is a partial isometry in $\mathcal{M}$. By Thm. 3.9(viii), a^* is also a partial isometry.

Now let $p = a^* a$ and $q = aa^*$, then $p, q \in \text{Proj}(\mathcal{M})$ with $p \sim q$.

Moreover, $\widehat{p} = \widehat{a^* a} = a$ and $\widehat{q} = \widehat{aa^*} = b$

The other direction is trivial by Thm. 3.22(vi) and (vii).

(iii): For the nontrivial direction, by Thm. 3.22(vi), let $p, q_0 \in \text{Proj}(\mathcal{M})$ such that $\widehat{p} = a \leq b = \widehat{q_0}$.

Then Prop. 3.24 implies that $\widehat{q_0}\,\widehat{p} = \widehat{p}$, *i.e.* $q_0 p \approx p$. By II.3.3.5 in [Blackadar (2006)] and saturation, there is $q \in \mathrm{Proj}(\mathcal{M})$ with the property that $p \le q$ and $q \approx q_0$, hence $\widehat{q} = b$. $\qquad\square$

Classification of C^*-algebras according to the types of projections has been studied. Such classification is better understood and well-developed in the case of von Neumann algebras. (See [Blackadar (2006)] and [Takesaki (2002)].)

We only mention two types here and their relation to the nonstandard hulls of C^*-algebras.

A unital C^*-algebra $\mathcal{M}$ is said to be **infinite** if

$$\exists p \in \mathrm{Proj}(\mathcal{M})\, \big((p \ne 1) \wedge (p \sim 1)\big).$$

(Called **finite**, if not infinite.)

$\mathcal{M}$ is said to be **properly infinite** if

$$\exists p, q \in \mathrm{Proj}(\mathcal{M})\, \big((pq = 0) \wedge (p \sim 1) \wedge (q \sim 1)\big).$$

Note that being properly infinite implies being infinite. Also, the above implies $qp = 0$ since $(qp)^* = p^* q^* = pq = 0$.

Theorem 3.27. *Let $\mathcal{M}$ be an unital C^*-algebra. Then*

(i) $\mathcal{M}$ *is infinite iff* $\widehat{\mathcal{M}}$ *is infinite.*
(ii) $\mathcal{M}$ *is properly infinite iff* $\widehat{\mathcal{M}}$ *is properly infinite.*

Proof. (i): Suppose $\mathcal{M}$ is infinite, let $p \in \mathrm{Proj}(\mathcal{M})$ be such that $p \sim 1$ but $p \ne 1$. Then, by Lem. 3.7, $\widehat{p} \sim 1$. Also, by Cor. 3.12, $p \not\approx 1$, *i.e.* $\widehat{p} \ne 1$. Hence $\widehat{\mathcal{M}}$ is infinite.

Conversely, suppose $\widehat{\mathcal{M}}$ is infinite and let $\widehat{p} \in \mathrm{Proj}(\widehat{\mathcal{M}})$ be such that $1 \ne \widehat{p} \sim 1$.

Then, by Lem. 3.7, we can assume $p \in \mathrm{Proj}(\mathcal{M})$ with some $q \in \mathrm{Proj}(\mathcal{M})$ such that $q \approx 1$ and $q \sim p$. Again, by Cor. 3.12, $q = 1$ and since $p \not\approx 1$, we have $p \ne 1$. Therefore $1 \ne p \sim 1$, so $\mathcal{M}$ is infinite.

(ii): One direction is straightforward. For the other, suppose

$$\widehat{p_0} \sim 1 \sim \widehat{q_0} \quad \text{and} \quad \widehat{p_0}\,\widehat{q_0} = 0,$$

where we can assume by Thm. 3.22(vi) that $p_0, q_0 \in \mathrm{Proj}(\mathcal{M})$.

First note that $p_0 \approx 1 \approx q_0$, so we have by Lem. 3.7(i) that

$$p_0 \sim 1 \sim q_0.$$

So $(1 - \widehat{p_0})\widehat{q_0} = \widehat{q_0}$, hence, as $(1 - \widehat{p_0}) \in \mathrm{Proj}(\widehat{\mathcal{M}})$, it follows from Prop. 3.24 that $\widehat{q_0} \le (1 - \widehat{p_0}) = \widehat{1 - p_0}$.

Since $(1 - p_0) \in \mathrm{Proj}(\mathcal{M})$, so, by Lem. 3.7, there are $r, q \in \mathrm{Proj}(\mathcal{M})$ such that $\widehat{q} = \widehat{q}_0$, $\widehat{r} = \widehat{1 - p_0}$ and $q \leq r$.

Let $p := (1 - r) \in \mathrm{Proj}(\mathcal{M})$, then

$$\widehat{p} = \widehat{1 - r} = \widehat{p}_0 \quad \text{and} \quad q \leq (1 - p).$$

Consequently, $q = (1 - p)q$, by Prop. 3.24, hence $pq = 0$.

From $\widehat{p} = \widehat{p}_0$ and $\widehat{q} = \widehat{q}_0$, we have $p \approx p_0$ and $q \approx q_0$. Therefore, by Lem. 3.7(i), $p \sim p_0$ and $q \sim q_0$. In particular, $p \sim 1 \sim q$ and $pq = 0$.

So $\mathcal{M}$ is properly infinite. $\square$

3.5.2 P^*-algebras

Our interest here is to have a quick look at a class of unital C^*-algebras with some features close to von Neumann algebras but is stable under the norm-nonstandard hull construction. This is necessary, as Cor. 3.25 shows that the norm-nonstandard hull of a von Neumann algebra can easily fail to be von Neumann.

A unital C^*-algebra is called a P^*-**algebra** if $\mathrm{Re}(\mathcal{M}) =$

$$\overline{\mathrm{Lin}}\Big(\Big\{ \sum_{i=0}^{n} \alpha_i p_i \,|\, \alpha_i \in \mathbb{R},\, p_i \in \mathrm{Proj}(\mathcal{M}),\, p_i p_j = 0 \text{ if } i \neq j,\, 0 \leq i, j \leq n \in \mathbb{N} \Big\} \Big).$$

That is, self-adjoint elements in $\mathcal{M}$ are the norm-limit of real linear combinations of mutually orthogonal sequences of projections. Note that this is not the same as saying the $\mathcal{M}$ is the norm-closure of the complex linear span of such elements.

Example 3.21.

- Finite dimensional unital C^*-algebras are P^*-algebras.
- By an **AF-algebra**, *i.e.* an *approximately finite dimensional algebra*, we mean a unital C^*-algebra generated as the norm-closure of an increasing sequence of finite dimensional unital C^*-algebras. (So AF-algebras are separable.) We can see that AF-algebras are P^*-algebras. (See [Davidson (1996)] for more on AF-algebras.)
- Von Neumann algebras $\mathcal{M}$ are P^*-algebras. This can be seen from the fact that for each $a \in \mathrm{Re}(\mathcal{M})$, a is given by the spectral integral $\int_{\sigma(a)} x\, dp(x)$, where p is the spectral measure from the Borel subsets of $\sigma(a)$ to $\mathrm{Proj}(\mathcal{M})$. (Example 3.19.) Since $\sigma(a) \subset \mathbb{R}$, by an argument similar to the approximation of an integral by simple functions, we see

that for any $\epsilon \in \mathbb{R}^+$, there are $k_0 < \cdots < k_n$ in $\sigma(a)$, $\in \mathbb{N}$ such that

$$\left\| a - \sum_{0 < i \leq n} k_i \, p\big((k_{i-1}, k_i] \cap \sigma(a)\big) \right\| < \epsilon.$$

Since the $p\big((k_{i-1}, k_i] \cap \sigma(a)\big) \in \mathrm{Proj}(\mathcal{M})$ are mutually orthogonal, we conclude that $\mathcal{M}$ is a P^*-algebra.

- As for an example of a unital C^*-algebra which is not a P^*-algebra, one simply looks for those having few projections: for example, Ex. 1 on p.275. □

The main interest in P^*-algebras is that if $\mathcal{M}$ is such then so is $\widehat{^*\mathcal{M}}$.

Theorem 3.28.   *Let $\mathcal{M}$ be an internal P^*-algebra. Then the norm-nonstandard hull $\widehat{\mathcal{M}}$ is also a P^*-algebra.*

Proof.   Consider $\widehat{a} \in \mathrm{Re}(\widehat{\mathcal{M}})$, where $a \in \mathrm{Fin}(\mathcal{M})$. By Thm. 3.22 (i), we can assume that $a \in \mathrm{Re}(\mathcal{M})$. Then by $\mathcal{M}$ being a P^*-algebra, there are mutually orthogonal $p_n \in \mathrm{Proj}(\mathcal{M})$, $0 \leq n \leq N$ for some $N \in {}^*\mathbb{N}$, such that $a \approx \sum_{0 \leq n \leq N} \alpha_n p_n$ for some $\alpha_n \in {}^*\mathbb{R} \setminus \{0\}$.

So we can replace a by $a := \sum_{0 \leq n \leq N} \alpha_n p_n$.

By Thm. 3.16(vii) and Spectral Mapping Theorem (Cor. 3.7), $\sigma(\alpha_n p_n) \subset \{0, \alpha_n\}$. Since the $\alpha_n p_n$, $0 \leq n \leq N$, are orthogonal with each other, we iterate Cor. 3.16(ii) and obtain

$$\sigma(a) \subset \bigcup_{0 \leq n \leq N} \sigma(\alpha_n p_n) \subset \big(\{0\} \cup \{\alpha_n\}_{0 \leq n \leq N}\big) \subset \big(\{0\} \cup \sigma(a)\big).$$

By Prop. 3.21(ii), ${}^\circ\sigma(a) = \sigma(\widehat{a})$, so

$$\{\alpha_n\}_{0 \leq n \leq N} \subset \mathrm{Fin}({}^*\mathbb{R}).$$

If $0 \in \sigma(a)$, we can add 0 to the list of the α_n's if necessary. So we can assume that

$$\sigma(a) = \{\alpha_n \mid 0 \leq n \leq N\}.$$

Furthermore, by re-ordering the α_n's if necessary, we can further assume that the α_n's are nondecreasing.

Fix $m \in \mathbb{N}^+$. Define $r_0 := \alpha_0$ and, for $0 < n \leq N$, define

$$r_n := \begin{cases} r_{n-1} & \text{if } |r_{n-1} - \alpha_n| \leq m^{-1} \\ \alpha_n, & \text{otherwise} \end{cases}.$$

Then set $a_m = \sum_{0 \leq n \leq N} r_n p_n$.

By the same argument as before, we get:

$$\sigma(a - a_m) \subset \left(\{0\} \cup \{(\alpha_n - r_n)\}_{0 \le n \le N}\right).$$

Note that $a_m \in \mathrm{Fin}(\mathrm{Re}(\mathcal{M}))$ and so is $(a - a_m)$. Therefore Thm. 3.9(vii) implies that

$$\|a - a_m\| = \rho(a - a_m) = \max_{0 \le n \le N} |\alpha_n - r_n| \le m^{-1}.$$

Consequently $\lim_{m \to \infty} \|\widehat{a} - \widehat{a}_m\| = 0$.

So it remains to show that $\widehat{a}_m$ is a linear combination of mutually orthogonal projections.

In the definition of the $\widehat{a}_m$, the distinct elements from the sequence $\{r_n\}_{0 \le n \le N}$ form an increasing sequence in $\sigma(a)$ with step size $> m^{-1}$. As $^\circ\sigma(a) = \sigma(\widehat{a})$ is bounded, $\{r_n\}_{0 \le n \le N}$ contains only finitely many distinct elements. Say there are k of them.

Define a strictly increasing function $\zeta : \{0, \ldots, k\} \to \{0, \ldots, N+1\}$ by $\zeta(0) := 0$, $\zeta(k) := N+1$ and inductively, for $0 < i < k$, let $\zeta(i)$ be the least j so that $r_j > r_{\zeta(i-1)}$.

Next define $q_0 := p_0$ and $q_i := \displaystyle\sum_{j=\zeta(i)}^{\zeta(i+1)-1} p_j$ for $0 < i < k$.

Therefore $a_m = \displaystyle\sum_{0 \le i < k} r_{\zeta(i)} q_i$ and so $\widehat{a}_m = \displaystyle\sum_{0 \le i < k} {}^\circ r_{\zeta(i)} \widehat{q}_i$.

By mutual orthogonality of the p_n's, $q_i \in \mathrm{Proj}(\mathcal{M})$, hence $\widehat{q}_i \in \mathrm{Proj}(\widehat{\mathcal{M}})$. Moreover, since the q_i's are mutually orthogonal, so are the $\widehat{q}_i$'s.

Consequently, $\widehat{a}_m$ is a linear combination of mutually orthogonal projections in $\widehat{\mathcal{M}}$.

Therefore we conclude that $\widehat{\mathcal{M}}$ is a P^*-algebra. $\square$

By Example 3.21, von Neumann algebras are P^*-algebras, so we have:

Corollary 3.30. *Let $\mathcal{M}$ be an internal von Neumann algebra. Then $\widehat{\mathcal{M}}$ is a P^*-algebra.* $\square$

However, by Prop. 3.22, unless an internal P^*-algebra is finite dimensional, $\widehat{\mathcal{M}}$ is never a von Neumann algebra. In particular, together with Example 3.21, the class of P^*-algebras properly extends the class of von Neumann algebras.

As mentioned in Example 3.21, finite dimensional C^*-algebras are P^*-algebras. Hence

Corollary 3.31. *Let $\mathcal{M}$ be a hyperfinite dimensional C^*-algebra. Then $\widehat{\mathcal{M}}$ is a P^*-algebra.* $\square$

3.5.3   *Notes and exercises*

See [Dunford and Schwartz (1988c)] XV.2 and XVII for more on Boolean algebras of projections in the Banach algebra setting and connection to spectral integrals.

Thm. 3.27 and Thm. 3.28 were proved in [Baratella and Ng (2009)], where the notion of P^*-algebra was introduced and more properties about projections in the nonstandard hull of an internal unital C^*-algebra can be found.

We remark that if τ is a tracial state, then for Murray-von Neumann equivalent projections $p \sim q$, we have $\tau(p) = \tau(q)$. In fact the connection between tracial states and projections develops into a dimension theory and a classification of von Neumann algebras. See [Takesaki (2002)] or [Blackadar (2006)] for detail.

EXERCISES

(1) Given an example of an infinite dimensional unital C^*-algebra $\mathcal{M}$ such that $\mathrm{Proj}(\mathcal{M}) = \{0, 1\}$.
(2) Prove Prop. 3.26.
(3) Let $\mathcal{M}$ be a von Neumann algebra, show that $\big(\mathrm{Proj}(\mathcal{M}), \leq \big)$ forms a complete lattice, *i.e.* the supremum and infimum of any subset of $\mathrm{Proj}(\mathcal{M})$ exist in $\mathrm{Proj}(\mathcal{M})$.
(4) Check that the Murray-von Neumann equivalence is indeed an equivalence relation on projections.
(5) A projection p is called *infinite* if there is a projection q in the same algebra such that $p \sim q \lneq p$. Given an internal unital C^*-algebra $\mathcal{M}$, show that $p \in \mathrm{Proj}(\mathcal{M})$ is infinite iff $\widehat{p} \in \mathrm{Proj}(\widehat{\mathcal{M}})$ is infinite.

Chapter 4

Selected Research Topics

4.1 Hilbert space-valued integrals

We begin by defining Bochner integration, which at a first glance appears to be a *verbatim ac litteratim* copy of the definition of Lebesgue integration except with real-valued functions replaced by Banach space-valued functions. However Bochner integral is a powerful tool that reveals subtle connection between Banach space-valued vector measures and the geometry of Banach spaces. See [Dunford and Schwartz (1988a)], [Diestel and Uhl (1977)], [Dinculeanu (2000)] and [Benyamini and Lindenstrauss (2000)] for more details, especially topics concerning the Radon-Nikodým property.

Let X be a Banach space and $(\Omega, \mathcal{B}, \mu)$ a σ-finite measure space, *i.e.* $\mathcal{B}$ is a σ-algebra of some subsets of Ω, $\mu : \mathcal{B} \to \mathbb{F}$ a σ-additive measure such that for some $\{\Omega_n\}_{n \in \mathbb{N}} \subset \mathcal{B}$, $|\mu(\Omega_n)| < \infty$ and $\Omega = \cup_{n \in \mathbb{N}} \Omega_n$.

As in p.45, a simple function is a function $f : \Omega \to X$ of the form

$$f(t) = \sum_{k=0}^{n} a_k \chi_{A_k}(t)$$ where $n \in \mathbb{N}$, $a_k \in X$ and $A_k \in \mathcal{B}$ with $\mu(A_k) < \infty$.

The **Bochner integral** of the above simple function f is just defined as $\sum_{k=0}^{n} \mu(A_k)a_k$, with notation given by $\int_{\Omega} f(t)d\mu(t)$ or simply $\int_{\Omega} f d\mu$.

A function $f : \Omega \to X$ is called **Bochner-measurable** if it has an approximating sequence of simple functions, *i.e.* there are simple functions $f_n : \Omega \to X, n \in \mathbb{N}$, such that $\lim_{n \to \infty} \|f(t) - f_n(t)\| = 0$ a.e. μ.

Notice that for a Bochner-measurable function f, the function $\|f\|$ is always Borel-measurable. In fact it can be shown that f is Bochner-measurable iff there is $\Omega_0 \subset \Omega$ of full measure (*i.e.* $\Omega_0 \in \mathcal{B}$ and $\mu(\Omega \setminus \Omega_0) = 0$) so that $f{\restriction}_{\Omega_0}$ is Borel-measurable and has a separable range. (See [Dinculeanu (2000)].) In other words, f is Bochner-measurable iff there is a Borel-measurable g having a separable range such that $f = g$ a.e. μ. The separability property enables one to deal with a countable base and thus removes some technical difficulties involving the construction of an integral.

Given a Bochner-measurable $f : \Omega \to X$, suppose there are simple functions $f_n : \Omega \to X$, $n \in \mathbb{N}$, such that

$$\lim_{n\to\infty} \int_\Omega \|f(t) - f_n(t)\| \, d\mu(t) = 0,$$

we then say f is **Bochner-integrable**.

In such case; as $\left\| \int_\Omega (f_n - f_m)d\mu \right\| \leq \int_\Omega \|f_n - f_m\| \, d\mu$, $n, m \in \mathbb{N}$, $\left\{ \int_\Omega f_n d\mu \right\}_{n\in\mathbb{N}}$ is a Cauchy sequence in X and the **Bochner integral** of f *w.r.t.* μ is defined to be the limit:

$$\int_\Omega f(t)d\mu(t) := \lim_{n\to\infty} \int_\Omega f_n(t)d\mu(t).$$

Similar to the Lebesgue integral case, it can be checked that the definition is independent of the choice of the approximating sequence of the simple functions $\{f_n\}_{n\in\mathbb{N}}$.

For $A \in \mathcal{B}$, $\int_A f(t)d\mu(t)$ is defined similarly.

An important result that characterizes Bochner-integrability is the following. See [Diestel and Uhl (1977)] or [Benyamini and Lindenstrauss (2000)] for a proof.

Lemma 4.1. *Let X be a Banach space and $(\Omega, \mathcal{B}, \mu)$ a σ-finite measure space. Suppose $f : \Omega \to X$ is Bochner-measurable.*

Then f is Bochner-integrable iff $\int_\Omega \|f(t)\| \, d\mu(t) < \infty$. $\square$

Note in particular that bounded Bochner-measurable functions are Bochner-integrable.

Although not needed here, it is worthwhile mentioning that $\phi(\int_\Omega f d\mu) = \int_\Omega \phi(f)d\mu$ holds for any Bochner-integrable f and all $\phi \in X'$.

For a given Hilbert space X and a σ-finite measure space $(\Omega, \mathcal{B}, \mu)$, consider the following set of functions

$$Y := \left\{ f : \Omega \to X \,\middle|\, f \text{ is Bochner-measurable and } \int_\Omega \|f(t)\|^2 \, d\mu(t) < \infty \right\}.$$

That is, by Lem. 4.1, Y consists of Bochner-integrable functions $f : \Omega \to X$ such that the functions $\Omega \ni t \mapsto \|f(t)\| \in [0, \infty)$ are $L^2(\mu)$-functions. It is easily seen that Y forms a pre-inner product space under the pre-inner product $(f, g) \mapsto \left(\int_\Omega \langle f(t), g(t) \rangle d\mu(t) \right)^{1/2}$.

Let $Y_0 := \{f \in Y \mid \int_\Omega \|f(t)\|^2 \, d\mu(t) = 0.\}$.

Then we let $L_2(\Omega, X)$ denote the quotient space Y/Y_0 with each element identified with a function $\Omega \to X$ belonging to the equivalence class given by the element.

It is straightforward to see that $L_2(\Omega, X)$ is a Hilbert space. The inner product and norm are just denoted by $\langle \cdot, \cdot \rangle$ and $\|\cdot\|$ with no confusion with the same symbols used for the Hilbert space X.

For the following, we simply take $(\Omega, \mathcal{B}, \mu)$ to be a probability space with $\Omega = [0,1]$, $\mathcal{B}$ the Lebesgue measurable subsets of $[0,1]$ and $\mu = $ Leb. We write dt for $d\mathrm{Leb}(t)$.

So $L_2(\Omega, X)$ is now $L_2([0,1], X)$.

Our main result is the following isometric identities for Hilbert space-valued Bochner integrals:

Theorem 4.1. *Let X be a Hilbert space. Then the following isometric identities hold for every $f \in L_2([0,1], X)$:*

$$\left\| \int_0^1 f(t) dt \right\|^2 = \int_0^1 \int_0^t \|f(t) + f(s)\|^2 \, ds \, dt - \int_0^1 \|f(t)\|^2 \, dt \qquad (4.1)$$

and

$$\left\| \int_0^1 f(t) dt \right\|^2 = \int_0^1 \|f(t)\|^2 \, dt - \int_0^1 \int_0^t \|f(t) - f(s)\|^2 \, ds \, dt. \qquad (4.2)$$

$\square$

Example 4.1. Let $(\Omega, \mathcal{B}, \nu)$ be a probability space and take X to be $L_2(\Omega, \mathcal{B}, \nu)$. Then any element $f \in L_2([0,1], X)$ can be thought of as an L^2-stochastic process $(\Omega \times [0,1]) \ni (\omega, t) \mapsto (f(t))(\omega)$.

In this context, Thm. 4.1 deals with the isometric identities for the time-integration of an L^2-stochastic process as a Bochner integral and Thm. 4.1 is applicable when the process is L^2 *w.r.t.* $\nu \times$ Leb.

In particular, consider the case where $(\Omega, \mathcal{B}, \nu)$ is the Wiener space, where ν is the Wiener measure on the space of continuous functions, *i.e.* $\Omega = C([0,1], \mathbb{R})$, whose paths at $t \in [0,1]$ correspond to the 1-dimensional Brownian motion b_t (*i.e.* $b(\cdot, t)$).

Then we have from (4.1) of Thm. 4.1 that

$$\left\| \int_0^1 b_t dt \right\|^2_{L_2(\nu)} = \int_0^1 \int_0^t \|b_t + b_s\|^2_{L_2(\nu)} \, ds \, dt - \int_0^1 \|b_t\|^2_{L_2(\nu)} \, dt = \frac{5}{6} - \frac{1}{2} = \frac{1}{3}$$

and from (4.2) of Thm. 4.1 that

$$\left\| \int_0^1 b_t dt \right\|^2_{L_2(\nu)} = \int_0^1 \|b_t\|^2_{L_2(\nu)} \, dt - \int_0^1 \int_0^t \|b_t - b_s\|^2_{L_2(\nu)} \, ds \, dt = \frac{1}{2} - \frac{1}{6} = \frac{1}{3}.$$

Of course this can be computed from the left side directly.

The isometric identities for the special case for L^2-stochastic processes can be proved in a more complicated and tedious way by the stochastic method of the chaos decomposition of L^2-Wiener functionals.

For nonstandard approach to Brownian motion and the chaos decomposition see respectively [Albeverio *et al.* (1986)] and [Cutland and Ng (1991)].   □

Thm. 4.1 and some extensions will be proved with the following tools.

First fix an $N \in {}^*\mathbb{N} \setminus \mathbb{N}$ and write $\Delta t = N^{-1}$. We will work with the hyperfinite timeline

$$\mathbb{T} := \{ n\Delta t \mid n = 0, 1, \ldots, N \}$$

with μ denoting the internal normalized counting measure on $\mathbb{T}$ that assigns $\mu(\{t\}) = \Delta t$ for every $0 < t \in \mathbb{T}$ and $\mu(\{0\}) = 0$. The Loeb measure is denoted by $L(\mu)$. It turns out that $\mathbb{T}, \mu$ are often preferable than $[0,1]$, Leb.

The surjection from $\mathbb{T}$ to $[0,1]$ given by the standard part is denoted by

$$^{\circ} : \mathbb{T} \to [0,1] \quad \text{or} \quad \text{st} : \mathbb{T} \to [0,1].$$

By the Loeb measure theory (§ 1.5.3), st is measure-preserving between $L(\mu)$ and Leb. $\text{st}^{-1}(A)$ is $L(\mu)$-measurable for any Lebesgue measurable $A \subset [0,1]$ and $\text{Leb}(A) = L(\mu)(\text{st}^{-1}(A))$. Also, for every $L(\mu)$-measurable $S \subset \mathbb{T}$, $\text{st}(S)$ is Lebesgue measurable and $L(\mu)(S) = \text{Leb}(\text{st}(S))$. Moreover, if $A \subset [0,1]$ is Lebesgue-measurable, there is an internal $S \subset \mathbb{T}$ such that $L(\mu)(S \triangle \text{st}^{-1}(A)) = 0$ (Cor. 1.3).

We now define the notion of lifting (§ 1.5.3) in our setting. For convenience, instead of speaking of a lifting of $f \circ \text{st}$ we simply speak of a lifting of f.

As a natural extension from the real-valued case, we call a function $F : \mathbb{T} \to {}^*X$ a **lifting** of $f : [0,1] \to X$ if

$$L(\mu)\Big(\{t \in \mathbb{T} \mid F(t) \approx f(^{\circ}t)\}\Big) = 1.$$

Note the above is a Loeb-measurable subset of $\mathbb{T}$. Using the measure-preserving mapping st, an equivalent condition is:

$$\text{Leb}\Big(\{ ^{\circ}t \mid t \in \mathbb{T} \wedge F(t) \approx f(^{\circ}t)\}\Big) = 1.$$

Lemma 4.2. *Let X be a Hilbert space. Let $f \in L_2([0,1], X)$ be a simple function. Then there is an internal simple $F : \mathbb{T} \to {}^*X$ taking finitely many values such that*

(i) F lifts f.

(ii) $\sum_{t \in \mathbb{T}} F(t)\Delta t \approx \int_0^1 f(t)dt$.

(iii) The function $\mathbb{T} \ni t \mapsto {}^*\|F(t)\| \in {}^*[0,\infty]$ is an SL^2-integrable lifting of the function $[0,1] \ni t \mapsto \|f(t)\| \in [0,\infty)$.

Proof. Write f in the form $f(t) = \sum_{k=0}^n a_k \chi_{A_k}(t)$ where $n \in \mathbb{N}$, $a_k \in X$ and $A_k \subset [0,1]$ is Lebesgue measurable. By Cor. 1.3, choose internal $S_k \subset \mathbb{T}$ such that $L(\mu)\big(S_k \triangle \mathrm{st}^{-1}(A_k)\big) = 0$ and $\mathrm{Leb}(A_k) = L(\mu)\big(S_k\big)$.

Then let $F : \mathbb{T} \to {}^*X$ be given by $F(t) = \sum_{k=0}^n {}^*a_k \chi_{S_k}(t)$.

(i) : As $L(\mu)\Big(\bigcup_{k=1}^n \big(S_k \triangle \mathrm{st}^{-1}(A_k)\big)\Big) = 0$, it is clear that F lifts f.

(ii) : Note that

$$\sum_{t \in \mathbb{T}} F(t)\Delta t = \sum_{k=0}^n \mu(S_k)\, {}^*a_k \approx \sum_{k=0}^n \mathrm{Leb}(A_k)a_k = \int_0^1 f(t)dt.$$

(iii) : Since F takes finitely many values and is bounded, the function $t \mapsto {}^*\|F(t)\|$ is SL^2. Moreover, since F lifts of f, it follows from the continuity of the norm that $t \mapsto {}^*\|F(t)\|$ lifts $t \mapsto \|f(t)\|$. $\square$

Proof of Thm. 4.1. For (4.1):

Let $f \in L_2\big([0,1], X\big)$. So f is Bochner-integrable.

Let $f_n : [0,1] \to X$, $n \in \mathbb{N}$, be an approximating sequence of simple functions so that $\lim_{n \to \infty} \int_\Omega \|f(t) - f_n(t)\| \, d\mu(t) = 0$ and therefore we have $\int_0^1 f(t)dt = \lim_{n \to \infty} \int_0^1 f_n(t)dt$.

For each $n \in \mathbb{N}$, let $F_n : \mathbb{T} \to {}^*X$ be a lifting of f_n as given by Lem. 4.2.

Now transfer and apply (2.16) of Thm. 2.26 to the sequence $\{F_n(t)\}_{t \in \mathbb{T}}$ (assume without loss that $F(0) = 0$), we have

$$\sum_{\substack{0 < s < t \leq 1 \\ s, t \in \mathbb{T}}} {}^*\|F_n(t) + F_n(s)\|^2 = (N-2) \sum_{t \in \mathbb{T}} {}^*\|F_n(t)\|^2 + {}^*\Big\| \sum_{t \in \mathbb{T}} F_n(t) \Big\|^2.$$

The equation can be re-written as ${}^*\Big\| \sum_{t \in \mathbb{T}} F_n(t)\Delta t \Big\|^2 =$

$$\sum_{t \in \mathbb{T}} \Big(\sum_{0 < s < t} {}^*\|F_n(t) + F_n(s)\|^2 \Delta t \Big) \Delta t - (1 - 2\Delta t) \sum_{t \in \mathbb{T}} {}^*\|F_n(t)\|^2 \Delta t.$$

By the SL^2-integrable lifting of $t \mapsto \|f_n(t)\|$, $\sum_{t \in \mathbb{T}} {}^*\|F_n(t)\|^2 \Delta t \approx \int_0^1 \|f_n(t)\|^2 dt$, which is finite, so

$$
{}^*\left\| \sum_{t \in \mathbb{T}} F_n(t)\Delta t \right\|^2 \approx \sum_{t \in \mathbb{T}} \left(\sum_{0 < s < t} {}^*\|F_n(t) + F_n(s)\|^2 \Delta t \right) \Delta t - \sum_{t \in \mathbb{T}} {}^*\|F_n(t)\|^2 \Delta t.
$$

Then by the lifting properties again,

$$
{}^*\left\| \sum_{t \in \mathbb{T}} F_n(t)\Delta t \right\|^2 \approx \int_0^1 \int_0^t \|f_n(t) + f_n(s)\|^2 ds\, dt - \int_0^1 \|f_n(t)\|^2 dt.
$$

In particular $\left(\sum_{t \in \mathbb{T}} F_n(t)\Delta t \right) \in \mathrm{Fin}({}^*X)$.

Let $I(F_n) := \left(\sum_{t \in \mathbb{T}} F_n(t)\Delta t \right)^\wedge \in \widehat{{}^*X}$.

Then by Lem. 4.2, $I(F_n) = \int_0^1 f_n(t) dt \in X$.

For $n, m \in \mathbb{N}$, by transferring and applying (2.16) of Thm. 2.26 for the sequence $\{F_n(t) - F_m(t)\}_{t \in \mathbb{T}}$ and repeating the same argument again, we obtain

$$
{}^*\left\| \sum_{t \in \mathbb{T}} F_n(t) - \sum_{t \in \mathbb{T}} F_m(t) \right\|^2 \approx \int_0^1 \int_0^t \|((f_n - f_m)(t)) + (f_n - f_m)(s))\|^2 ds\, dt
$$

$$
- \int_0^1 \|f_n(t) - f_m(t)\|^2 dt \to 0, \quad \text{as } n, m \to \infty.
$$

Hence, for all small infinite $N \in \mathbb{N}$, $I(F_N) = \int_0^1 f(t) dt$ and ${}^*\|F_N(\cdot)\|$ is a SL^2-lifting of $\|f(\cdot)\|$. For any such small infinite $N \in \mathbb{N}$, by (2.16) of Thm. 2.26,

$$
\left\| \int_0^1 f(t) dt \right\|^2 = I(F_N) \approx {}^*\left\| \sum_{t \in \mathbb{T}} F_N(t)\Delta t \right\|^2
$$

$$
\approx \sum_{0 < s < t \in \mathbb{T}} {}^*\|F_N(t) + F_N(s)\|^2 \Delta t^2 - \sum_{t \in \mathbb{T}} {}^*\|F_N(t)\|^2 \Delta t
$$

$$
\approx \int_0^1 \int_0^t \|f(t) + f(s)\|^2 ds\, dt - \int_0^1 \|f(t)\|^2 dt.
$$

For (4.2): the proof follows the same line of argument by using (2.17) of Thm. 2.26 instead. $\qquad \square$

Orbiter dictum, a more straightforward lifting useful in the above proof can be found:

Proposition 4.1. *Let* $f \in L_2([0,1], X)$. *Define* $F : \mathbb{T} \to {}^*X$ *by* $F(0) = 0$ *and* $F(t) := N \displaystyle\int_{t-\Delta t}^{t} {}^*f(s)d\,{}^*\mathrm{Leb}(s)$ *for* $0 < t \in \mathbb{T}$.

Then F *lifts* f *and* $t \mapsto {}^*\|F(t)\|$ *is an* SL^2-*integrable lifting of the function* $t \mapsto \|f(t)\|$. $\square$

We just mention that Prop. 4.1 can be proved by applying the fact that $f(t) = \lim_{h \to 0^+} h^{-1} \int_{t-h}^{t} f(s)dt$ a.e. Leb. (See [Diestel and Uhl (1977)] Thm.9 on p.49.)

What the proof of Thm. 4.1 indicates is that the result can be generalized. Now we are no longer concerned with integration over $[0,1]$ and will work with an internal probability space $(\Omega, \mathcal{B}, \mu)$, where $\mathcal{B} = {}^*\mathcal{P}(\Omega)$, and its Loeb space $(\Omega, L(\mathcal{B}), L(\mu))$.

Suppose X is an internal Banach space. We now make the following obvious and straightforward generalization of similar notions of real-valued functions.

A **hyper-measurable** function $f : \Omega \to \widehat{X}$ is defined to be one such that for some internal $F : \Omega \to X$, we have $\widehat{F(t)} = f(t)$ a.e. $L(\mu)$. For convenience, such F is called a **lifting** of f.

By a **hyper-integrable** function $f : \Omega \to \widehat{X}$ we mean one such that for some internal $F : \Omega \to X$, F lifts f as above and F is S-integrable. Here the latter means that $\int_{\Omega} F(t)d\mu(t) \in \mathrm{Fin}(X)$ and

$$\forall S \in {}^*\mathcal{P}(\Omega) \left(\mu(S) \approx 0 \Rightarrow \int_S \|F(t)\| \, d\mu(t) \approx 0. \right)$$

In such case, we have $\left(\int_{\Omega} F(t)d\mu(t) \right)^{\wedge} \in \widehat{X}$ and the **hyper-integral** of f w.r.t. $L(\mu)$ is unambiguously defined (*i.e.* independent of a particular choice of S-integrable lifting) by

$$\int_{\Omega} f(t)dL(\mu)(t) := \left(\int_{\Omega} F(t)d\mu(t) \right)^{\wedge} \in \widehat{X}.$$

When X is a Banach space, $f : \Omega \to X$ is naturally called hyper-measurable/hyper-integrable if it is so as a function $f : \Omega \to \widehat{{}^*X}$.

Proposition 4.2. *Let* X *be an internal Banach space and* $(\Omega, \mathcal{B}, \mu)$ *an internal probability space. Suppose* $f : \Omega \to \widehat{X}$.

(i) *If* f *is Bochner-measurable, then* f *is hyper-measurable.*

(ii) *If* f *is Bochner-integrable, then* f *is hyper-integrable. In this case, the Bochner integral coincides with the hyper-integral.*

Proof.   For a simple function of the form $f(t) = \sum_{k=0}^{n} \widehat{a}_k \chi_{A_k}(t)$ where $a_k \in \text{Fin}(X)$ and $A_k \in L(\mathcal{B})$, let $F(t) := \sum_{k=0}^{n} a_k \chi_{S_k}(t)$, where $S_k \in {}^*\mathcal{P}(\Omega)$ is such that $L(\mu)\big(S_k \triangle A_k\big) = 0$. Then F is an S-integrable lifting of f.

To prove (i) and (ii), given a sequence of simple functions $\{f_n\}_{n \in \mathbb{N}}$ approximating f, we let F_n be as defined above, then a limit argument shows that for any small infinite $N \in {}^*\mathbb{N}$, F_N is a lifting of f and if f is Bochner-integrable, the Bochner integral is the same as $\left(\displaystyle\int_{\Omega} F(t) d\mu(t) \right)^{\wedge}$.

$\square$

Example 4.2.   Take $\Omega = \mathbb{T}$, the hyperfinite timeline with the internal normalized counting probability measure μ as before. Let $X = {}^*L_2(\mu)$.

Define $F : \mathbb{T} \to X$ by $F(t) := \sqrt{N} \chi_{\{t\}}$. So $\forall t \in \mathbb{T} \left({}^*\|F(t)\| = 1 \right)$ and it is clear that F is S-integrable.

Now let $f : \mathbb{T} \to \widehat{X}$ be given by $f(t) = \widehat{F(t)}$. Then F lifts f, so f is hyper-integrable.

However, the range of f is non-separable, therefore f is not Bochner-measurable.

Observe that $\int_{\Omega} f(t) dL(\mu)(t) = 0$, because

$$\left\| \int_{\Omega} f(t) dL(\mu)(t) \right\| \approx {}^* \left\| \sum_{0 < t \in \mathbb{T}} F(t) \Delta t \right\| = N^{-1/2} \approx 0.$$

However, for ${}^*L_1(\mu)$, we have

$$\forall t \in \mathbb{T} \left({}^* \|N \chi_{\{t\}}\|_{L_1} = 1 \right) \quad \text{and} \quad {}^* \left\| \sum_{0 < t \in \mathbb{T}} N \chi_{\{t\}} \Delta t \right\|_{L_1} = 1.$$

So, in the case $X = {}^*L_1(\mu)$, if we define $f(t) := \widehat{N \chi_{\{t\}}}$, $t \in \mathbb{T}$, then, as before, f is non-Bochner-integrable but hyper-integrable, although $\left\| \int_{\Omega} f(t) dL(\mu)(t) \right\| = 1$.

$\square$

Let an internal Hilbert space X be given, then $HL_2(\Omega, \widehat{X})$ denotes the collection of hyper-integrable functions $f : \Omega \to \widehat{X}$ with an additional condition imposed on their S-integrable liftings F by requiring that

$$\int_{\Omega} \|(F(t)\|^2 \, d\mu(t) < \infty.$$

As for the case of $L_2(\Omega, \widehat{X})$, $HL_2(\Omega, \widehat{X})$ also forms a Hilbert space under $(f, g) \mapsto \left(\int_{\Omega} \langle f(t), g(t) \rangle dL\mu(t) \right)^{1/2}$.

Note that $L_2(\Omega, \widehat{X})$, is a closed subspace of $HL_2(\Omega, \widehat{X})$, but the above example shows that the it could be a proper subspace.

When X is a Hilbert space, $HL_2(\Omega, X)$ is the set of functions of $HL_2(\Omega, {}^*\widehat{X})$ having range in X. It is not hard to check that $HL_2(\Omega, X)$ is a closed subspace of $HL_2(\Omega, {}^*\widehat{X})$.

With these notions, we have a generalized version of Thm. 4.1.

Theorem 4.2. *Let X be an internal Hilbert space and $(\Omega, \mathcal{B}, \mu)$ a hyperfinite probability space and μ a normalized counting probability measure on Ω. Then for every $f \in HL_2(\Omega, \widehat{X})$,*

$$\left\| \int_\Omega f(t) dL(\mu)(t) \right\|^2$$
$$= \frac{1}{2} \iint_{\Omega^2} \|f(t) + f(s)\|^2 \, dL(\mu)(s) dL(\mu)(t) - \int_\Omega \|f(t)\|^2 \, dL(\mu)(t)$$

and

$$\left\| \int_\Omega f(t) dL(\mu)(t) \right\|^2$$
$$= \int_\Omega \|f(t)\|^2 \, dL(\mu)(t) - \frac{1}{2} \iint_{\Omega^2} \|f(t) - f(s)\|^2 \, dL(\mu)(s) dL(\mu)(t)$$

Hence the same equations also hold for $f \in HL_2(\Omega, X)$, where X is a Hilbert space.

Proof.   We only prove the first equality, for the second one is similar, as in the proof of Thm. 4.1.

Let $F : \Omega \to X$ be an SL^2-lifting of f.

Let $|\Omega| = N \in {}^*\mathbb{N}$. Then as μ is a normalized counting probability measure on Ω, it assigns $\Delta t := N^{-1}$ to each singleton $\{t\}$, $t \in \Omega$.

By applying (2.16) of Thm. 2.26 to the sequence $\{F(t)\}_{t \in \Omega}$, we have

$$\left\| {}^* \sum_{t \in \Omega} F(t) \Delta t \right\|^2 \approx \frac{1}{2} \sum_{s,t \in \Omega} {}^* \|F(t) + F(s)\|^2 \, \Delta t^2 - \sum_{t \in \Omega} {}^* \|F(t)\|^2 \, \Delta t.$$

The rest of the proof is similar to that of Thm. 4.1.

Note however that Keisler's Fubini's Theorem is needed here, see Ex. 8 on p.52. $\square$

For other topics related to Loeb measure and integration theory in Banach spaces setting, *Cf.* [Sun (1992)], [Osswald and Sun (1991)], [Osswald (1995)] and [Zimmer (1998)].

Question 1. Does Thm. 4.2 (or Thm. 4.1) hold for some Banach spaces which are not Hilbert spaces?

Question 2. Are there equalities or inequalities generalizing those in Thm. 4.1 for functions in $L_p([0,1], X)$, $1 \leq p < \infty$?

Problem 3. Generalize Thm. 4.2 for other internal measures other than the normalized counting probability measures.

Question 4. For a hyper-integrable f that takes values in a Hilbert space, is it always possible to decompose f as $f_0 + f_1$ such that f_0 is Bochner-integrable and f_1 has zero hyper-integral? (See Example 4.2.)

4.2 Reflexivity and fixed points

Recall §2.7.4. We say that a subset X of a Banach space has the **fixed point property** if for every nonexpansive self-mapping on a nonempty b.c.c. (bounded closed convex) subset of X has a fixed point.

Question 5. Does every reflexive (or even superreflexive) Banach space have the fixed point property?

Recall that in a reflexive space (Cor.2.22), b.c.c. sets coincide with w.c.c. (weakly compact convex) sets. By an example in [Alspach (1981)], it is known that in a nonreflexive Banach space nonempty w.c.c. sets need not have the fixed point property.

A more general one is the following.

Question 6. Is it true that a nonempty convex subset of a reflexive Banach space has the fixed point property iff it is weakly compact?

An even more general but vague question is the following.

Question 7. Given a class of subsets of a topological space with possibly additional structures and a class of self-mappings on such subsets, does the guarantee of possessing fixed points corresponds to a notion of compactness?

In the absence of a positive answer to Question 5, whether nonempty b.c.c. sets have the fixed point property, one may try to obtain some weaker results. One approach is the following:

Given a nonexpansive $f : C \to C$, where $C \neq \emptyset$ is b.c.c. (or even w.c.c.), can one find a b.c.c. D with a nonexpansive function $g : D \to D$ having a fixed point and such that $C \cap D \neq \emptyset$ and $f \restriction_{C \cap D} = g \restriction_{C \cap D}$?

Note that if there is such D and g in the above, f must be invariant on $C \cap D$, i.e. $f[C \cap D] \subset (C \cap D)$.

We need some notions before stating a result related to the above.

In the theory of convex metric spaces, one study the notion of a **metric segment**. i.e. given a metric space (X, d), and $x, y \in X$ a metric segment from x to y is defined to be an isometry

$$\gamma : [0, d(y - x)] \to X \quad \text{such that } \gamma(0) = x \text{ and } \gamma(d(y - x)) = y.$$

So we have then $\forall t \in [0, d(y - x)] \, (d(\gamma(t), x) = t)$.

We also call the set $\{\gamma(t) \mid t \in [0, d(y - x)]\}$ a metric segment joining x to y.

By Menger's Theorem (see for example [Goebel and Kirk (1990)]), in a complete metric space (X, d) with the property that for distinct $x, y \in X$ there is z distinct from both x, y such that $d(x, y) = d(x, z) + d(z, y)$, any two points are joint by a metric segment. In general, there may be more than one metric segment between them.

In a normed linear space X and $x, y \in X$, the set

$$[x, y] := \big\{ (1 - t)x + ty \,|\, t \in [0, 1] \big\}$$

is obviously a metric segment joining x to y.

The technical lemma next basically produces an invariant b.c.c. subset of a nonexpansive function that includes a fixed point and a certain w.c.c. subset whose elements are of equal distance to it.

Lemma 4.3. *Let X be a normed linear space, $C \subset X$ be b.c.c. with a nonexpansive $f : C \to C$ having a fixed point $c \in C$.*

(i) *Suppose there is a nonempty w.c.c. $A \subset C$ such that f is invariant on A and $\forall x, y \in A \,(\, \|c - x\| = \|c - y\| \,)$. Let*

$$D := \bigcup \big\{ L \subset C \,|\, L \text{ is a metric segment joining a point } a \in A \text{ to } c \big\}.$$

Then $(A \cup \{c\}) \subset D$ and D is b.c.c.
Moreover, f is invariant on D.

(ii) *Let $\emptyset \neq C_0 \subset (C \setminus \{c\})$ be w.c.c. such that f is invariant on it.*
Then there is nonempty w.c.c. $A \subset C_0$ on which f is invariant and $\forall x, y \in A \,(\, \|c - x\| = \|c - y\| \,)$.

Proof.   (i) : For any $a \in A$, $[a, c] \subset D$ by C being convex, so $(A \cup \{c\}) \subset D$. By scaling, we can assume that $\|c - a\| = 1$ for any $a \in A$.

Note that $D \subset \bar{B}(c, 1)$, so D is bounded.

Now let $b \in X$ such that $b \approx b_0$ for some $b_0 \in {}^*D$. Then there is an internal metric segment in ${}^*C$ joining some $a_0 \in {}^*A$ to c. By Cor. 2.25, there is $a \in A$ such that $\|b - a\| = \text{dist}(b, A)$. But

$$\text{dist}(b, A) \approx {}^*\text{dist}(b_0, {}^*A) \leq {}^*\|b_0 - a_0\|,$$

so we have

$$1 = \|c - a\| \leq \|c - b\| + \|b - a\| \lesssim {}^*\|c - b_0\| + {}^*\|b_0 - a_0\| = {}^*\|c - a_0\| = 1,$$

so $[a, b] \cup [b, c]$ is a metric segment in C containing b and joining a to c, hence $b \in D$.

Therefore D is closed.

Now let $b_1, b_2 \in D$. So there are $a_1, a_2 \in A$ such that b_1, b_2 are on some metric segments joining a_1, a_2 to c respectively. Then we have

$$\|c - b_1\| + \|b_1 - a_1\| = 1 \quad \text{and} \quad \|c - b_2\| + \|b_2 - a_2\| = 1.$$

Hence, by A being convex $2^{-1}(a_1 + a_2) \in A$ and

$$1 = \left\| c - \frac{a_1 + a_2}{2} \right\| \leq \left\| c - \frac{b_1 + b_2}{2} \right\| + \left\| \frac{b_1 + b_2}{2} - \frac{a_1 + a_2}{2} \right\|$$

$$\leq \frac{1}{2} \|c - b_1\| + \frac{1}{2} \|c - b_2\| + \frac{1}{2} \|b_1 - a_1\| + \frac{1}{2} \|b_2 - a_2\| = 1,$$

$\left[2^{-1}(a_1 + a_2), \, 2^{-1}(b_1 + b_2)\right] \cup \left[2^{-1}(b_1 + b_2), \, c\right]$ is a metric segment joining $2^{-1}(a_1 + a_2)$ to c and contains $2^{-1}(b_1 + b_2)$, $i.e.$ $2^{-1}(b_1 + b_2) \in D$, therefore D is convex.

Now let $\gamma : [0, 1] \to X$ be a metric segment joining some $a \in A$ to c. By assumption, $f(a) \in A$.

Let $t \in [0, 1]$. Then by f nonexpansive and $f(c) = c$,

$$1 = \|c - f(a)\| \leq \|c - f(\gamma(t))\| + \|f(\gamma(t)) - a\| \leq \|c - \gamma(t)\| + \|\gamma(t) - a\| = 1,$$

$i.e.$, since this holds for all $t \in [0, 1]$, $(f \circ) \gamma : [0, 1] \to X$ is a metric segment joining $f(a)$ to c.

Therefore $f[D] \subset D$.

Finally, note that $c \in D$, so f has a fixed point in D.

(ii) : By Cor. 2.25, the set given by $A := \left\{ x \in C_0 \mid \|c - x\| \doteq \text{dist}(c, C) \right\}$ is nonempty. Clearly, $\forall x, y \in A \left(\|c - x\| = \|c - y\| \right)$.

Let $a \in X$ such that $a \approx a_0$ for some $a_0 \in {}^*A$. Then $\|c - a\| \approx {}^*\|c - a_0\| = \text{dist}(c, C)$, $i.e.$ $\|c - a\| = \text{dist}(c, C)$. Moreover, C_0 is closed (Prop. 2.26), $a \in C_0$, therefore $a \in A$ and hence A is closed.

A is clearly convex. So by Cor. 2.20, A is w.c.c..

Now let $a \in A$. Then $f(a) \in C_0$. By f nonexpansive, we have

$$\|c - f(a)\| = \|f(c) - f(a)\| \leq \|c - a\| = \text{dist}(c, C),$$

hence $f(a) \in A$. $i.e.$ $f[A] \subset A$. $\qquad\square$

We say that a subset X of a topological space has density $< \kappa$, the cardinality of saturation, if there is $X_0 \subset X$ such that $|X_0| < \kappa$ and $\overline{X}_0 = X$. So separable sets are of density $< \kappa$.

The following is a weak solution to the problem of getting a fixed point for a nonexpansive self-mapping on a b.c.c. set (Thm. 2.4. in [Baratella and Ng (1998)], with a minor inaccuracy in the statement corrected).

Theorem 4.3. *Let X be an internal normed linear space. Suppose $C \subset \widehat{X}$ is a nonempty, w.c.c. with density $< \kappa$ and $f : C \to C$ is nonexpansive.*

Then for some b.c.c. $D \subset \widehat{X}$, having the same bound as C does and $C \cap D \neq \emptyset$, there is some nonexpansive $g : D \to D$ having a fixed point and satisfying $f \!\restriction_{C \cap D} = g \!\restriction_{C \cap D}$.

Proof.   Without loss of generality, we assume that $C \subset \bar{B}_{\widehat{X}}$ and it suffices to find the required $D \subset \bar{B}_{\widehat{X}}$.

Let $C_0 \subset \bar{B}_X$ such that $|C_0| < \kappa$ and $\overline{\widehat{C_0}} = C$. (Recall that $\widehat{C_0}$ denotes $\{\widehat{x} \mid x \in C_0\}$.) Let $\rho : C_0 \to \bar{B}_X$ be a function satisfying $f(\widehat{x}) = \widehat{\rho(x)}$ for all $x \in C_0$. Note that, except in trivial cases, both C_0 and ρ are external.

Let $Y \subset C_0$ be finite and $n \in \mathbb{N}$. So $\mathrm{conv}(Y)$ is compact. Apply Thm. 1.22, choose a small enough $\epsilon \in \mathbb{R}^+$ and finite $H \subset \mathrm{conv}(Y)$ such that $\mathrm{conv}(Y) \subset H^\epsilon$, we see that there is an internal function $h : \mathrm{conv}(Y) \to \bar{B}_X$ such that

$$\forall y \in Y \left(\|h(y) - \rho(y)\| < n^{-1} \right)$$

and $\forall y_1, y_2 \in \mathrm{conv}(Y) \left(\|h(y_1) - h(y_2)\| < (1+n^{-1}) \|y_1 - y_2\| \right).$

(*i.e.* of Lipschitz constant $(1 + n^{-1})$.)

Then, by saturation, there is an *convex $K \subset \bar{B}_X$ with $C_0 \subset K$ and an internal $h : K \to \bar{B}_X$ of Lipschitz constant $(1 + \epsilon)$ for some $0 \approx \epsilon \in {}^*\mathbb{R}^+$ such that $\forall y \in C_0 \left(h(y) \approx \rho(y) \right)$.

Since h has Lipschitz constant $(1 + \epsilon)$, h is S-continuous. Note also $K \subset \bar{B}_X \subset \mathrm{Fin}(X)$. (So $\widehat{K} \subset \bar{B}_{\widehat{X}}$.) Therefore the function

$$\widehat{h} : \widehat{K} \to \widehat{X} \quad \text{given by} \quad \widehat{x} \mapsto \widehat{h(x)}, \ x \in K,$$

is well-defined.

By $C_0 \subset K$ and $\overline{\widehat{C_0}} = C$, it follows from $\widehat{K}$ being closed that $C \subset \widehat{K}$.

By h having Lipschitz constant $(1 + \epsilon)$, where $\epsilon \approx 0$, $\widehat{h}$ is nonexpansive.

By $\forall x \in C_0 \left(h(x) \approx \rho(x) \right)$ and saturation, $\forall x \in C \left(\widehat{h}(x) = f(x) \right)$.

That is, $\widehat{h}$ is a nonexpansive function extending f, although it needs not be a self-mapping.

Since $f = \widehat{h} \!\restriction_C$ is a nonexpansive self-mapping, $\widehat{h}$ has a fixed point $c \in \widehat{K}$ by Thm. 2.35.

Now define $A := \left\{ x \in C \mid \|c - x\| = \mathrm{dist}(c, C) \right\}$. By Cor. 2.25, $A \neq \emptyset$.

It is easily seen that A is also b.c.c. and $\forall x, y \in A \left(\|c - x\| = \|c - y\| \right)$.

Moreover, for any $a \in A$, by $\widehat{h}$ being nonexpansive

$$\left\| c - \widehat{h}(a) \right\| = \left\| \widehat{h}(c) - \widehat{h}(a) \right\| \leq \|c - a\| = \mathrm{dist}(c, C),$$

hence $\left\| c - \widehat{h}(a) \right\| = \mathrm{dist}(c, C)$ since $\widehat{h}(a) = f(a) \in C$. So $\widehat{h}(a) \in A$. Therefore $\widehat{h}[A] \subset A$.

With all assumptions in Lem. 4.3 satisfied by $\widehat{h}$ and $\widehat{K}$, we define

$$D := \bigcup \{ L \subset \widehat{K} \mid L \text{ is a metric segment joining a point } a \in A \text{ to } c \},$$

let $g = \widehat{h} \upharpoonright_D$ and conclude that D is a nonempty, b.c.c. and $g : D \to D$ has a fixed point.

Moreover, $(C \cap D) \supset A \neq \emptyset$ and $D \subset \widehat{K} \subset \bar{B}_{\widehat{X}}$.

Furthermore $f \upharpoonright_{C \cap D} = g \upharpoonright_{C \cap D}$, since $C, D \subset \widehat{K}$ and the f, g are restrictions of $\widehat{h}$ to C, D respectively.   □

Observe that in the above theorem if f already has a fixed point, then D is simply a singleton containing that point.

Question 8. Does the conclusion in Thm. 4.3 still hold if the density requirement is dropped?

The main problem of course is that we want f, g to be compatible in a strong sense, *i.e.* they must agree on their common domain $C \cap D$. If we fix any $c \in C$, define

$$C_0 := \{c\} \quad \text{and} \quad C_{n+1} := \mathrm{conv}\big(C_n \cup f[C_n]\big), \ n \in \mathbb{N},$$

and let $\tilde{C} := \overline{\bigcup_{n \in \mathbb{N}} C_n}$, then $\tilde{C}$ is a nonempty separable b.c.c. subset of C on which f is a nonexpansive self-mapping. Therefore, by Thm. 4.3, there is a fixed-point-possessing nonexpansive self-mapping g on a b.c.c. D that agrees with f on $\tilde{C} \cap D$. However f, g need not agree on $C \cap D$.

In the proof of Thm. 4.3, g was obtained from a "lifting" of f which exists by saturation and the small density.

Recall from Cor. 2.26 that the nonstandard hull $\widehat{{}^*X}$ of a superreflexive space X is superflexive. Moreover, $X \subset \widehat{{}^*X}$ as a closed subspace, so such X has the fixed point property iff $\widehat{{}^*X}$ does. Therefore, to settle Question 5 in the superreflexive case, it suffices to consider nonstandard hulls only.

Moreover, on these nonstandard hulls, the problem can be reduced to nonexpansive functions on w.c.c. subsets of the unit sphere, as the following corollary to Thm. 4.3 shows.

Corollary 4.1. *Let X be an internal normed linear space such that $\widehat{X}$ is reflexive (hence superreflexive, by Thm. 2.24).*

Then $\widehat{X}$ has the fixed point property iff $S_{\widehat{X}}$ does.

Proof. $(\Rightarrow)$ is trivial as $S_{\widehat{X}}$ is closed.

$(\Leftarrow)$: Suppose $\widehat{X}$ fails to satisfy the fixed point property.

Let $f : C \to C$ be a fixed-point-free nonexpansive self-mapping on a nonempty w.c.c. $C \subset \widehat{X}$. We now show that $S_{\widehat{X}}$ also has a nonempty w.c.c. subset supporting a fixed-point-free nonexpansive self-mapping.

By the remark above, we can assume that C is separable. Let g and D be obtained from Thm. 4.3. Let $c \in D$ denotes the fixed point of g in Thm. 4.3. Note that, by assumption $c \notin C$. Then by applying Lem. 4.3(ii) to g and the w.c.c. $C \cap D$, there is a nonempty w.c.c. $A \subset (C \cap D)$ such that $f[A] \subset A$ and $\forall x, y \in A \left(\|c - x\| = \|c - y\| \right)$.

By scaling if necessary, we can assume without loss of generality that $\forall x \in A \left(\|c - x\| = 1 \right)$. Hence $(A - c) \subset S_{\widehat{X}}$.

Now let $h : (A - c) \to (A - c)$ be $(x - c) \mapsto (f(x) - c)$, $x \in A$. By assumption, f is fixed-point-free on A hence h is fixed-point-free on $(A-c)$. Clearly, h is nonexpansive. Moreover, $(A - c)$ is still nonempty and w.c.c..

Therefore there exists a fixed-point-free nonexpansive self-mapping on a nonempty w.c.c. subset of $S_{\widehat{X}}$. $\square$

See [Wiśnicki (2002)] for other applications of nonstandard techniques for fixed point results.

4.3 Arens product on a bidual

Let $\mathcal{M}$ be a normed algebra, say over $\mathbb{C}$. So $\mathcal{M}$ is canonically identified with a normed linear subspace of its bidual $\mathcal{M}''$ through the evaluation mappings (Prop. 2.23).

By the work of R. Arens, two products can be defined on $\mathcal{M}''$, both are extensions of the product on $\mathcal{M}$ and turn $\mathcal{M}''$ into a Banach algebra. (See §1.4 in [Palmer (1994)] for more detail.)

The two products, called **Arens products**, are defined by the following steps.

First, for $c \in \mathcal{M}$ and $\omega \in \mathcal{M}'$, we define $\omega_c \in \mathcal{M}'$ and $_c\omega \in \mathcal{M}'$ by:

$$\omega_c : \mathcal{M} \ni x \mapsto \omega(cx) \in \mathbb{C} \quad \text{and} \quad _c\omega : \mathcal{M} \ni x \mapsto \omega(xc) \in \mathbb{C}.$$

Next, for $a \in \mathcal{M}''$ and $\omega \in \mathcal{M}'$, we define $\omega_a \in \mathcal{M}'$ and $_a\omega \in \mathcal{M}'$ by;

$$\omega_a : \mathcal{M} \ni x \mapsto a(_x\omega) \in \mathbb{C} \quad \text{and} \quad _a\omega : \mathcal{M} \ni x \mapsto a(\omega_x) \in \mathbb{C}.$$

Finally, for $a, b \in \mathcal{M}''$, we define $(a \ltimes b) \in \mathcal{M}''$ and $(a \rtimes b) \in \mathcal{M}''$ by:

$$(a \ltimes b) : \mathcal{M}' \ni \omega \mapsto a(_b\omega) \quad \text{and} \quad (a \rtimes b) : \mathcal{M}' \ni \omega \mapsto b(\omega_a).$$

It is easily seen that for $a, b \in \mathcal{M}$ (as canonically identified in $\mathcal{M}''$), $(a \ltimes b) = ab = (a \rtimes b)$.

We call $\mathcal{M}$ **Arens-regular** if both Arens products are the same on $\mathcal{M}''$.

It can be checked that $\ell_1(\mathbb{Z})$ under the convolution product is an example of a unital C^*-algebra which is not Arens-regular.

Suppose $\mathcal{M}$ is a unital C^*-algebra. Let $\pi : \mathcal{M}'' \to \mathcal{B}(H)$ be the embedding given by Thm. 3.26. Then a product on $\mathcal{M}''$ can be defined by

$$\widehat{ab} := \pi^{-1}\big(\pi(\widehat{a})\pi(\widehat{b})\big), \quad \text{where} \quad a, b \in \mathrm{Fin_w}(\,^*\mathcal{M}).$$

Moreover, this product coincides with both Arens products by Thm. II.1 in [Godefroy and Iochum (1988)].

In other words, any unital C^*-algebra $\mathcal{M}$ is Arens-regular, and both Arens products are the same as the product on the universal enveloping von Neumann algebra of $\mathcal{M}$.

The following shows that the Arens products for the bidual of a unital C^*-algebra can be represented by the ordinary product in the nonstandard extension.

Theorem 4.4. Let $\mathcal{M}$ be a unital C^*-algebra and $a, b \in \mathrm{Fin_w}(\,^*\mathcal{M})$.

Let H be the Hilbert space given by the universal representation of $\mathcal{M}$ and $\pi : \mathcal{M}'' \to \mathcal{B}(H)$ be given by Thm. 3.26. Then

(i) *There is $a_0 \in \mathrm{Fin}_w(\,^*\mathcal{M})$ such that $\widehat{a}_0 = \widehat{a}$ and $\pi(\widehat{a})\pi(\widehat{b}) = \pi(\widehat{a_0\,b})$.*

(ii) *There is $b_0 \in \mathrm{Fin}_w(\,^*\mathcal{M})$ such that $\widehat{b}_0 = \widehat{b}$ and $\pi(\widehat{a})\pi(\widehat{b}) = \pi(\widehat{a\,b_0})$.*

Proof. By applying the (i) to $b^*\,a^*$, one obtains (ii).

To prove (i), let $a, b \in \mathrm{Fin}_w(\,^*\mathcal{M})$ and write $c = a^*$.

By Thm. 2.13 and Thm. 3.26, all $\overline{\mathcal{M}^{\mathrm{WOT}}}$, $\mathcal{M}''$ and $\widehat{\,^*\mathcal{M}^w}$ are identified with each other.

Let $\{c_i\}_{i \in I} \subset \mathcal{M}$ be such that $c_i \xrightarrow{\mathrm{WOT}} \pi(\widehat{c})$.

Then for any $\epsilon \in \mathbb{R}^+$ and $\{\xi_j, \zeta_j\}_{j \in J} \subset H$, where J is finite, the following holds for all large enough $i \in I$:

$$\left| \langle c_i(\xi_j), \pi(\widehat{b})(\zeta_j) \rangle - \langle \pi(\widehat{c})(\xi_j), \pi(\widehat{b})(\zeta_j) \rangle \right| < \epsilon.$$

i.e.

$$\left| \langle c_i(\xi_j), \eta_j \rangle - \langle \pi(\widehat{c})(\xi_j), \eta_j \rangle \right| < \epsilon,$$

where we write $\eta_j := \pi(\widehat{b})(\zeta_j)$. Since

$$\langle c_i(\xi_j), \eta_j \rangle \approx \langle c_i(\xi_j), b(\zeta_j) \rangle$$

and

$$\langle \pi(\widehat{c})(\xi_j), \eta_j \rangle \approx \langle c(\xi_j), \eta_j \rangle,$$

we have for all large $i \in I$ and all $j \in J$ that

$$\left| \langle c_i(\xi_j), b(\zeta_j) \rangle - \langle c(\xi_j), \eta_j \rangle \right| < \epsilon.$$

By saturation, for some $k \in \,^*I$

$$\langle c_k(\xi), b(\zeta) \rangle \approx \langle c(\xi), \pi(\widehat{b})(\zeta) \rangle \quad \text{for all } \xi, \zeta \in S_H. \tag{4.3}$$

On the left side of (4.3), we have

$$\langle c_k(\xi), b(\zeta) \rangle = \langle \xi, c_k^* b(\zeta) \rangle \approx \langle \xi, \pi(\widehat{c_k^*\,b})(\zeta) \rangle.$$

While on the right side of (4.3), we have

$$\langle c(\xi), \pi(\widehat{b})(\zeta) \rangle = \langle a^*(\xi), \pi(\widehat{b})(\zeta) \rangle = \langle \xi, a(\pi(\widehat{b})(\zeta)) \rangle \approx \langle \xi, (\pi(\widehat{a})\pi(\widehat{b}))(\zeta) \rangle.$$

Hence, as a consequence of (4.3), $\pi(\widehat{a})\pi(\widehat{b}) = \pi(\widehat{c_k^*\,b})$

Now let $a_0 := c_k^*$.

Then since $c_k \approx_w c = a^*$, we get $a_0 \approx_w a$ and (i) is proved. $\qquad \square$

In general it is necessary to make a particular choice from the equivalence classes in order to satisfy the statements in Thm. 4.4.

For example, consider $H = \ell_2$ and $\mathcal{M} = \mathcal{B}(H)$. Let $N \in {}^*\mathbb{N} \setminus \mathbb{N}$ and $a \in {}^*\mathcal{M}$ be the operation that interchanges the 1st coordinate with the Nth coordinate in each $\xi \in H$. Then $a \in \text{Fin}({}^*\mathcal{M})$, with $a^2 = 1$ hence $\widehat{a^2} = 1$. But $\widehat{a}$ is the operation that replaces the 1st coordinates in $\xi \in H$ by 0, therefore $(\widehat{a})^2 \neq \widehat{a^2}$.

Question 9. Suppose one replaces $\{\widehat{a}, \widehat{b}\}$ in the statements in Thm. 4.4 by an arbitrary $X \subset \widehat{{}^*\mathcal{M}^{\text{w}}}$. For which X are the statements still valid?

So one needs to find $\rho : X \to \text{Fin}({}^*\mathcal{M})$ such that

$$\forall x \in X \left(x = \widehat{\rho(x)}\right) \quad \text{and} \quad \forall x, y \in X \left(xy = \widehat{\rho(x)\rho(y)}\right).$$

If Thm. 4.4 is valid for $X = \widehat{{}^*\mathcal{M}^{\text{w}}}$, one may even want the choice function ρ be such that $\rho\left[\widehat{{}^*\mathcal{M}^{\text{w}}}\right]$ extends to an internal subalgebra of ${}^*\mathcal{M}$, *i.e.* ρ must satisfy some internal isometric conditions.

Problem 10. Characterize the class of Arens-regular normed linear algebras $\mathcal{M}$ for which the representation of the Arens products using the ordinary product from ${}^*\mathcal{M}$, as given in Thm. 4.4, is possible.

294   *Nonstandard Methods in Functional Analysis*

4.4   Noncommutative Loeb measures

As remarked on p.265-266, von Neumann algebras can be viewed as non-commutative measure spaces.

More generally, let $\mathcal{M}$ be an unital C^*-algebra identified as a C^*-subalgebra of $\mathcal{B}(H)$, where H is the Hilbert space given by the universal representation. By Prop. 3.23, $\mathcal{M}'$ is the span of positive functionals on H.

The rough idea here is that each positive functional $\phi \in \mathcal{M}'$ can be though of as a noncommutative integral operator, each $a \in \mathcal{M}^+$ as a positive integrand and $\phi(a)$ as the integral; moreover, there is a common noncommutative positive finite measure w.r.t. which the measures for the ϕ's under some natural conditions are absolutely continuous. Of course this oversimplifies the matter, as it requires great care to obtain the noncommutative version of the Radon-Nikodým Theorem.

When $\mathcal{M}$ is a commutative unital C^*-algebra, we identify by Thm. 3.14 $\mathcal{M}$ with $C(\Omega)$ for some compact space Ω. Then $\mathcal{M}'$ is just $\mathbb{M}(\Omega)$ (the space of σ-additive complex Borel measures on Ω), by the Riesz Representation Theorem (Thm. 2.14). In the case when $\mathcal{M}$ is von Neumann algebra, $\mathcal{M}'$ is simply the predual of $\mathcal{M}$ (Thm. 3.26).

Now take as an example the commutative von Neumann algebra $\mathcal{M}$ given by $\{m_f \mid f \in L_\infty(\mu)\}$, where μ is some positive measure, the m_f's are multiplication operators and $\phi \in \mathcal{M}'$ is positive with a cyclic element $\xi \in L_2(\mu)$, as in the notes on p.265-266. Then for $m_f \in \mathcal{M}$, we have

$$\phi(m_f) = \langle m_f(\xi), \xi \rangle = \int_\Omega m_f(\xi)\, d\mu = \int_\Omega f\, \xi d\mu.$$

Here the positive functional ϕ corresponds to the integral operator w.r.t. a positive measure ν with the Radon-Nikodým derivative $f = d\nu/d\mu$ being a positive function. We would like to do all these in a noncommutative manner.

Therefore, back to the case of a general unital C^*-algebra $\mathcal{M}$, we regard elements in $\mathcal{M}'$ as integral operators w.r.t. noncommutative complex measures and positive functionals as those for noncommutative finite real measure. In particular, states correspond to noncommutative probability measures.

More ambitiously we would like to extend this analog to possibly infinite measures. Hence we need the following notion. As in the case with infinite measure, to avoid problems like $\infty - \infty$, one restricts to positive infinite measures and deals with positive integrands for convenience.

Given a unital C^*-algebra $\mathcal{M}$, an additive positively homogeneous $[0,\infty]$-valued function on $\mathcal{M}^+$ is called a *weight*.

That is, a function $\phi : \mathcal{M}^+ \to [0,\infty]$ such that

$$\forall x, y \in \mathcal{M}^+ \; \forall r \in [0,\infty) \left(\big(\phi(x+y) = \phi(x) + \phi(y)\big) \wedge \big(\phi(rx) = r\phi(x)\big) \right).$$

We adopt the convention that $0 \cdot \infty = 0$. Consequently $\phi(0) = 0$ is always satisfied.

Note that a positive functional forms a weight by restricting to $\mathcal{M}^+$.

A finite weight ϕ on $\mathcal{M}^+$ extends canonically to an element in $\mathcal{M}'$ as follows: Let $a \in \mathcal{M}$. Decompose a according to Thm. 3.9(vi) and Cor. 3.21 as $(a_1 - a_2) + i(a_3 - a_4)$ for some unique $a_1, a_2, a_3, a_4 \in \mathcal{M}^+$. Then define the extension at a as $\phi(a_1) - \phi(a_2) + i\phi(a_3) + i\phi(a_4)$.

Some simple properties:

Proposition 4.3. *Let ϕ be a weight on $\mathcal{M}^+$, where $\mathcal{M}$ is a unital C^*-algebra. Then*

(i) *If $a, b \in \mathcal{M}^+$ and $a \le b$, then $\phi(a) \le \phi(b)$.*
(ii) *If $a \in \mathcal{M}^+$, then $\phi(a) \le \|a\| \, \phi(1)$.*
(iii) *$\phi = 0$ iff $\phi(1) = 0$.*

Proof. (i) follows from the additive condition in the definition of a weight. (iii) follows from (ii).

As for (ii), by Cor. 3.20, $(\|a\| - a) \in \mathcal{M}^+$, hence

$$\phi(a) \le \phi(a) + \phi(\,\|a\| - a) = \phi(\,\|a\|\,) = \|a\| \, \phi(1). \qquad \square$$

A weight ϕ on $\mathcal{M}^+$ is called *normal* if whenever $\{a_i\}_{i \in I} \subset \mathcal{M}^+$ is a bounded net such that $\sup_{i \in I} a_i$ exists in $\mathcal{M}^+$, then

$$\phi\left(\sup_{i \in I} a_i \right) = \sup_{i \in I} \phi(a_i).$$

If the above holds only with an additional condition $|\mathcal{F}| < \kappa$, where κ is the cardinality of saturation, then we say that ϕ is κ-*normal*.

These notions are meant to impose some continuity on the integration process. In the context of von Neumann algebras, normal weights can be characterized by very strong additive properties. (See [Haagerup (1975)].) For more applications, see Chap. 4 in [Arveson (2003)].

Given an internal weight $\phi : \mathcal{M} \to {}^*[0,\infty]$, where $\mathcal{M}$ is an internal unital C^*-algebra, the question is how to convert it to a weight on some unital C^*-algebra. Therefore it is natural to call any construction of such weight as the **noncommutative Loeb integral** operator corresponding to some **noncommutative Loeb measure**. It turns out that this is easier than one expects, at least for the S-continuous case.

We call an internal weight to be S-*continuous* if

$$\forall x \in \mathcal{M}^+ \left((x \approx 0 \Rightarrow (\phi(x) \approx 0)) \right).$$

There is an abundant supply of S-continuous weights:

Lemma 4.4. *Let $\mathcal{M}$ be an internal unital C^*-algebra and consider an internal weight $\phi : \mathcal{M}^+ \to {}^*[0, \infty]$. Then the following are equivalent.*

(i) ϕ *is S-continuous.*

(ii) $\phi(1) < \infty$.

(iii) $\forall x, y \in \mathcal{M}^+ \left(\phi(x) \approx \phi(y) \right)$.

Proof. ((i) $\Rightarrow$ (ii)) : If $\phi(1)$ is infinite, then $(\phi(1))^{-1} \cdot 1 \approx 0$ and $\phi((\phi(1))^{-1} \cdot 1) = 1$, hence ϕ is not S-continuous.

((ii) $\Rightarrow$ (i)) : If $\phi(1) < \infty$, then by Prop. 4.3(ii), whenever $a \approx 0$, $\phi(a) \leq \|a\| \, \phi(1) \approx 0$, *i.e.* ϕ is S-continuous.

((i) $\Rightarrow$ (iii)) : Let $a, b \in \mathcal{M}^+$ be such that $a \approx b$. By Prop. 3.12 there is $c \in \mathcal{M}^+$, such that $c \approx a \approx b$ and $a, b \leq c$. Moreover, $0 \approx (c - a) \in \mathcal{M}^+$ and $0 \approx (c - b) \in \mathcal{M}^+$, so, by S-continuity, $\phi(c - a) \approx 0 \approx \phi(c - b)$. Therefore $\phi(a) \approx \phi(c) \approx \phi(b)$.

((iii) $\Rightarrow$ (i)) : Follows trivially from the definition. $\qquad\qquad\square$

In some ways, S-continuous weights resemble finite positive atomless measures. An internal S-continuous weight can be converted to a weight on the norm-nonstandard hull by the following result.

Theorem 4.5. *Let an internal unital C^*-algebra $\mathcal{M}$ and an S-continuous internal weight $\phi : \mathcal{M}^+ \to {}^*[0, \infty]$ be given. Define $\widehat{\phi} : \left(\widehat{M} \right)^+ \to [0, \infty)$ by $\widehat{\phi}(\widehat{a}) = {}^\circ\phi(a)$, $a \in \mathrm{Fin}(\mathcal{M}^+)$.*

Then $\widehat{\phi}$ is well-defined and forms a weight.

Moreover, if ϕ is normal and $\mathcal{M}$ is commutative, then $\widehat{\phi}$ is κ-normal.

Proof. Two things to recall: Thm. 3.22(iv) that $\left(\widehat{M} \right)^+ = \left(\mathrm{Fin}(\mathcal{M}^+) \right)^\wedge$ and Thm. 3.17(i) that $\mathcal{M}^+$ forms a closed cone.

Let $a, b \in \mathrm{Fin}(\mathcal{M}^+)$ with $a \approx b$. By Cor. 3.21, we have decomposition $a - b = c_1 - c_2$ for some $c_1, c_2 \in \mathcal{M}^+$ such that $\|c_1\|, \|c_2\| \leq \|a - b\| \approx 0$.

By Prop. 4.3(ii) and Lem. 4.4, $\phi(c_1) \approx 0 \approx \phi(c_2)$, hence

$$\phi(a) \approx \phi(a) + \phi(c_2) = \phi(a + c_2) = \phi(b + c_1) = \phi(b) + \phi(c_1) \approx \phi(b).$$

Therefore $\widehat{\phi}$ is well-defined.

Also, by S-continuity, Lem. 4.4 and Prop. 4.3(ii), the range of $\widehat{\phi}$ is contained in $[0, \infty)$.

It is easily seen that $\widehat{\phi}$ is additive and positively homogeneous, so $\widehat{\phi}$ is a weight on $(\widehat{M})^+$.

Now suppose ϕ is normal and $\mathcal{M}$ is commutative. Then, by Thm. 3.14, $\mathcal{M}$ is *isomorphic to $C(\Omega)$ for some compact space Ω. Therefore every finite subset $A \subset \mathcal{M}^+$ has a unique supremum $\sup A \in \mathcal{M}^+$ with $\|\sup A\| = \max_{x \in A} \|x\|$. In particular, $\mathcal{M}^+$ forms a lattice.

To show κ-normality, let $\{\widehat{a}_i\}_{i \in I} \subset (\widehat{\mathcal{M}})^+$ be a bounded net with $\sup_{i \in I} \widehat{a}_i \in (\widehat{\mathcal{M}})^+$, where $\{a_i\}_{i \in I} \subset \mathrm{Fin}(\mathcal{M})$ and $|I| < \kappa$.

Without loss of generality, we assume that $\{\widehat{a}_i\}_{i \in I} \subset \bar{B}_{\widehat{\mathcal{M}}}$. We can further assume that $\{a_i\}_{i \in I} \subset \mathrm{Fin}(\mathcal{M}^+)$.

Let $r := \sup_{i \in I} \widehat{\phi}(\widehat{a}_i)$. By Prop. 4.3(i), clearly $r \leq \widehat{\phi}\Big(\sup_{i \in I} \widehat{a}_i\Big)$, so we only need to show that $\geq$ holds.

For each finite $J \subset I$ and $\epsilon \in \mathbb{R}^+$, if we let $a = \sup_{i \in J} a_i \in \mathcal{M}^+$, then, by ϕ being normal and $\{\widehat{a}_i\}_{i \in J} \subset \bar{B}_{\widehat{\mathcal{M}}}$, the following internal conditions are satisfied by such $a \in \mathcal{M}^+$:

$$\|a\| \leq 1 + \epsilon, \quad \phi(a) = \sup_{i \in J} \phi(a_i) \leq r + \epsilon \quad \text{and} \quad a \geq a_i, \quad i \in J.$$

Hence, by an application of saturation, there is $a \in \mathcal{M}^+$ such that

$$\|a\| \lessapprox 1, \quad \phi(a) \lessapprox r \quad \text{and} \quad \forall i \in I \, (a \geq a_i).$$

Subsequently, $\widehat{\phi}(\widehat{a}) = r$ and $\widehat{a} \geq \sup_{i \in I} \widehat{a}_i$. By the later and $\widehat{\phi}$ being a weight, Prop. 4.3(ii) gives $r \geq \widehat{\phi}\Big(\sup_{i \in I} \widehat{a}_i\Big)$ as required. $\qquad \square$

Question 11. In the second part of Thm. 4.5, can one show that $\widehat{\phi}$ is normal? Does the conclusion still hold for noncommutative $\mathcal{M}$?

Question 12. Thm. 4.5 gives finite weights from S-continuous weights. How to generalize this and produce infinite weights from some internal weights?

Problem 13. Develop noncommutative stochastic processes (quantum processes) based on noncommutative Loeb measures.

4.5   Further questions and problems

Here is a list of questions and problems of various degree of interest and
difficulty. The relevant results and pages are indicated in the brackets $[\cdots]$.

Problem 14. [Thm. 1.11, p.88, [Ziman and Zlatoš (2006)].] Extend the
Loeb measure construction for internal complex-valued measures of possibly infinite total variation. This would be important for analyzing the
nonstandard hull of an internal $\mathbb{M}(\Omega)$, *i.e.* an internal space of σ-additive
complex Borel measures.

Problem 15. [§2.3.2.] Characterize normed linear spaces Y such that, for
any normed linear spaces with $X_1 \subset X_2$, every $f \in \mathcal{B}(X_1, Y)$ extends to
some $\bar{f} \in \mathcal{B}(X_2, Y)$ with $\|f\| = \|\bar{f}\|$.

Problem 16. [Thm. 2.13.] Investigate whether a sequence of families of
internal seminorms $\{W_n\}_{n \in \mathbb{N}}$ can be found for a given normed linear space
X so that the nonstandard hull of $^*X$ *w.r.t.* W_n is isometric isomorphic to
$X^{(2n)}$, the higher dual of X of order $2n$. Note that implicit in our requirement is that all these nonstandard hulls are constructed within the same
nonstandard universe. Of course, by repeated applications of Thm. 2.13,
one can get such representation of $X^{(2n)}$ as a nonstandard hull, but that
requires changing to a different nonstandard universe each time.

Question 17. [Thm. 2.31.] (*The invariant subspace problem.*) Does every
bounded linear operator on a Hilbert space of dimension > 1 has a nontrivial
closed subspace?

   Put in another from, let H be a Hilbert space of dimension > 1 and
$a \in \mathcal{B}(H)$. Is there $p \in \mathrm{Proj}(\mathcal{M}) \setminus \{0, 1\}$ such that $pap = a$?

Problem 18. [Prop. 3.2.] Let X be a unital Banach algebra. Define
functions $\phi_\mathrm{R}, \Phi_\mathrm{R}, \phi_\mathrm{L}, \Phi_\mathrm{L} : X \to [0, \infty]$ by

$$\phi_\mathrm{R}(x) = \inf \big\{ \, \|xy^{-1} - 1\| \; \big| \; x \approx y \in {}^*X^{-1} \big\},$$
$$\Phi_\mathrm{R}(x) = \sup \big\{ \, \|xy^{-1} - 1\| \; \big| \; x \approx y \in {}^*X^{-1} \big\},$$

with $\phi_\mathrm{L}, \Phi_\mathrm{L}$ given similarly by using $y^{-1}x$ instead. Note that $\phi_\mathrm{R}(0) =
\Phi_\mathrm{R}(0) = \phi_\mathrm{L}(0) = \Phi_\mathrm{L}(0) = 1$ and $\forall x \in X^{-1}\big(\phi_\mathrm{R}(x) = \phi_\mathrm{L}(x) = 0\big)$.
   Investigate general properties about these functions.

Problem 19. [§2.4.1, §3.2.2 and §3.2.3.] Study the class of nonstandard
hulls of unital Banach algebras $\mathcal{M}$ constructed *w.r.t.* the seminorms given
by $\mathrm{hom}(\mathcal{M})$.

Do the same for the nonstandard hull construction of unital C^*-algebra $\mathcal{M}$ w.r.t. the seminorms given by states $\mathbb{S}(\mathcal{M})$.

Question 20. [§3.3.1.] Does the nonstandard extended functional calculus given in Thm. 3.23 include the Borel functional calculus?

Problem 21. [§3.3.2.] Classify elements a in an internal unital C^*-algebra $\mathcal{M}$ such that $\widehat{\mathcal{M}_a} = \mathcal{M}_{\widehat{a}}$.

Problem 22. [§3.2 and §3.4.] Develop a GNS representation based on the nonstandard hulls of hyperfinite dimensional matrix algebras, relate the spectrum to eigenvalues, normal elements to normal matrices *etc.*

Problem 23. [§1.5.4 and Example 3.20.] In a unital C^*-algebra $\mathcal{M}$, consider Boolean algebras $\mathbb{B} \subset \mathrm{Proj}(\mathcal{M})$. Using Stone's Theorem, represent $\mathbb{B}$ as the Boolean algebra of clopen subsets of a compact totally disconnected Hausdorff space and formulate Kelley's Theorems (Thm. 1.15 or Thm. 1.16) in this setting.

Is there a Kelley-style characterization of subsets $\mathbb{B} \subset \mathrm{Proj}(\mathcal{M})$ that generates a von Neumann algebra? If so, one could regard such $\mathbb{B}$ as a noncommutative measure algebra and such result as noncommutative Kelley's Theorem.

Problem 24. [§3.5.2.] Investigate more thoroughly the class of P^*-algebras. In particular, find out more about properties that it shares with the class of von Neumann algebras.

By Cor. 3.31, it should be worthwhile to use the nonstandard hull of hyperfinite dimensional C^*-algebras as a guidance for studying P^*-algebras. Moreover hyperfinite dimensional C^*-algebras are in a sense nonseparable version of AF-algebras.

Problem 25. [§3.3.] Find a natural identification of the minimal class of C^*-algebras that includes von Neumann algebras and is closed under the nonstandard hull construction.

300   *Nonstandard Methods in Functional Analysis*

Suggestions for Further Reading

Nonstandard Analysis:

S. Albeverioo, R. Høegh-Krohn, J. E. Fenstad and T. Lindstrøm. **Nonstandard methods in stochastic analysis and mathematical physics**. [Albeverio *et al.* (1986)]

Banach Spaces:

J.B. Conway. **A course in functional analysis**. [Conway (1990)]

R. E. Megginson. **An introduction to Banach space theory**. [Megginson (1998)]

Banach Algebras:

T.W. Palmer. **Banach algebras and the general theory of ∗-algebras** *vol.* I, II. [Palmer (1994)], [Palmer (2001)]

C*-*algebras*:

B. Blackadar. **Operator algebras**. [Blackadar (2006)]

K.R. Davidson. **C*-algebras by example**. [Davidson (1996)]

Spectral Analysis:

B. Aupetit. **A primer on spectral theory**. [Aupetit (1991)]

Noncommutative Mathematics:

A. Connes. **Noncommutative geometry**. [Connes (1994)]

Other Related Topics:

Y. Benyamini and J. Lindenstrauss. **Geometric nonlinear functional analysis**. *vol.* 1. [Benyamini and Lindenstrauss (2000)]

G. Pisier. **Introduction to operator space theory**. [Pisier (2003)]

W. Arveson. **Noncommutative dynamics and E-semigroups**. [Arveson (2003)]

N.P. Brown and N. Ozawa. **C^*-algebras and finite-dimensional approximations**. [Brown and Ozawa (2008)]

General References:

N. Dunford and J. T Schwartz. **Linear operators**. *vol.* I, II, III. [Dunford and Schwartz (1988a)], [Dunford and Schwartz (1988b)], [Dunford and Schwartz (1988c)]

Bibliography

Aksoy, A. G. and Khamsi, M. A. (1990). *Nonstandard methods in fixed point theory*, Universitext (Springer-Verlag, New York), ISBN 0-387-97364-8, with an introduction by W. A. Kirk.

Albeverio, S., Høegh-Krohn, R., Fenstad, J. E. and Lindstrøm, T. (1986). *Nonstandard methods in stochastic analysis and mathematical physics*, Pure and Applied Mathematics, Vol. 122 (Academic Press Inc., Orlando, FL), ISBN 0-12-048860-4; 0-12-048861-2, also reprinted by Dover Publications, ISBN: 0-48-646899-2.

Alspach, D. E. (1981). A fixed point free nonexpansive map, *Proc. Amer. Math. Soc.* **82**, 3, pp. 423–424.

Anderson, R. M. (1976). A non-standard representation for Brownian motion and Itô integration, *Israel J. Math.* **25**, 1-2, pp. 15–46.

Anderson, R. M. (1982). Star-finite representations of measure spaces, *Trans. Amer. Math. Soc.* **271**, 2, pp. 667–687.

Aronszajn, N. and Smith, K. T. (1954). Invariant subspaces of completely continuous operators, *Ann. of Math. (2)* **60**, pp. 345–350.

Arveson, W. (2003). *Noncommutative dynamics and E-semigroups*, Springer Monographs in Mathematics (Springer-Verlag, New York), ISBN 0-387-00151-4.

Ash, R. B. (2000). *Probability and measure theory*, 2nd edn. (Harcourt/Academic Press, Burlington, MA), ISBN 0-12-065202-1.

Aupetit, B. (1991). *A primer on spectral theory*, Universitext (Springer-Verlag, New York), ISBN 0-387-97390-7.

Baratella, S. and Ng, S.-A. (1998). Fixed points in the nonstandard hull of a Banach space, *Nonlinear Anal.* **34**, 2, pp. 299–306.

Baratella, S. and Ng, S.-A. (2003). A nonstandard proof of the Eberlein-Šmulian theorem, *Proc. Amer. Math. Soc.* **131**, 10, pp. 3177–3180 (electronic).

Baratella, S. and Ng, S.-A. (2009). Nonstandard hulls of C^*-algebras, Preprint, http://arxiv.org/abs/0910.5058.

Benyamini, Y. and Lindenstrauss, J. (2000). *Geometric nonlinear functional analysis. Vol. 1*, American Mathematical Society Colloquium Publications, Vol. 48 (American Mathematical Society, Providence, RI), ISBN 0-8218-

0835-4.

Bernstein, A. R. and Robinson, A. (1966). Solution of an invariant subspace problem of K. T. Smith and P. R. Halmos, *Pacific J. Math.* **16**, pp. 421–431.

Blackadar, B. (2006). *Operator algebras, Encyclopaedia of Mathematical Sciences*, Vol. 122 (Springer-Verlag, Berlin), ISBN 978-3-540-28486-4; 3-540-28486-9, theory of C^*-algebras and von Neumann algebras, Operator Algebras and Non-commutative Geometry, III.

Brown, N. P. and Ozawa, N. (2008). C^*-*algebras and finite-dimensional approximations, Graduate Studies in Mathematics*, Vol. 88 (American Mathematical Society, Providence, RI), ISBN 978-0-8218-4381-9; 0-8218-4381-8.

Chang, C. C. and Keisler, H. J. (1990). *Model theory, Studies in Logic and the Foundations of Mathematics*, Vol. 73, 3rd edn. (North-Holland Publishing Co., Amsterdam), ISBN 0-444-88054-2.

Connes, A. (1994). *Noncommutative geometry* (Academic Press Inc., San Diego, CA), ISBN 0-12-185860-X.

Conway, J. B. (1990). *A course in functional analysis, Graduate Texts in Mathematics*, Vol. 96, 2nd edn. (Springer-Verlag, New York), ISBN 0-387-97245-5.

Conway, J. B. (1991). *The theory of subnormal operators, Mathematical Surveys and Monographs*, Vol. 36 (American Mathematical Society, Providence, RI), ISBN 0-8218-1536-9.

Cutland, N. and Ng, S.-A. (1991). On homogeneous chaos, *Math. Proc. Cambridge Philos. Soc.* **110**, 2, pp. 353–363.

Dacunha-Castelle, D. and Krivine, J.-L. (1970). Ultraproduits d'espaces d'Orlicz et applications géométriques, *C. R. Acad. Sci. Paris Sér. A-B* **271**, pp. A987–A989.

Davidson, K. R. (1996). C^*-*algebras by example, Fields Institute Monographs*, Vol. 6 (American Mathematical Society, Providence, RI), ISBN 0-8218-0599-1.

Davidson, K. R. and Szarek, S. J. (2001). Local operator theory, random matrices and Banach spaces, in *Handbook of the geometry of Banach spaces, Vol. I* (North-Holland, Amsterdam), pp. 317–366.

Day, M. M. (1973). *Normed linear spaces*, 3rd edn. (Springer-Verlag, New York), Ergebnisse der Mathematik und ihrer Grenzgebiete, Band 21.

Diestel, J. and Uhl, J. J., Jr. (1977). *Vector measures* (American Mathematical Society, Providence, R.I.), with a foreword by B. J. Pettis, Mathematical Surveys, No. 15.

Dinculeanu, N. (2000). *Vector integration and stochastic integration in Banach spaces*, Pure and Applied Mathematics (New York) (Wiley-Interscience, New York), ISBN 0-471-37738-4.

Dunford, N. and Schwartz, J. T. (1988a). *Linear operators. Part I*, Wiley Classics Library (John Wiley & Sons Inc., New York), ISBN 0-471-60848-3, general theory, With the assistance of William G. Bade and Robert G. Bartle, Reprint of the 1958 original, A Wiley-Interscience Publication.

Dunford, N. and Schwartz, J. T. (1988b). *Linear operators. Part II*, Wiley Classics Library (John Wiley & Sons Inc., New York), ISBN 0-471-60847-5, spectral theory. Selfadjoint operators in Hilbert space, With the assistance of William G. Bade and Robert G. Bartle, Reprint of the 1963 original, A Wiley-Interscience Publication.

Dunford, N. and Schwartz, J. T. (1988c). *Linear operators. Part III*, Wiley Classics Library (John Wiley & Sons Inc., New York), ISBN 0-471-60846-7, spectral operators, With the assistance of William G. Bade and Robert G. Bartle, Reprint of the 1971 original, A Wiley-Interscience Publication.

Enderton, H. B. (1977). *Elements of set theory* (Academic Press [Harcourt Brace Jovanovich Publishers], New York).

Fabian, M., Habala, P., Hájek, P., Montesinos Santalucía, V., Pelant, J. and Zizler, V. (2001). *Functional analysis and infinite-dimensional geometry*, CMS Books in Mathematics/Ouvrages de Mathématiques de la SMC, 8 (Springer-Verlag, New York), ISBN 0-387-95219-5.

Fajardo, S. and Keisler, H. J. (1996). Neometric spaces, *Adv. Math.* **118**, 1, pp. 134–175.

Gass, S. I. (2003). *Linear programming*, 5th edn. (Dover Publications Inc., Mineola, NY), ISBN 0-486-43284-X, methods and applications.

Godefroy, G. and Iochum, B. (1988). Arens-regularity of Banach algebras and the geometry of Banach spaces, *J. Funct. Anal.* **80**, 1, pp. 47–59.

Goebel, K. and Kirk, W. A. (1990). *Topics in metric fixed point theory, Cambridge Studies in Advanced Mathematics*, Vol. 28 (Cambridge University Press, Cambridge), ISBN 0-521-38289-0.

Haagerup, U. (1975). Normal weights on W^*-algebras, *J. Functional Analysis* **19**, pp. 302–317.

Heinrich, S. (1980). Ultraproducts in Banach space theory, *J. Reine Angew. Math.* **313**, pp. 72–104.

Henson, C. W. (1979). Unbounded Loeb measures, *Proc. Amer. Math. Soc.* **74**, 1, pp. 143–150.

Henson, C. W. (1988). Infinitesimals in functional analysis, in *Nonstandard analysis and its applications (Hull, 1986)*, London Math. Soc. Stud. Texts, Vol. 10 (Cambridge Univ. Press, Cambridge), pp. 140–181.

Henson, C. W. and Moore, L. C., Jr. (1983). Nonstandard analysis and the theory of Banach spaces, in *Nonstandard analysis—recent developments (Victoria, B.C., 1980)*, Lecture Notes in Math., Vol. 983 (Springer, Berlin), pp. 27–112.

Hewitt, E. (1948). Rings of real-valued continuous functions. I, *Trans. Amer. Math. Soc.* **64**, pp. 45–99.

Hewitt, E. and Ross, K. A. (1963). *Abstract harmonic analysis. Vol. I: Structure of topological groups. Integration theory, group representations*, Die Grundlehren der mathematischen Wissenschaften, Bd. 115 (Academic Press Inc., Publishers, New York).

Hinokuma, T. and Ozawa, M. (1993). Conversion from nonstandard matrix algebras to standard factors of type II_1, *Illinois J. Math.* **37**, 1, pp. 1–13.

Hochstadt, H. (1979/80). Eduard Helly, father of the Hahn-Banach theorem, *Math. Intelligencer* **2**, 3, pp. 123–125.

Jech, T. (2003). *Set theory*, Springer Monographs in Mathematics (Springer-

Verlag, Berlin), ISBN 3-540-44085-2, the third millennium edition, revised and expanded.

Kanovei, V. and Reeken, M. (2004). *Nonstandard analysis, axiomatically*, Springer Monographs in Mathematics (Springer-Verlag, Berlin), ISBN 3-540-22243-X.

Keisler, H. J. (1986). *Elementary Calculus: An Infinitesimal Approach*, 2nd edn. (Prindle, Weber and Schmidt Inc., Boston), ISBN 0-871-50911-3, http://www.math.wisc.edu/~keisler/calc.html.

Kelley, J. L. (1959). Measures on Boolean algebras, *Pacific J. Math.* **9**, pp. 1165–1177.

Kunen, K. (1983). *Set theory, Studies in Logic and the Foundations of Mathematics*, Vol. 102 (North-Holland Publishing Co., Amsterdam), ISBN 0-444-86839-9, an introduction to independence proofs, Reprint of the 1980 original.

Loeb, P. A. (1975). Conversion from nonstandard to standard measure spaces and applications in probability theory, *Trans. Amer. Math. Soc.* **211**, pp. 113–122.

Luxemburg, W. A. J. (1969). A general theory of monads, in *Applications of Model Theory to Algebra, Analysis, and Probability (Inte rnat. Sympos., Pasadena, Calif., 1967)* (Holt, Rinehart and Winston, New York), pp. 18–86.

McDuff, D. (1970). Central sequences and the hyperfinite factor, *Proc. London Math. Soc. (3)* **21**, pp. 443–461.

Megginson, R. E. (1998). *An introduction to Banach space theory, Graduate Texts in Mathematics*, Vol. 183 (Springer-Verlag, New York), ISBN 0-387-98431-3.

Moore, G. H. (1982). *Zermelo's axiom of choice, Studies in the History of Mathematics and Physical Sciences*, Vol. 8 (Springer-Verlag, New York), ISBN 0-387-90670-3.

Munkres, J. R. (2000). *Topology: a first course*, 2nd edn. (Prentice-Hall Inc., Englewood Cliffs, N.J.), ISBN 0-131-81629-2.

Nelson, E. (1977). Internal set theory: a new approach to nonstandard analysis, *Bull. Amer. Math. Soc.* **83**, 6, pp. 1165–1198.

Ng, S.-A. (1991). A new proof of Kelley's theorem, *Fund. Math.* **140**, 1, pp. 63–67.

Osswald, H. (1995). A nonstandard approach to the Pettis integral, in *Advances in analysis, probability and mathematical physics (Blaubeuren, 1992)*, Math. Appl., Vol. 314 (Kluwer Acad. Publ., Dordrecht), pp. 75–90.

Osswald, H. and Sun, Y. (1991). On the extensions of vector-valued Loeb measures, *Proc. Amer. Math. Soc.* **111**, 3, pp. 663–675.

Palmer, T. W. (1994). *Banach algebras and the general theory of *-algebras. Vol. I, Encyclopedia of Mathematics and its Applications*, Vol. 49 (Cambridge University Press, Cambridge), ISBN 0-521-36637-2.

Palmer, T. W. (2001). *Banach algebras and the general theory of *-algebras. Vol. 2, Encyclopedia of Mathematics and its Applications*, Vol. 79 (Cambridge University Press, Cambridge), ISBN 0-521-36638-0, *-algebras.

Pisier, G. (2003). *Introduction to operator space theory, London Mathematical*

Society Lecture Note Series, Vol. 294 (Cambridge University Press, Cambridge), ISBN 0-521-81165-1.

Reed, M. and Simon, B. (1980). Methods of modern mathematical physics. I, 2nd edn. (Academic Press Inc. [Harcourt Brace Jovanovich Publishers], New York), ISBN 0-12-585050-6.

Render, H. (1993). Pushing down Loeb measures, Math. Scand. 72, 1, pp. 61–84.

Rudin, W. (1987). Real and complex analysis, 3rd edn. (McGraw-Hill Book Co., New York), ISBN 0-07-054234-1.

Rudin, W. (1991). Functional analysis, 2nd edn., International Series in Pure and Applied Mathematics (McGraw-Hill Inc., New York), ISBN 0-07-054236-8.

Salbany, S. and Todorov, T. (2000). Nonstandard analysis in topology: nonstandard and standard compactifications, J. Symbolic Logic 65, 4, pp. 1836–1840.

Sherman, S. (1951). Order in operator algebras, Amer. J. Math. 73, pp. 227–232.

Simpson, S. G. (2009). Subsystems of second order arithmetic, 2nd edn., Perspectives in Logic (Cambridge University Press, Cambridge), ISBN 978-0-521-88439-6.

Stroyan, K. D. and Bayod, J. M. (1986). Foundations of infinitesimal stochastic analysis, Studies in Logic and the Foundations of Mathematics, Vol. 119 (North-Holland Publishing Co., Amsterdam), ISBN 0-444-87927-7.

Sun, Y. (1992). On the theory of vector valued Loeb measures and integration, J. Funct. Anal. 104, 2, pp. 327–362.

Takesaki, M. (2002). Theory of operator algebras. I, Encyclopaedia of Mathematical Sciences, Vol. 124 (Springer-Verlag, Berlin), ISBN 3-540-42248-X, reprint of the first (1979) edition, Operator Algebras and Non-commutative Geometry, 5.

van Dalen, D. (2004). Logic and structure, 4th edn., Universitext (Springer-Verlag, Berlin), ISBN 3-540-20879-8.

Wiśnicki, A. (2002). Neocompact sets and the fixed point property, J. Math. Anal. Appl. 267, 1, pp. 158–172.

Ziman, M. and Zlatoš, P. (2006). Hyperfinite dimensional representations of spaces and algebras of measures, Monatsh. Math. 148, 3, pp. 241–261.

Zimmer, G. B. (1998). An extension of the Bochner integral generalizing the Loeb-Osswald integral, Math. Proc. Cambridge Philos. Soc. 123, 1, pp. 119–131.

Živaljević, B. (1992). Uniqueness of unbounded Loeb measure using Choquet's theorem, Proc. Amer. Math. Soc. 116, 2, pp. 529–533.

308   *Nonstandard Methods in Functional Analysis*

Index

Index   311

314   *Nonstandard Methods in Functional Analysis*